MODERN AMERICAN PROTESTANTISM AND ITS WORLD

Modern American Protestantism And Its World
Historical Articles on Protestantism in American Religious Life

Edited by Martin E. Marty

Series ISBN 3-598-41530-3

MODERN AMERICAN PROTESTANTISM AND ITS WORLD
Historical Articles on Protestantism in American Religious Life

10. Fundamentalism and Evangelicalism

Edited with an Introduction by

Martin E. Marty
The University of Chicago

K · G · Saur
Munich · London · New York · Paris 1993

BT
82.2
. F834
1993

Library of Congress Cataloging-in-Publication Data
Fundamentalism and evangelicalism / edited with an
   introduction by Martin E. Marty.
      p. cm. -- (Modern American Protestantism and its world ; 10)
     Includes bibliographical references and index.
     ISBN 3-598-41541-9 (alk. paper)
     1. Fundamentalism--United States. 2. Evangelicalism--United States.
   3. United States--Church history--20th century.
   I. Marty, Martin E., 1928-   . II. Series.
   BT82.2.F834 1992
   277.3'082--dc20   92-3091
                                          CIP

Die Deutsche Bibliothek - CIP - Einheitsaufnahme

Modern American Protestantism and its world: historical
articles on Protestantism in American religious life / ed. with
an introduction by Martin E. Marty. - Munich ; London ; New York ;
Paris : Saur.
   ISBN 3-598-41530-3
NE: Marty, Martin E. (Hrsg.)
Vol. 10. Fundamentalism and evangelicalism. - 1993
   ISBN 3-598-41541-9

♾

Printed on acid-free paper/Gedruckt auf säurefreiem Papier

Printed in the United States of America

Printed/Bound by Edwards Brothers Incorporated, Ann Arbor

ISBN 3-598-41541-9 (vol. 10)
ISBN 3-598-41530-3 (series)

Contents

Series Preface

Protestantism is America's "only national religion and to ignore that fact is to view the country from a false angle." Andre Siegfried, an astute visitor from France, made that observation in 1927. One did not have to be very astute to see America as the French journalist did, for reasons that are beginning to be lost among Americans today.

Wherever one looked, there was Protestantism. It had come with the Anglicans of Virginia and the other southern colonies, which established their Episcopal church and did what they could to suppress Catholics who had settled in Maryland. Protestantism in its Puritan form motivated the Pilgrims and Puritans who settled the New England states and established Congregationalism there. They did what they could to quarantine Rhode Island, because it welcomed people of many faiths--Jews settled in early--but New England was firmly Protestant. As for the middle colonies, where no one church was supported by law, Protestants prevailed.

At the time of the birth of America, there were probably somewhat more than four million people in the colonies. Many estimate that only about 30,000 of these were Catholics and 3,000 were Jews. Native Americans, most of them resistant to Protestantism, were being pushed away; when slave owners started caring about the souls of slaves, they Protestantized them, and black America remains overwhelmingly Protestant.

The birth of America meant the gradual ending of state support for Protestantism. But the power that Protestants lost in the legal realm they recouped in other ways. They formed strong competitive church bodies, headed out to win the West for their religion and ways of life, developed voluntary organizations for reform and good works, and tried to pass laws favoring their outlooks. They developed what I have called a "religious empire."

What about everyone else? A year after Andre Siegfried wrote, Protestant America helped defeat Governor Al Smith's bid for the United States Presidency; Smith's Catholicism offended them, and not until 1960 was a Catholic elected President. Catholicism made up the largest single church for a century, and Catholics had enormous power in northern cities. But Catholic were often not part of the American elite. Jews also were not.

Today it is hard to picture how strong the Protestant hold was on the whole nation. Nowadays we speak of America as "pluralist," not a Protestant society. There are as many Muslims as Episcopalians. The heirs of established Congregationalism are few, while the Asian population grows. Judaism has unimaginably more influence than it did a century ago. Catholics are at home and at ease in all elements of American life, exerting influence where and as they will. What the Protestants of old called "infidelity" and of more recent times called "secularism" has a power they feared but did not really believe could establish itself.

How Protestantism came from its position of near-monopoly to mere hegemony to its place as a set of faiths among other sets of faiths is central to the drama of American religious life, indeed, American life as a whole. Yielding power as it did is an act which normally one would expect would have issued in bloodshed and produced revolutions. Every morning's newspaper and television bring stories of interreligious, intertribal, and intercultural clashes that exact terrible tolls on human life: one thinks of Lebanon, the Asian subcontinent, Northern Ireland, and elsewhere for examples. In America the change was made without loss of life--but not without drama.

The drama of adjustment by Protestants to the world they had made and was now moving beyond them makes up the background to the articles chosen to depict Protestantism's changing place in American life in this series of articles and books.

What is this Protestantism that exercised so much influence and still has to be reckoned with in most dimensions of American life? The Protestant churches represent that element in Western Christianity which is not obedient to the Pope at Rome. That negative definition seems to be as far as one can go in finding something which all Protestants held and hold in common. But Protestantism is also a cultural force of great influence in the culture, as a glance at American history shows.

Protestants, for instance, believed in exalting the individual. They formed churches and other collectives, but they insisted that the conscience and reason of the individual mattered supremely in the plan of God. While Protestantism is not the only source of American individualism, to overlook it would be to misunderstand America. But individualism was by no means a unique stress. The early Protestants came with a sense of "covenant," a claim that God had reached out and, in effect, made an agreement with them. God would favor them if the Americans would be faithful. So citizens of the United States started thinking of themselves as "chosen" people with a "mission" and a "destiny."

Many of the extensions of these themes of chosenness, mission, and destiny helped produced the best and worst times in American life. They have led to moral productivity, as in support of human rights, just as they have led to pride and folly, as when Americans tried to impose their will and ways on other nations.

Protestant themes course through American literature, through the novels of Hawthorne and Melville and more. They are present in economic thought, beginning with a concept of one's "vocation" or "calling" in daily life, in the use of the world's resources. Americans do their debating of vital issues: defense policy, attitudes toward Israel, a welfare society, abortion, and the like, employing themes planted by Protestants.

The Protestant movement, even though it no longer has hegemony, is by no means weak. Taking black and white Protestantism together, the churches of this movement attract the loyalty and active participation of well over a hundred million Americans. They give billions of dollars and hours to the cause, providing places of worship, shelters for the homeless, agencies for involvement in political life, centers for training the young, participating in institutions of human care, keeping contact with people around the globe, and marking the stages of peoples' individual lives. They share "clout," but that does not mean they do not have "clout."

Protestantism has passed through many stages in America. This series concentrates on those that occurred in a period code-named "modern." Making sense of modernity is a full-time job for individuals, for a nation, for the churches, and for the scholars. In this collection, we take the modern period to represent roughly the past century. Decades after the Civil War and toward the end of the century, Americans saw that they were turning urban, industrial, pluralist. Village life, a natural home for Protestants, was turning metropolitan, where Protestants were being overwhelmed by non-Protestant immigrants. Yet, instead of growing weaker and seeing their churches empty, as in the European experience, American Protestants built more and filled the houses of God. If one Protestant movement weakens--as the mainstream seems to have done in recent decades--others, such as black, fundamentalist, evangelical, and pentecostal fill the void and Protestantism finds new ways to prosper.

If Protestants once "ran the show" in an America that no longer is ruled by a single force, they are still not folding their tents, turning their churches into museums, or ceasing in efforts to influence the whole society. For that reason, there is a strong interest in the near-contemporary, the turmoil of the recent past, which helps shape American political and social discourse as the century ends. The articles in these volumes are alert to changes as America comes to a period called "post-modern." These are times when many pay attention, if only eclectically, to traditions. When Americans rummage in their collective attics looking for what they might make use of in a new day, much of what they find has "Protestant" on the label. Understanding Protestantism has become urgent.

In comparison to other movement--Mormonism, "cults," the Asian presence, the New Age, and the like--Protestantism suffers. Because it is so identified with the wallpaper and woodwork of the mental furnished apartment in which Americans live, many take it for granted. It does not seem arcane, exotic, alluring to a public which seeks novelty. Yet that very bonding with the environment which makes Protestantism hard to study is part of the secret of its power. It remains a sort of insider in a pluralist and secular and modern world. To overlook its power and the changes it has experienced in the past century is to miss the point of much of American Life. These articles help a new generation find the point, as a step toward helping them then make up their mind what to do about it.

Introduction

Modern American Protestantism has not one world but at least two. When the polltaker asks Americans what religion they prefer or how they would characterize their preference, one out of three will say "moderate Protestant" or "liberal Protestant" (while about one out of eleven or twelve will identify with "black Protestant.") But one out of six will say "conservative Protestant." While that means only half as many as there are moderates and liberals, in actual churchly vitality, cultural force, and visibility, the two major camps are probably of equal weight.

Moderates and conservatives make up what is often called mainline or mainstream Protestantism, names given them just as they were losing their hegemony after mid-century. Meanwhile, evangelicals, fundamentalists, and pentecostals have worked their cohort well. If many others "prefer" Protestantism but do little about it, the evangelicals and fundamentalists back their preference more readily with commitment. They are "born again," and like to say so. Church plays a bigger part in the lives of more of them than of their more adapted counterparts. They make up what Peter Berger calls a "cognitive minority," which means that they look at everything somewhat differently, because of their religious faith, than do those Protestants who have accommodated more blithely to the secular and pluralist environment.

It can be seen from even that brief description that "there's a war on" in Protestantism, or that there was. Today many moderates and liberals have gotten tired of the battles; they see little at stake in the theological critiques of evangelicals and find no reason at all to share company with fundamentalists, who seem to be a different religion to them. They only pay attention when the polemics of the "ultras" disrupt their lives, as in efforts to limit free expression on television or in humanistic scholarship, and when fundamentalists try to legitimate "creation research" in textbooks; when they sense political encroachments. The theology holds no lures.

Many evangelicals are so sufficiently moderate that they find company with mainstream Protestants, and in fact they have often become the mainstream as much as anyone else. They tend to stay in the classic denominations from which fundamentalists have gone in frustration and rage or from which pentecostalists were called by their revivalists. And it should be said that many mainstream

Protestants feel sympathies with the religious professions of moderate evangelicals and work in tandem with them. Fundamentalists represent, then, the opposite pole of people who at times explicitly used the language of being "come-outers," separatists, who would have nothing to do with other Protestants, and who rarely have engaged themselves religiously with anyone but their own kind.

This is a collection on history and not polemics, so the fire and fury of anti-modernism appears more regularly through quotations in the essays than in the essays themselves. Yet a goodly number of them have been written by scholars of evangelical Protestant disposition. It happens that biblical and theological scholarship, where more strictures are imposed or readily accepted, have had less hearing in the larger academy. But the historians of religion have not only made their way but excelled in the company of historians, and are, as a group, looked to with as much expectation and appreciated as richly as any other set of historians.

Being historians, they have to give an accounting of how American Protestantism got to where it is. Most complex religious movements have polarities and contentions and are marked by something like political two-party systems. When these polarities and contentions are held in tension, they can bring vitality to a movement. Yet the twentieth century has seen political metaphors yield to military ones, and has seen breaches, disruptions, and militancy across the lines in Protestantism.

These essays describe the long roots of the schisms. In the middle of the nineteenth century, there was more relative homogeneity, as evangelists and social reformers stayed in the same camp. By the early twentieth century, liberal theology, the Social Gospel, and open commitments to the surrounding progressive culture led liberalism one way. Meanwhile, conservative reaction, the desire to win souls, and the impulse to prophecy against the culture and accommodation to it marked the conservative parties.

The adaptation to Darwinian evolution by modernist Protestants and their introduction of German-style Higher Criticism of the Bible were catalysts for counter-liberal movements. On the other hand, they generated viewpoints and doctrines of their own, most notably the premillennial dispensationalism which led them to part with all progressivisms and to look for the second coming of Christ, apart from reform of the world.

In the course of time, as the gathering fundamentalist party looked for a way to counter liberalism, it sought and found authority in a peculiar understanding of the Bible. Not without precedent in seventeeth century scholastic Protestantism, the doctrine of biblical inerrancy (of the original autographs of the Bible, before copyists got at them), was not developed until the turn of the century period of militancy.

With the insistence that the Bible was true in all its historical, geographical, and scientific propositions, came an insistence on the use of the word "literal" to describe the attachments to scriptural passages by the fundamentalists and evangelicals. They claimed to find "literal" prophesying about the end time that was applicable to the present in the prophesies of Ezekiel and Daniel and the Book of Revelation. William Jennings Bryan, the politically progressivist lay person who is the subject of two essays here, kept insisting that when liberals allegorized, spiritualized, or symbolized a teaching, they sucked all the truth out of it.

Many of the essays included here are designed to shatter stereotypes. Nathan Hatch shows that evangelicalism was not an anti-democratic movement, as some of its critics believed it was when they saw certain sub-movements within it. Ernest Sandeen goes to the roots to show how fundamentalism took shape out of Princeton scholastic theology and premillennialism, while George Marsden, the most consistent of the historians of fundamentalism, shows its international scope. Here is an essay by Joel Carpenter which complements Robert T. Handy's article on the American religious depression (in another volume in this series). They provide contexts out of which the others write.

Most of the articles, then, feature local, personal (J. Frank Norris), or particularized elements in fundamentalism and evangelicalism. At the end, Ronald L. Numbers in an essay on Creationism, Robert G. Clouse on New Christian Right politics, and Jeffrey K. Hadden on religious broadcasting, bring the story down to the mid-1980s, when evangelicalism had come into its political prime. Together the essays prepare readers to follow the still unfolding plots of these vital movements.

* * *

Peter D'Agostino and Gil Waldkoenig, successively my research assistants while this project was being conceived and executed, made major contributions to the selecting and editing, and I thank them for their creative and persistent participation.

In order to assure representationality and variety, for the choice of articles in this set of volumes we consulted with dozens of scholars who specialize in American religion. In some cases, several of them nominated the same articles. We wish to acknowledge the cooperation of Josef L. Altholz, Robert G. Clouse, Brian Clarke, Jacob H. Dorn, James Findley, Robert C. Fuller, Samuel S. Hill, Richard Kern, John Kloos, Bill J. Leonard, Mark Noll, James B. North, Ronald B. Numbers, Richard V. Pierard, George Shriver, John G. Stackhouse, Jr., Ferenc M. Szasz, Grant Walker, and Louis Weeks.

FUNDAMENTALISM AND EVANGELICALISM

Evangelicalism as a Democratic Movement

Nathan O. Hatch

> *But in a democracy organized on the model of the United States there is only one authority, one source of strength and of success, and nothing outside of it.*
> Alexis de Tocqueville

The vengeance with which religious issues have reentered the public arena in our time draws attention to what samplers of public opinion long have known: the United States stands at the top of Western industrial societies in the importance religion plays in the lives of its citizens. After surveying much of this evidence, historian Laurence Veysey has concluded that at least two out of three adults in America still maintain fairly bedrock religious beliefs. In a recent Gallup Poll that asked how important religion should be in life, 41 percent of young Americans (ages 18–24) answered "very important." In France, Germany, and Great Britain, fewer than ten percent of the young people gave the same response. On any given Sunday morning, over 40 percent of the population in the United States attends church. In Canada and Australia this number tails off to about 25 percent; in England to about 10 percent; and in Scandinavia to around 5 percent—despite the fact that 95 percent of the population is confirmed in the church.

Nathan O. Hatch is Associate Professor of History and Associate Dean of the College of Arts and Sciences at the University of Notre Dame. This article appears as a chapter in Evangelicalism and Modern America, a collection edited by George Marsden and forthcoming from Eerdmans in October.

How does one begin to explain the religious continuity and vitality found in America? What kind of long-term cultural mores have allowed the roots of Christianity to sink so deeply within popular culture? It is certainly not that Americans have developed a genius for ecclesiastical organization. While the United States Army or General Motors may attribute their success to a well-honed bureaucracy, centrally directed and tied together from top to bottom, American Christianity has muddled along in a state of anarchic pluralism, a sort of free-market religious economy. Dietrich Bonhoeffer once commented that it had been granted to America, less than to any nation on earth, to realize the visible unity of the church of God. Nor can the success of American Christianity be attributed to the prestige of its clergy. While some religious leaders such as Billy Graham or Theodore Hesburgh may *win* respect, American clergymen as a group enjoy less prestige than do their colleagues in other Western democracies. The shadow of Elmer Gantry still lingers. Neither can Christianity here attribute its strength to an agility in making faith plausible to the modern world. European Protestant churches for at least a century have made it possible for their parishioners to embrace modernity without a twinge of conscience, while American churchmen who attempt to make peace with this age may be surprised to find parishioners still fretting about issues such as evolution, secular humanism, prayer in the schools, and the kinds of books that libraries should own.

What then is the driving force behind American Christianity if it is not the quality of its organization, the status of its clergy, or the power of its intellectual life? I would suggest that a central dynamic has been its democratic orientation. In America the principal mediator of God's voice has not been state, church, council, confession, ethnic group, university, college, or seminary; it has been, quite simply, the people. American Christianity, particularly its evangelical varieties, has not been something held aloof from the rank and file, a faith to be appropriated on someone else's terms. Instead, the evangelical instinct for two centuries has been to pursue people wherever they could be found, to embrace them without regard to social standing, to challenge them to think, interpret Scripture, and organize the church for themselves, and to endow their lives with the ultimate meaning of knowing Christ personally, being filled with the Spirit, and knowing with assurance the reality of eternal life. These democratic yearnings are among the oldest and deepest impulses in American religious life. Given this fact, what is surprising is not the continued dynamism of evangelical Christianity in this century. What is surprising is that analysts of religion and culture have paid so little attention to its democratic foundations and thus too readily assumed its demise.

What do I mean when I say that American Christianity is characteristically democratic? I mean that in the century and a half from the Great Awakening to the dawn of the twentieth century America underwent a profound Christianization. Describing this culture as an "Evangelical Empire," Martin Marty has written that "the first half century of national life saw the development of evangelicalism as a kind of national religion." While many factors contributed to this "Golden Day of

---

The evangelical instinct has been to pursue people wherever they could be found.

Democratic Evangelicalism"—to use Sydney Ahlstrom's phrase—none is more striking than the impulse to rework Christianity into forms that were unmistakably popular. Noting in 1840 that the United States appeared to be more religious than European nations, the Frenchman Alexis de Tocqueville reasoned that in Europe the spirit of liberty and of religion had marched in opposite directions. In America the two had become "one undivided current," so much so that people had trouble distinguishing between them. The style of Christianity which Tocqueville described in *Democracy in America* was democratic in at least three respects: it was audience centered, intellectually open to all, and organizationally pluralistic and innovative.

THE SOVEREIGN AUDIENCE

The genius of evangelicals long has been their firm identification with people. While others may have excelled in defending and elaborating the truth and in building institutions to weather the storms of time, evangelicals in America have been passionate about communicating a message. The enduring legacy of the first Great Awakening, Harry Stout has suggested, was a new mode of persuasion. Defying a church callous to its common folk, John Wesley thundered that he would preach nothing but "plain truth for plain people." The primary emphasis of Wesley and George Whitefield was not orthodoxy per se, as important as both thought theology to be. The end of religion was that each person would know a profound personal experience with God. This required an idiom in touch with people. By the time of the American Revolution, the warmth of such evangelical appeals and their ability to draw the unchurched into cohesive fellowships made evangelicalism a major social force on both sides of the Atlantic.

John Wesley, George Whitefield, and Jonathan Edwards certainly were not democrats, but whatever concessions they failed to make to a broader public were swept away by the tidal wave of democracy that engulfed America in the first decades of the nineteenth century. While staid graduates of Harvard, Yale, and Princeton continued to serve solidly respectable congregations, they watched as bystanders to the real winning of America's soul—by relentless Methodist circuit riders, roughhewn Baptist preachers, and no-nonsense elders from the Disciples of Christ. The utterly unpre-

While others may have excelled in building institutions, evangelicals have been passionate about communicating a message.

tentious message of these groups, like that of Joseph Smith, founder of the Latter-day Saints, or William Miller, the Adventist, stripped away the power of creed and confession, the authority of staid institutional forms, and the inherent power of the clergy. What they promised in return was that people could make their own religious commitments rather than obeying those handed down to them.

Charles G. Finney and Dwight L. Moody stand as the finest exemplars of this "proclamation theology." Both did espouse coherent theologies, but both despised the formal study of divinity. This was precisely because it produced such dull and ineffective communication. Attuned to the needs and concerns of average people, both discarded hidebound forms for new methods that would awaken the unconcerned and reawaken the complacent. "What do the politicians do?" queried Finney.

> They get up meetings, circulate handbills and pamphlets, blaze away in the newspapers, send their ships about the streets on wheels with flags and sailors, send coaches all over town, with handbills, to bring people up to the polls, all to gain attention to their cause and elect their candidate. . . . The object of our measures is to gain attention and you *must have* something new.
> [*Lectures on Revivals of Religion*]

Finney despised sermons that were read because the preacher remained content oriented rather than sensitive to the audience's reaction.

> The preacher preaches right along just as he has it written down and cannot observe whether he is understood or not. . . . If a minister has his eye on the people he is preaching to, he can commonly tell by their looks whether they understand him. And if he sees they do not understand any particular point, let him stop and illustrate it. If they do not understand one illustration, let him give another.   [*Lectures on Revivals of Religion*]

Finney noted elsewhere that the gospel had to be preached *to* men, not *about* them.

Dwight L. Moody shared the same instincts about communication. For him, nothing was gained by mere allegiance to creed, church, or churchmen; the goal of preaching was to expose men and women to God, not merely to talk about him. Recalling his experience with Moody as a college student, Henry Sloane Coffin, then president of Union Theological Seminary, commented on Moody's personal style: "The gospel was never to him something to be discussed . . . but to be tried and passed on." Coffin remembered Moody as a peerless storyteller, but what stood out was his effectiveness in driving home the story's main point: "There was nothing bizarre, nothing spectacular, nothing theatrical, nothing irreverent," Coffin said of Moody telling the story of Daniel. "This was the Word of God, but it was so vivid to him that he made us feel that we were right on the spot."

A FREE MARKET OF IDEAS

Such audience orientation had profound effects on the way American evangelicals came to organize and carry out religious thinking. As the common man rose in power in the early republic, the inevitable consequence was the displacement from power of the uncommon man, the man of ideas. Democratic America would never produce another theologian to rank with Edwards, just as it would never elect statesmen of the caliber of Adams, Jefferson, and Madison. In the main, evangelicals did not simply become anti-intellectual; what they did was destroy the monopoly that classically educated and university trained clergymen had enjoyed. They threw theology open to any serious student of Scripture, and they considered the "common sense" intuition of people at large more reliable, even in the realm of theology, than the musing of an educated few.

This shift involved new faith in public opinion as an arbiter of truth. Common folk were no longer mistrusted as irresponsible and willful; they now came to be seen as ready to embrace truth if only it was unclouded from the visionary speculations of the academic and uncoupled from the heavy hand of the past. Arguing against the Standing Order in the 1790s, the Massachusetts reformer Benjamin Austin insisted that common people had always possessed an instinct for truth and virtue: "The multitude had received Christ with great acclaim," he suggested, while "the monarchical, aristocratical and priestly authorities cried crucify him." By thus admitting the sovereignty of the audience, evangelicals, knowingly or not, undercut the structure that could support critical theological thinking on the level of a Jonathan Edwards or a John Wesley. Not only did theology proper recede in importance before the task of proclaiming the gospel; the new ground rules for theology, opening it to all, meant that the measure of theology would be its acceptability in the marketplace of ideas. This meant that uncomfortable complexity would be flattened out, that issues would be resolved by a simple choice of alternatives, and that, in many cases, the fine distinctions from which truth alone can emerge were lost in the din of ideological battle. In this process, few evangelicals would admit that further reduction to popularity could at times involve downright falsification.

In such a free market of ideas, Tocqueville feared a "tyranny of the majority," that serious thinking would be hooted down in the marketplace before it could mature. He objected to the American penchant for trusting the people to judge issues of awesome moment—as when Alexander Campbell debated Robert Owen for five days on the truth of Christianity, and then wanted to settle the issue by popular vote. One should not underestimate the high degree of theological literacy and self-education fostered in this culture. But neither can one dismiss the reality that Tocqueville found: expecting that great freedom of thought would generate great ideas, he found instead that Americans easily became "slaves of slogans."

RADICAL PLURALISM

Orienting Christianity so profoundly around the free will of the individual also had implications for religious organizations. "Protestants in general," said H. Richard Niebuhr, "but American Protestants in particular, seem to have developed its institutions and orders not as ends in view or as representative of its purposes but as the necessary instruments and pragmatic devices of a movement that could not come to rest in any structure." Whatever common spirit bound evangelicals together in the nineteenth century, it assumed few institutional manifestations. Recurrent dissentings blasted any semblance of organizational coherence. The array of denominations, mission boards, reform agencies, newspapers and journals, revivalists, and colleges is at best an amorphous collectivity, an organizational smorgasbord. Power, influence, and authority were radically dispersed, and most came by way of democratic means: popular appeals to the good will of the audience. "No minister can be forced upon his people without their suffrage and voluntary support," said Lyman Beecher. "Each pastor stands upon his own character and deeds, without anything to break the force of his responsibility to his people." In this climate, Tocqueville said, where he expected to find priests, he found politicians.

The democratic winnowing of the church produced not just pluralism but also striking diversity. The flexibility and innovation involved in American religious organizations meant that, within certain broad limits, an American could find an amenable group no matter what his or her preference in belief, practice, or institutional structure. Churches ranged from the most egalitarian to the most autocratic and included all degrees of organizational complexity. One could be a Presbyterian who favored or opposed the freedom of the will, a Methodist who promoted or denounced democracy in the church, a Baptist who advocated or condemned foreign missions, and a member of virtually any denomination who upheld or opposed slavery. One could revel in Christian history with John W. Nevin or wipe the slate clean with Alexander Campbell. The range of options within evangelical communions seemed virtually unlimited: one could choose to worship on Saturday, practice footwashing, ordain women, advocate pacifism, prohibit alcohol, or practice health reform. Or, like Abraham Lincoln, one could simply choose a biblical form of Christianity without the slightest ecclesiastical encumbrance.

The English sociologist David Martin, writing about secularization, comments that in America dissent has become universalized. While this theme has certainly not escaped church historians in the twentieth century, the implications Martin draws are strikingly different. Historians generally treat the atomistic structure of American Christianity as a maladaptive characteristic, a weakness, a sign of deficiency. The general impression given is that Christianity has somehow remained vital *despite* its fragmentation. Turning such thinking

One could choose to worship on Saturday, practice footwashing, ordain women, advocate pacifism, prohibit alcohol, or practice health reform.

on its head, Martin argues in *A General Theory of Secularization* that the vitality of American Christianity can be correlated precisely with the degree of pluralism and dissent present here.

> Disassociate religion from social authority and high culture, let religion adapt to every status group through every variety of pulsating sectarianism. The result is that nobody feels ill at ease with his religion, that faith is distributed along the political spectrum, that church is never *the* axis of dispute. [p. 36]

Luther P. Gerlach and Virginia H. Hine (*People, Power, Change: Movements of Social Transformation,* 1970) have made a similar argument, namely, that the success of a religious movement can actually spring from organizational fission and lack of cohesion. They argue for the inherent dynamism of certain reticulate religious movements—those that are weblike, with parts tied together not through a central point but through intersecting sets of personal relations and intergroup linkages. Splitting, combining, and proliferating can be seen as clear signs of health. Such movements are superb at recruiting new members and offering the in-

Whatever their foibles, evangelicals have not lost the common touch.

dividual a keen sense of personal access to knowledge, truth, and power.

From these vantage points, the democratic structure of evangelicalism—audience centered, intellectually open to all, organizationally fragmented—takes on new import. One can more easily understand how popular culture in nineteenth-century America became thoroughly Christianized while the English laboring classes were becoming alienated from the church. Tocqueville understood well the difference between Christianity here, firmly linked to democracy, and that of Europe, which after the French Revolution dismissed any taint of liberty, equality, and fraternity. "By allying itself with political power," he said of Europe,

> religion increases its strength over some but forfeits the hope of reigning over all. . . . Unbelievers in Europe attack Christians more as political than as religious enemies; they hate the faith as the opinion of a party much more than as a mistaken belief, and they reject the clergy less because they are representatives of God than because they are the friends of authority.

Twentieth-century evangelicals have not relinquished their grip on democracy, and they have certainly not withered away, despite being largely alienated from the culture's intellectual centers, losing control of the mainline denominations, and experiencing recurrent fragmentation. Whatever their foibles, evangelicals have not lost the common touch. While Harry Emerson Fosdick or Shailer Mathews could rejoice that good theological education was solidly in the liberal camp by the time of the Depression, evangelicals were busy building Bible colleges by the score—250 in the course of this century. The magna charta of the Bible school movement was to make Christian education possible for the widest possible audience. Liberal prophets like Walter Rauschenbusch or neo-orthodox ones like Reinhold Niebuhr could bemoan the plight of urban America, but who was it that actually brought Christianity to Detroit's autoworkers, Gary's steelworkers, and migrant blacks in Chicago? It was black revivalists, black and white Pentecostals, and fundamentalists like Jack Hyles. Mainline Protestants in the twentieth century have been hardly more effective in deepening the layers of their audience than were old lights like

Charles Chauncy and Jonathan Mayhew in the eighteenth century. For all its intellectual sophistication, liberal theology has always been a bit like deism of an earlier day in forgetting that God is a person who communicates, and that he delights to do so with the meek and the lowly.

Evangelicals in this century, on the other hand, have virtually organized their faith around the issue of communicating the gospel. Evangelism and missions were the principal burden of leaders such as A. B. Simpson, A. J. Gordon, R. A. Torrey, Charles E. Fuller, V. Raymond Edman, Harold Ockenga, and Billy Graham; and it is no accident that when evangelicals have gathered for major concerted purpose it is the subject of world evangelization that has preoccupied their attention.

Evangelical thinking over the last fifty years reflects this audience orientation in at least four respects. In the first place, evangelicals have sustained the conviction that religious knowledge is not an arcane science to be mediated by an educated elite. "This is a layman's age," declared C. I. Scofield, whose notes to the King James Bible became a best seller for Oxford University Press. They offered to tens of thousands the hope of understanding the Bible for themselves. A recent advertisement for another study Bible offered the same promise: "study notes that they [everyday people] can understand without having to be theologians." Where but in America could a converted seaman named Hal Lindsey make bold to undertake the same kind of rational study of biblical prophecy that had taxed the mind of a Jonathan Edwards or an Isaac Newton? The popularity of *The Late Great Planet Earth*—the best-selling paperback book in America during the 1970s—is testimony to the confidence American evangelicals still have that truth is simple and open to all.

A second tendency is that evangelical scholars are far more likely to speak and write to a popular evangelical audience than to pursue serious scholarship. This pattern is quite evident in the extraordinary contingent of evangelicals who first did graduate work in theology at Harvard in the 1940s and 1950s. The collective experience of Merrill Tenney, Samuel Schultz, Roger Nicole, Kenneth Kantzer, John Gerstner, Harold Kuhn, Paul Jewett, Glenn Barker, and E. J. Carnell demonstrates a much greater concern with instructing the evangelical rank and file than in engaging in serious theological scholarship. The career choices of these men were, no doubt, shaped partly by pastoral concern for the church and partly by pressing responsibilities of leadership. But that is just the point. Evangelicals characteristically subordinate the task of first-order thinking to tasks that seem to affect more tangibly the lives of people at large.

Even when facing the most serious and complex intellectual issues, the instinct of evangelicals is to play them out before a popular audience. Professor William Hutchison has made the fascinating point that *The Fundamentals*, published between 1910 and 1915, were more a warning to the general Christian public than they were a scholarly grappling with the roots of modernism. It seemed far more important to those backing the project to distribute 300,000 of *The Fundamentals* free of charge than to enter into serious doctrinal discussion with liberals. The effect was that the work was virtually ignored in those sectors of the theological community that might have been expected to respond. Similarly, issues such as evolution, inerrancy, and "secular humanism" are much more likely to be treated in ways that rally a large constituency than in ways that admit of the genuine complexities involved and that allow scholars to retreat from ideological battle to carefully weigh and clarify the issues at hand. In this sense, it is the people who are the custodians of orthodoxy, and persons who tamper with these issues, as they are popularly understood, might well receive the same fate as would a greenhorn politician seeking to revamp the entire Social Security system. The public simply will not stand for it.

Because of its democratic coloring, evangelical thinking also manifests a third tendency: it measures

The people are the custodians of orthodoxy.

the importance of an issue by its popular reception. By this logic, any position worth its salt will command a significant following. A best seller by definition becomes a "classic"; to be read is to deserve to be read. A good case in point is the recent revamping of the magazine *Christianity Today* to become less an evangelical voice within the larger theological world and more a popular medium catering to a solidly evangelical audience. Such an appeal to a mass readership may well reflect financial necessity, but that has always been the dilemma of evangelical concerns. Radically dependent on popular support, they easily drift into merely reinforcing rather than improving popular taste. Flushed with tangible success, they allow the issue of quality to become a moot point. Evangelicals take pride that they are the defenders of orthodoxy in the twentieth century. Yet they have now abandoned their one real attempt to represent evangelical thinking in broader arenas of the modern world, among those who do not find evangelical convictions self-evident. It is a telling commentary on evangelical priorities.

The keen sensitivity of evangelicals to public opinion also has a fourth implication: a tendency for the values of the audience to color the substance of thinking. If mainline Protestants in this century have taken their cues from secular and academic culture, evangelicals have remained in tune with popular mores. And if leadership in the former is a function either of academic renown or bureaucratic skill, leadership in the latter has been bestowed on those who best embody the hopes of mass audiences—the Charles E. Fullers, Billy Grahams, Jerry Falwells, and Robert Schullers. Evangelicals spoke the language of peace of mind in the 1950s, developed a theology of "body-life" and community in the wake of the 1960s, and are currently infatuated with a gospel of self-esteem that correlates precisely with the contemporary passion for self-fulfillment.

But it is not only charismatic leaders like Robert Schuller and Jerry Falwell who seem to embody widely shared values. This talent is a much broader phenomenon among evangelicals. More progressive figures, such as Ron Sider, Jim Wallis, John Alexander, and

Donald Dayton, are equally given to proclaiming an evangelical message premised on other self-evident values—in this case a progressive agenda of social action. I have often thought that the Moral Majority and the Sojourners Community, for all their differences, are in many ways birds of a feather: both are equally adept at offering moral solutions to an agenda drawn up by someone else. If this penchant for adaptation gives evangelical thinking a derivative quality, it is a problem not likely to go away: influential voices within the church growth movement on the right and among evangelical anthropologists on the left are both making forceful arguments that, if it is to survive, the church must develop theologies that are even *more* "receptor oriented."

To say that evangelicals have certain widely shared assumptions about communications and certain common approaches to religious thinking is not to suggest that uniformity is the hallmark of the movement. Pluralism among evangelicals over the last fifty years has grown more intense, even as the burgeoning number of organizations manifest more of a cooperative spirit. But in the final analysis, power, authority, and influence are more widely dispersed than ever. There are many denominations that range from Orthodox Presbyterian to the Assembly of God, organizations— generally lay controlled—within mainline denominations, a hundred or so liberal arts colleges, twice that many Bible colleges, scores of seminaries, foreign mission boards, evangelistic organizations, campus ministries, magazines, publishing firms, hundreds of radio and television stations and independent programs (with a combined weekly audience of some 129 million), professional groups for doctors, lawyers, businessmen, the military, Bible study organizations, Bible conferences and camps. The evangelical world also includes powerful, magnetic personalities with driving visions: Pat Robertson, Robert Schuller, Bill Gothard, Jimmy Swaggart, Oral Roberts, Charles Colson, Chuck Swindoll, John Perkins, James Kennedy, James Dobson, and the late Francis Schaeffer. The range of evangelical belief and practice is so broad that few conceivable inter-

For all their success at a popular level, evangelicals have failed notably in sustaining serious intellectual life.

est groups are left unattended. One can join an evangelical church of thousands or a house church; enjoy music that ranges from Andre Crouch to Bob Dylan to Bill Gaither to J. S. Bach; take social ethics from John Howard Yoder or Harold Criswell; learn psychology from Robert Schuller, Jay Adams, or Paul Vitz; learn about the role of woman from Phyllis Schlafley, Elisabeth Elliot, or Virginia Mollenkott; imitate the lifestyle of the Fellowship of Christian Athletes or the Sojourners Community; vote with the politics of Jesse Helms, Mark Hatfield, or Jesse Jackson. Such radical pluralism involves a healthy measure of entrepreneurial activity and is especially adapted to the task of spreading a movement across class and cultural boundaries. This enables evangelicals to meet a broad range of ideological, psychological, and social needs and to draw adherents from the widest possible backgrounds.

Yet for all this success at a popular level, evangelicals have failed notably in sustaining serious intellectual life. They have nourished millions of believers in the simple verities of the gospel, but have abandoned the universities, the arts, and other realms of "high" culture. Even in its more progressive wing, evangelicalism has little intellectual muscle. Feeding the hungry, living simply, and banning the bomb are tasks not inherently more intellectual than winning souls. Evangelicals have only begun to span the yawning chasm between modern modes of thinking and the safe and comfortable world view of evangelical congregations— mores that have changed little in the last century.

Part of the reason is institutional. Evangelicals spend enormous sums of money on higher education. But the diffusion of resources among hundreds of colleges and seminaries means that almost none can begin to afford a research faculty, theological or otherwise.

The problem is compounded by the syndrome of the reinvented wheel. Democratic authority figures such as Bill Bright, Oral Roberts, Jerry Falwell, and Jim Bakker all assume that no education that has gone before is capable of meeting the demands of the hour. Despite the absence of any formal educational credentials, each man presumes to establish a genuinely Christian university. Small wonder that evangelical thinking, which once was razor-sharp and genuinely profound, now seems dull, rusty, even banal.

The net effect is that evangelical experience over the last fifty years has been schizophrenic, knowing extremes of strength and weakness. Within their own walls evangelicals have never seemed stronger; yet outside those walls the juggernaut of secularism rolls on, conquering position after strategic position—outposts abandoned by evangelicals years ago. When aroused by these reverses, evangelicals can only react by rallying the troops already within the walls.

By continuing to exploit what they do best, reaching people at large, and continuing to abdicate what they pursue awkwardly, the life of the mind, evangelicals must sooner or later face the specter of Pyrrhic victory. The vitality of evangelical life does little to reverse the pervasive secularization of American thought—a current that undercuts the foundation while evangelicals are admiring the fine job of decorating being done on the third floor. "The problem is not only to win souls but to save minds," Charles Malik said prophetically to evangelicals at the dedication of the Billy Graham Center. "If you win the whole world and lose the mind of the world, you will soon discover you have not won the world. Indeed it may turn out that you have actually lost the world." ∎

TOWARD A HISTORICAL INTERPRETATION OF
THE ORIGINS OF FUNDAMENTALISM

ERNEST R. SANDEEN, *Assistant Professor of History*
Macalester College

The fate of Fundamentalism in historiography has been worse than its lot in history. The whirlwind of the twenties, after twisting through the denominations, ended by tearing even the name to shreds. Who were the Fundamentalists? Few today will use the name, and there seems to be no unity among those that do.

There have been only two book-length attempts to trace the history of this movement. Stewart G. Cole, who wrote his *History of Fundamentalism* in the middle of the controversy (1931), has remained the standard authority although he was writing too close to the events to put them in proper perspective. The late Norman F. Furniss published in 1954 a monograph limited to one decade of the controversy (1918-1931).[1] Both of these books are not so much discussions of Fundamentalism as investigations of one aspect of denominationalism during the twenties. Both Cole and Furniss were interested in Fundamentalism primarily as a negative force, interested in it because it impinged upon denominational machinery. This approach shaped their understanding of the movement. Fundamentalism was described as a political controversy within denominationalism, which it was, of course; but in their accounts it never appears to have been a religious movement at all.

In addition to its treatment as a political controversy, Fundamentalism has been discussed as a psychological and sociological phenomenon. In fact, this is the context in which discussions of the origin and nature of the movement most commonly appear. Even when we exclude those who treat Fundamentalism as one among many "varieties of religious experience" and concentrate upon those who are specifically interested in the history of this one movement, sociological and psychological explanations still predominate. The factors that explain the Fundamentalists' brash behavior, most historians have argued, can be discovered in the economic and intellectual forces which alarmed and agitated the churches so terribly in the last decades of the nineteenth century and in the psychological states of those whose lot it was to live through those days.[2]

But no matter what tack scholars have taken in approaching Fundamentalism—political, sociological, psychological—they have all

1. Stewart G. Cole, *The History of Fundamentalism* (New York, 1931). Norman F. Furniss, *The Fundamentalist Controversy, 1918-1931* (New Haven, 1954).
2. The most famous advocate of this position has been H. Richard Niebuhr. He has written, ". . . fundamentalism was closely related to the conflict between rural and urban cultures in America. . . .Furthermore, fundamentalism in its aggressive forms was most prevalent in those isolated communities in which the traditions of pioneer society had been most effectively preserved and which were least subject to the influence of modern science and industrial civilization" (H. Richard Niebuhr, "Fundamentalism," *Encyclopaedia of Social Sciences*, V, 527).

tacitly assumed or flatly asserted that theologically and dogmatically Fundamentalism was indistinguishable from nineteenth-century Christianity.[3] The Fundamentalists themselves always proclaimed the same theme. As James M. Gray once wrote, ". . . there is nothing new in Fundamentalism except it may be its name. It is the same old 'offense of the cross'."[4]

Thus, though historical interest in Fundamentalism has not diminished over the years, historians have not found it necessary to study this group within its religious and theological context.[5] The aura of intellectual disrepute surrounding the Scopes trial has discouraged serious consideration of the faith of the Fundamentalists. Most American historians have felt that the Fundamentalists were mistaken and seem to have concluded that they cannot have been serious—that their theology must have been only a cloak to hide their socio-economic or psychological nakedness. This paper will attempt to prove that it was these neglected theological affirmations which gave structure and identity to Fundamentalism and that only through the understanding of this aspect of American intellectual history can we lay the foundation for a historical interpretation of Fundamentalism. Therefore, this paper will not proceed any further with the analysis of Fundamentalist historiography but will instead suggest a thesis for the understanding of Fundamentalism as a religious movement. The thesis of this article is that Fundamentalism was comprised of an alliance between two newly-formulated nineteenth-century theologies, dispensationalism and the Princeton Theology which, though not wholly compatible, managed to maintain a united front against modernism until about 1918.

* * * * * * *

Dispensationalism stems principally from a small British sect, usually called Plymouth Brethren, which sprang up in Ireland and England during the 1820's and from one man in particular, John Nelson Darby. Dissatisfied with what they felt to be the dead hand of tradition and legalism in the Church of England, the Brethren admitted any professing Christian to their informal weekly communion services, refused to acknowledge or create any special caste of clergy, conducted their meetings without order of service in order to allow the Holy Spirit to lead their worship, and in everything at-

3. "The Bible conference . . . represented fifty years of conservatives' effort to maintain their Christian witness in a cultural situation that was slipping from their control" (Cole, *op. cit.*, p. 35). "The principal cause for the rise of the fundamentalist controversy was the incompatibility of the nineteenth-century orthodoxy cherished by many humble Americans with the progress made in science and theology since the Civil War" (Furniss, *op. cit.*, p. 14).
4. James M. Gray, "The Deadline of Doctrine Around the Church," *Moody Monthly* (November, 1922), p. 2.
5. Happily there are a few exceptions which ought not to be passed over in silence: Winthrop S. Hudson, *Religion in America* (New York: Scribner's, 1965), and H. Shelton Smith, Robert T. Handy, and Lefferts A. Loetscher, *American Christianity* (New York: Scribner's, 1963), Vol. II.

tempted to recreate the New Testament pattern of church government and worship.[6]

Dispensationalism refers primarily to the division of history into periods of time, dispensations, seven of which are usually named. The *Scofield Reference Bible,* the most influential dispenser of dispensationalism in America, named them Innocence (the Garden of Eden), Conscience (Adam to Noah), Human Government (Noah to Abraham), Promise (Abraham to Moses), Law (Moses to Christ), Grace (Christ through the present to the judgment of the world), and the Kingdom or Millennium.[7] Proponents argued that God judged man not on an absolute and unchanging standard but according to ground rules especially devised for each dispensation. For example, under the dispensation of Grace, men are required to repent and turn in faith to Christ, while under Moses they were commanded to obey the law.[8]

For the dispensationalist, the earthly people of Israel and the spiritual community of the Church must be sharply distinguished. The one is entered by natural birth, the other by conversion—the "new birth." Both have promises and prophecies given to them which must be distinguished and separated. In the millennium, the Church will reign as the "bride of Christ," while Israel will be restored to its ancestral land and inherit the earthly kingdom forecast by the prophets. This particular emphasis within dispensationalism accounts for the enthusiastic Zionism manifested by many Fundamentalists.[9]

The dispensationalist accepted an intensely pessimistic view of the world's future combined with a hope in God's imminent and direct intervention in his own life. God has established covenants which have always been broken by virtually all those involved in them. God has waited, restraining judgment, but eventually punished the disobedient while saving out of the destruction a little band, a remnant of just men such as Noah, Joshua, or Ezra. This pattern of past events was projected into the future through the interpretation of prophecy. Dispensationalists became prophets themselves, predicting the speedy end of their own era in an act of God's cataclysmic judgment. They looked for a literal, imminent second coming of Christ as the next event before God judged the world and brought in the next dispensation, the millennium, and, therefore, referred to their eschatology as pre-millennialism.[10] In their view the religious lead-

6. W. Blair Neatby, *The History of the Plymouth Brethren* (London, 1901); and W. G. Turner, *John Nelson Darby* (London, 1944).
7. *Scofield Reference Bible,* ed. by C. I. Scofield (New York: Oxford University Press, 1909). For a systematic discussion by one of Scofield's disciples, see Lewis Sperry Chafer, *Systematic Theology* (Dallas, 1948).
8. C. Norman Kraus, *Dispensationalism in America* (Richmond), 1958), p. 67-8.
9. Charles C. Ryrie, *Dispensationalism Today* (Chicago: Moody Press, 1965), p. 44 ff. See also Daniel P. Fuller, "*The Hermeneutics of Dispensationalism*" (unpublished Th.D. dissertation, Northern Baptist Seminary, 1957).
10. Many historians have become quite familiar with references to premillennialism without ever becoming acquainted with the dispensational system of which it frequently

ership has always been the chief center of apostasy (as in the case of Israel and the golden calf) while the righteous remnant has been neglected, overlooked and even despised. In nineteenth-century America as in Europe, the apostates were quickly identified as liberal theologians.

Thus a doctrine of the Church emerged from a philosophy of history: The church was made up of God's elect who were always only a handful, seldom if ever the possessors of power. The true church could not possibly be identified with any of the large denominations, which were riddled with heresy, but could only be formed by individual Christians who could expect to be saved from the impending destruction.[11] It is impossible to overestimate the importance of this ecclesiology for the history of Fundamentalism. Most protest groups within American Protestantism turn into denominations themselves. Yet Fundamentalism has not so solidified, and one of the unappreciated factors in this anomalous situation is certainly the retarding influence of dispensationalism. According to their teaching, the true Church can never be an organization but must remain a spiritual fellowship of individual Christians.[12]

As has been intimated, the interpretation of biblical prophecy played a large role in dispensationalism. A glance at the history of other nineteenth century religious groups—Millerites, Irvingites, Mormons, Campbellites, or Shakers—would show them to be concerned with prophetic interpretation as well. Millennial expectations are woven into the fabric of early nineteenth century life in both Europe and America.[13] One factor which differentiates the dispensationalists is their concern for biblical literalism. To speak of a concern for biblical literalism may seem redundant in the context of the evangelical tradition in which dependence upon scripture alone had become a shibboleth. But care must be taken to differentiate between the common evangelical belief in biblical inspiration, the effect of which was to distinguish the Bible from other books, and the prin-

formed a part. Not all premillennialists were dispensationalists, but every dispensationalist was a premillennialist. Some of the best-known works of dispensationalists have been tracts on the premillennial return of Christ. Three of the most influential were William E. Blackstone, *Jesus Is Coming* (2d. ed., New York, 1886), James H. Brookes, *Maranatha* (5th ed., New York, 1878), and Adoniram J. Gordon, *Ecce Venit* (New York, 1889).

11. The ecclesiology of dispensationalism is so individualistic that each individual becomes his own church; his own sanctification is the only holiness the church can know. Through this emphasis holiness teachings became linked to the Fundamentalist movement. The English Keswick movement, which entered the U. S. through Moody's Northfield Conferences, made a great impact upon dispensationalism and Fundamentalism generally.

12. The very character of dispensationalism has thus made the identification of its adherents difficult. It is obvious, with this type of group, that one must work very carefully in identifying individuals as dispensationalists. There is a great temptation to label men on less than adequate evidence. But despite the difficulties, dispensationalists *are* identifiable. Other than outright avowals of dispensationalism, I have loooked for remarks in correspondence and published materials in which the subject expresses his debt to known dispensationalist authors, gives evidence of dispensational theology or falls into the characteristic vocabulary of the dispensationalist.

13. David E. Smith, "Millenarian Scholarship in America," *American Quarterly*, XVII (Fall, 1965), 535 ff.

ciples of hermeneutics which guide the interpretation of the Bible itself. Literalism, in the early nineteenth century, usually refers quite specifically to the interpretation of prophecy and contrasts with the figurative or symbolic manner of interpretation. This literalistic approach puzzled many scholars although they themselves might not have any doubts concerning the inspiration of the scriptures. Writing in the *Princeton Review,* one American noted:

> Millenarianism has grown out of a new "school of Scripture interpretation" and its laws of interpretation are so different from the old, that the Bible may almost be said to wear a new visage and speak with a new tongue—a tongue not very intelligible, in many of its utterances, to the uninitiated. The central law by which millenarians profess always to be guided, is that of giving the literal sense.[14]

It is not difficult to see how some of the characteristic doctrines of dispensationalism arose from this hermeneutic. The second coming of Christ, the restoration of the Jews to the land of Israel, the Great Tribulation, and the 1000 years of peace and justice—dispensationalists believed these prophecies would be fulfilled quite as literally as Christ had fulfilled prophecy during his first advent. When the verbal inspiration of the Bible became a matter of theological dispute later in the century, the dispensationalists were able to win many converts to their cause by arguing that only dispensationalism really took the Bible seriously. Dispensational theology was based upon hermeneutical principles which required, in fact presupposed, a frozen biblical text in which every word was supported by the same weight of divine authority.[15]

Dispensationalism was being taught in the United States and Canada as early as the 1840's, and not the least important apostle of the new movement was John Nelson Darby himself, who travelled to North America on seven occasions from 1862 through 1877, frequently traveling around the continent for as long as a year. During this sixteen-year period (1862-1877), Darby resided in North America at least forty percent of the time. The great bulk of his time was spent working in large cities, mostly on the eastern seaboard. During his last trip to the U.S., Darby ministered for at least 15 months in Boston and New York City.[16] Darby did not preach to the heathen of America, but primarily to the more committed Christians, particularly to the clergy. He accepted invitations to speak in the pulpit, but seemed to prefer meeting during week-day mornings for informal discussions and Bible readings.[17] Of course,

14. "Modern Millenarianism," *The Princeton Review,* XXV (January, 1853), 68.
15. For an early reference to the connection between inspiration and literalism, see "Inspired Literality of Scripture," *Quarterly Journal of Prophecy,* II (1850), 297-307; and for a contemporary reference to the same point, see Ryrie, *op. cit.,* p. 86 ff.
16. Darby, *Letters* (London: Stow Hill Tract Depot), *passim.*
17. One historian of the Brethren stated—without citing evidence—that Darby preached in the pulpit of James H. Brookes (H. A. Ironside, *A Historical Sketch of the Brethren Movement* [Grand Rapids, 1952], p. 196). Robert Cameron wrote that he first met Darby when he was conducting Bible readings in a "humble kitchen" in New York city (*Watchword and Truth,* XXIV [December, 1902], 327).

Darby's was not the only voice preaching dispensationalism. Other advocates of these views visited North America, dispensational publications found their way to the U. S., and a few were published in the U. S. as well.[18]

Many converts to dispensationalism were won during these years, but few of these would take the step of leaving their denominations. According to Darby, this was the main aim of his teaching. He once wrote, ". . . our real work . . . is to get Christians clear practically of a great corrupt baptized body. . . ."[19] But most converts to dispensational theology refused to abandon their denominations and pastoral posts. Darby complained in one of his letters,

> . . . There is a great effort to keep souls in the various systems while taking advantage of the light which brethren have and preaching their doctrines. They do not even conceal it. One. of the most active who has visited Europe told ministers that they could not keep up with the brethren unless they read their books, but he was doing everything he could to prevent souls leaving their various systems called churches.[20]

Thus the instrument of propagation for the dispensational system in the U. S. became the clergy and religious periodicals of American denominations and voluntary societies, who, without announcing their conversion to anything new or different, began to influence the evangelical churches.

To what kind of Christians did dispensationalism appeal?—particularly the Calvinists.[21] Most of the converts seem to be Presbyterians or Calvinistic Baptists. Very few Methodists were ever caught up in dispensationalism, nor were many U. S. Episcopalians, although many British and Canadian Anglicans became converts.[22] This alignment is significant for the later composition of Fundamentalism, for the Presbyterian and Baptist denominations were the two most racked by the Fundamentalist controversy in the 1920's.

[After dispensationalism had become an American movement, the institution most influential in its spread was the summer Bible con-

---

18. The works of other Plymouth Brethren, such as William Kelly and C. H. Mackintosh, were well known among U. S. millenarians. Probably even more influential were the works of Samuel Tragelles, B. W. Newton and George Müller, all of whom had at one time been associated with Darby but had broken with him over, among other issues, prophetic interpretation. Though these three rejected Darby's emphasis upon the imminent return of Christ and insisted that events predicted in prophecy, such as the return of the Jews to Israel, must precede the second advent, they nevertheless retained a dispensational stance. That the rejection of the "any-moment" return of Christ is not a sufficient grounds for discriminating between dispensationalist and non-dispensationalist premillennialists has not been well recognized. Kraus falls into this mistake (cf. *Dispensationalism in America*, p. 99 ff.).
19. Darby, *Letters*, II, 228.
20. Darby, *Letters*, II, 304.
21. Darby once wrote, ". . . one had to insist on the first principles of grace. No one will have it as a rule in the American churches. Old school Presbyterians, or some of them, have the most of it" (Darby, *Letters*, II, 193). Cf. also Kraus, *op. cit.*, pp. 57 ff.
22. Two qualifications ought to be noted. German-speaking Methodists seem to have been attracted to dispensationalism in undue proportion to their numbers within Methodism, and the nineteenth century secession from the Protestant Episcopal Church. **the Reformed Episcopal Church,** also seems to have been especially susceptible to dispensationalist penetration.

ference and the most significant of these was the Niagara Conference. From 1868 until 1900, a relatively small but stable group of pastors and laymen met for one or two weeks at a summer resort (during 1883-97 at Niagara Falls) for concentrated Bible study.[23] The men who led these conferences during the 1870's deserve to be known as the founding fathers of Fundamentalism: James H. Brookes, a Presbyterian and alumnus of Princeton Seminary, for many years pastor of the Walnut Street (now Memorial) Presbyterian Church in St. Louis and editor of his own periodical, *The Truth*,[24] William J. Erdman, at various times pastor of Presbyterian and Congregational churches as well as Moody's Chicago Avenue Church, one of the founders of the Moody Bible Institute, an editor of the *Scofield Bible* and father of Charles R. Erdman, Professor of Practical Theology at Princeton Seminary;[25] Adoniram Judson Gordon, a Baptist, pastor for most of his life of the Clarendon Street Church, Boston, founder of the Boston Missionary Training School (now Gordon College and Seminary), editor of the periodical *Watchword*, a close associate of Moody in the management of the Northfield Conferences;[26] William G. Moorehead, a Presbyterian, Professor of New Testament and President of Xenia Seminary, an editor of the *Scofield Bible*.[27]

The series of prophetic and premillennial conferences which began in 1878 were the direct outgrowth and offspring of this Niagara Group. But not only dispensationalists collaborated in the calling and direction of the First International Prophetic Conference in 1878. A group of conservative Calvinists closely related to Princeton Theological Seminary were drawn into this conference movement through their concern with the premillennial return of Christ and other prophetic themes. The 1878 Premillennial Conference marks the beginning of a long period of dispensationalist cooperation with Princeton-

23. Sources for this conference are extremely varied. Addresses from the conferences were published occasionally (James H. Brookes, *Bible Reading on the Second Coming of Christ* [Springfield, Illinois, 1877]; *Lakeside Studies, Proceedings of the 1892 Niagara Conference* [Toronto, N. D.] and *The Second Coming of Our Lord, Papers Read at a Conference Held at Niagara, July 14-17, 1885* [Toronto, N. D.]). An account of the origin of the conference can be found in George C. Needham, *The Spiritual Life* (Philadelphia, 1895), pp. 18 ff. James H. Brookes edited a periodical which made cryptic references to the conference regularly from 1876 on, and it is in this source that the Niagara creed was first published (*The Truth*, IV [1878], 452-8). The best place to catch a glimpse of the workings of the conference is the July, 1897, number of *The Watchword*, where narrative statements concerning the progress of the conference are combined with virtually a complete list of the sessions and sermons.
24. David R. Williams, *James H. Brookes: A Memoir* (St. Louis, 1897) and Kraus, *op. cit.*, p. 36.
25. *Alumni Catalogue of Union Theological Seminary* (New York, 1926), p. 101. He participated in virtually every Bible conference of the nineteenth century, including the Niagara, the Prophetic Conferences of 1878, 1886 and 1895, and the Northfield Conferences.
26. His obituary appears in *Northfield Echoes*, II, 8 ff. Adoniram J. Gordon, *How Christ Came To Church: A Spiritual Autobiography* (Philadelphia, 1895) makes clear Gordon's conversion to dispensationalism.
27. Moorehead was a close friend of both Erdman and Brookes, and acted as one of the corresponding editors of *The Truth*. He participated in the 1878, 1886 and 1895 Prophetic Conferences. See also Kraus, *op. cit.*, p. 101.

oriented Calvinists. The unstable and incomplete synthesis which is now known as Fundamentalism at this point first becomes visible to the historian.

* * * * * *

[The Princeton Theology was born with the founding of Princeton Seminary in 1812 and endured as a living force for about 100 years.]Inspired by its first professor, Archibald Alexander, and given its most complete formulation by Charles Hodge in his *Systematic Theology,*[the Princeton Theology was defended and modified throughout the nineteenth and early twentieth centuries by competent scholars such as A. A. Hodge, Benjamin B. Warfield and J. G. Machen.[28] [Princeton Seminary was very inbred, but its outreach was extensive, passing far beyond the bounds of Presbyterians into Episcopal, Congregational, Baptist and other denominations.] The Princeton faculty never admitted that they were teaching a unique theology, but staunchly insisted that they only intended to defend the system of John Calvin. In this belief they were deceived, both the methodology and the conclusions of their theology differing clearly from the work of Calvin himself and the standard of the Westminster Confession.[29]

The methodology of the Princeton Theology laid down by Archibald Alexander and Charles Hodge remains the most characteristic aspect of the school's teaching. [Insisting that theology must be pursued scientifically, the Princeton professors, completely ignoring the criticism of Hume and Kant, constructed a rationalistic method which was compared by Charles Hodge himself to Newtonian physics.

> As natural science was a chaos until the principle of induction was admitted and faithfully carried out, so theology is a jumble of human speculations, not worth a straw, when men refuse to apply the same principle to the study of the Word of God. . . . The Bible gives us not only the facts concerning God, and Christ, ourselves, and our relations to our Maker and Redeemer, but also records the legitimate effects of those truths on the minds of believers. So that we cannot appeal to our own feelings or inward experience, as a ground or guide, unless we can show that it agrees with the experience of holy men as recorded in the Scriptures.[30]

[Princeton thus took the position of the scientist who observes, arranges, and systematizes but does not participate in his experiment.] Furthermore, the world of epistemology for Princeton seemed to be divided between reason and mysticism, fact and inner light. Though they criticized their early deist rivals for mistreating right reason, they reserved their hardest words for mystics, hewing a rationalist

---

28. Charles Hodge, *Systematic Theology* (New York, 1874). The best general treatment of this subject is Lefferts A. Loetscher, *The Broadening Church* (Philadelphia, 1957). See also William D. Livingstone, ''The Princeton Apologetic as Exemplified by the Work of Benjamin B. Warfield and J. Gresham Machen: A Study in American Theology, 1880-1930'' (Unpublished Ph.D. dissertation, Yale University, 1948).
29. I have analyzed the accuracy of their assertion in my article, ''The Princeton Theology,'' *Church History,* XXXI (September, 1962).
30. Hodge, *Systematic Theology,* I, 14 ff.

line in their own theology. Their doctrine of inspiration, as it developed during the century, never wavered from this fundamental tenet—that if the Bible was to be proven to be God's inspired word, the demonstration must be made on the basis of reason through the use of external marks of authenticity—not inner convictions.[31]

Building upon this methodology, Charles Hodge, A. A. Hodge, and B. B. Warfield constructed what they considered a shock-proof doctrine of Biblical authority. Their fundamental assumption seems to have been that God would not reveal his truths through a fallible book. They tried to prove that God had so inspired the Biblical authors that their every word as recorded on the original autographs was inerrant—a term more specifically rationalistic than the word infallible. The frequency with which these aspects of the doctrine of inspiration occur in the Fundamentalist controversy seems largely due to the influence of the Princeton Theology. That the Bible was 1) verbally inspired, 2) inerrant in its every reference, statistic, and quotation, 3) when first written down on the original autographs— these phrases have become the shibboleth of the Fundamentalist doctrine of the Scriptures. This doctrine did not exist either in Europe or America prior to its formulation in the last half of the nineteenth century. It has become an essential ingredient in the theology of Fundamentalism.[32]

That dispensationalists and advocates of the Princeton Theology should find it possible to cooperate and accept each other as fellow Christians (when they rejected so many others) should not seem strange. They agreed with one another in general mood and in the elaboration of their central theme of Biblical authority. Both groups insisted upon an inerrant scripture, and, whether by accident or design, began at about the same time to defend their views by recourse to the "original autographs."[33] Both groups thought in pre-Kantian, pre-Schliermacherian, rationalistic terms. Over against the new theologies of immanence and social gospel, both stressed God's transcendence and supra-historical power and expressed themselves in very pessimistic terms when discussing social problems. The two movements were by no means completely compatible, but the common Modernist foe kept them at peace with one another throughout the nineteenth and early twentieth centuries. Attacks upon the dispensa-

---

31. J. Gresham Machen, illustrating the state to which this kind of rationalism was finally carried, once wrote, "'Christian doctrine, I hold, is not merely connected with the Gospel, but it is identical with the Gospel'" (Ned B. Stonehouse, *J. Gresham Machen* [Grand Rapids, 1957], p. 376).

32. I am not ignoring the Lutheran and Reformed dogmatic tradition of the sixteenth and seventeenth centuries. I have shown in my article in *Church History*, XXXI, that Princeton began as the offspring of. that tradition and developed from that point, in the course of the nineteenth century creating something unique. As will be illustrated when discussing *The Fundamentals* later in this paper, the Princeton doctrine of inspiration has become the common property of dispensationalists and Calvinists alike.

33. The first reference to the original autographs in the Princeton Theology occurs in 1879 (A. A. Hodge, *Outlines of Theology* [New York, 1879], pp. 66 and 75).

tionalists were occasionally heard from such a man as B. B. Warfield, but at the same time the books of the dispensationalists were
being regularly reviewed and recommended in the *Presbyterian and
Reformed Review.*[34]

*   *   *   *   *   *

Bible conferences seem to have provided the Niagara Group with
a good opportunity to widen its influence. This may have been their
chief purpose in calling the first International Prophetic Conference
in the Holy Trinity Episcopal Church in New York City, October
30, 1878.[35] The religious press treated the conference in sceptical
terms, labeling the participants Millerites and Adventists.[36] One
rather complacent Baptist editor, noting that pessimism about the future was a distinguishing mark of the pre-millennialists, prophesied
that only Anglicans and Roman Catholics were likely to be attracted
to it.[37] But the conference itself proved impressive. Whether by design or default, very little dispensationalism, as such, was taught.
Instead a considerable body of respected American and European
support was rallied behind pre-millennialist beliefs.[38] Although not
a dispensationalist, the able Samuel H. Kellogg, a Princeton Seminary graduate and at that time Professor of Theology at Western
Seminary, delivered a paper in support of premillennialism at the
conference and followed up his address with a long article in a
respected American theological journal, in which he defended premillennialism as an historic doctrine of the Church.[39] In the premillennial advent, the dispensationalists had apparently hit upon a
theme suitable for building an alliance with certain other Biblically-
oriented conservatives, particularly those following the path toward
Biblical inerrancy laid out by Princeton. No formal alliances were
ever drawn up. No official conferences were held in which the two

34. See Warfield's review of R. A. Torrey, *What the Bible Teaches* (Chicago, 1898) in
the *Presbyterian and Reformed Review*. (1899), 562, and his review of Nathaniel West,
Studies in Eschatology, in *Ibid.*, I, 513-4, Many of James H. Brookes' books were
reviewed in the *Presbyterian and Reformed Review*. *Chaff and Wheat*, a defense
of verbal inspiration, was reviewed without either much praise or blame (*Ibid.*, III.
369). *God Spake All These Words*, another defense of inspiration, was reviewed
quite favorably (*Ibid.*, VI, 573). *Mystery of Suffering* is described as a good book
badly printed (*Ibid.*, I, 705). *The Christ* received a very favorable review (*Ibid.*,
V, 554).
35. The proceedings of the conference were published (Nathaniel West [ed.], *Premillennial Essays* [Chicago, 1879]), The *New York Tribune* gave the conference good coverage on October 25, 28, 30, 31, and November 1, 1878.
36. "We confess that we look with some anxiety upon the spread of the view represented
at the so-called Prophetic Conference held in New York last week" (*Watchman*,
[November 7, 1878], p. 356). See also the *Christian Advocate* (October 31, 1878)
and the *Standard* (November 7, 1878).
37. *Watchman* (November 14, 1878), p. 364.
38. There was considerable name-dropping during the conference sessions, and messages of
greeting were read from foreign scholars (West, *Premillennial Essays, passim*). Samuel
H. Kellogg, in an article which appeared immediately after the conference, also cited
a great many European scholars in support of premillennialism, including Frederic
Godet, Franz Delitzch, Thomas R. Birks, Karl A. Auberlen and Johannes Van Oosterzee
(Samuel H. Kellogg, "Premillennialism, Its Relation to Doctrine and Practice,"
Bibliotheca Sacra, XLV [1888], 234-74).
39. *Ibid.*

groups publically declared their intention to cooperate. But during the last two decades of the nineteenth century, there were frequent occasions on which representatives of these two groups were found speaking on the same platforms and publishing their articles in the same books, all with an end to defeating the Modernist heresies.

The Niagara Group continued to hold its small but influential summer conferences. Four more International Prophetic Conferences were held before the outbreak of World War I—in 1886, 1895, 1901 and 1914.[40] A conference devoted exclusively to the problem of Biblical inspiration was held in Philadelphia in 1887 led by Arthur T. Pierson who had by that time joined the Niagara Group.[41] During 1890 two conferences were held, one devoted to the study of the Holy Spirit in Baltimore, and one especially for Baptists in Brooklyn— both led by dispensationalists.[42] In 1893 a long-time leader of the Niagara Group, L. W. Munhall, arranged a conference in Asbury Park, New Jersey, which, along with the usual large contingent of dispensationalist speakers, included two Princeton Seminary Professors, W. Henry Green, who delivered two papers on the problems of the Pentateuch, and Talbot W. Chambers, who spoke on "The Book of Psalms."[43]

[Another important conference of this period was D. L. Moody's Northfield Conference—important because of the size and continued duration of the series (from 1880 to 1902), and because of the importance that Moody's presence lent to the proceedings. Few men in American nineteenth-century Protestantism could equal Moody's influence. Comments about the broad, perhaps too inclusive nature of Moody's spirit have frequently been made. No historian of Moody's amazing career has noted, however, that his Northfield Conferences were virtually dominated by dispensationalists, particularly from 1880 through 1887 and again from 1894 to 1902.[44] Moody himself never appears to have become a convinced dispensationalist. But, whatever his own beliefs, he gave up control of the conference, when he himself was absent, to dispensationalist leaders such as A. J. Gordon or A. T. Pierson and invited so many dispensationalist speakers that they were referred to as "the usual war horses" in accounts of the

40. George C. Needham (ed.), *Prophetic Studies of the International Prophetic Conference* (Chicago, 1886). *Addresses on the Second Coming of the Lord Delivered at the Prophetic Conference, Allegheny, Pa., December 3-6, 1895* (Pittsburgh, N. D.), *Addresses of the International Prophetic Conference Held December 10-15, 1901, in the Clarendon Street Baptist Church, Boston* (Boston, N. D.). *Coming and Kingdom of Christ* (Chicago, 1914).

41. A. T. Pierson (ed.) *The Inspired Word* (New York, 1888).

42. For the Baltimore conference see A. C. Dixon (ed.), *The Person and Ministry of the Holy Spirit* (London, 1891). For the Brooklyn Baptist conference see George C. Needham (ed.), *Primitive Paths in Prophecy* (Chicago, 1891).

43. L. W. Munhall (ed.), *Anti-Higher Criticism* (New York, 1894).

44. A. T. Pierson, "The Story of the Northfield Conferences," *Northfield Echoes*, I (June, 1894), 1-13.

proceedings.[45] "Dispensational Truth" was the announced theme of the conference in 1886, the same year that the Student Volunteer Movement was initiated in Northfield. In that famous conference, in which the first 100 volunteers of the S.V.M. were recruited, the leadership was dispensationalist to a man.[46]

[By 1900 Fundamentalism, though still unchristened, was already a significant force in American life.] How influential is difficult to estimate, for, as has been indicated, the movement had not yet become divisive. The lines of battle were becoming clear by 1900, but (as was tragically true in Europe at just this time) no one could predict that a great war was imminent nor that this war would involve almost everyone in American Protestantism. The great majority of the pastors and laymen had not yet been forced—as they later would be—to choose sides between the Fundamentalists and the Modernists, and many Christians felt that they could live peaceably with both camps. It is clear that the Fundamentalists, though alarmed and dismayed by the teachings of the Modernists, were not ill-informed nor ignorant. Nor were they behaving like obscurantists or retreating from the world. Their movement at this time possessed great vigor, particularly in evangelism and world missions. The leadership was concentrated in urban centers, particularly in the Philadelphia-New York-Boston area with lesser centers in Chicago, St. Louis and Los Angeles. The South was almost unrepresented. Fundamentalist leaders were occasionally found in the pastorate, but more often held positions which allowed them wider influence—the editorship of a journal (Charles G. Trumbull or Arno C. Gaebelein), the deanship of a Bible school (James M. Gray or Reuben A. Torrey), a chair in a seminary (Wm. G. Moorehead or Melvin G. Kyle) or the calling of an evangelist (J. Wilbur Chapman or L. W. Munhall).

* * * * * *

[In the early twentieth century, the most commonly cited source of Fundamentalist teaching was the series of volumes entitled *The Fundamentals.*] This series of twelve pamphlets was published and distributed free, in numbers ranging from 175,000 to 300,000 copies by two brothers who preferred to be known only as "Two Christian Laymen."[47] That these two were Lyman and Milton Stewart, founders

45. *Northfield Echoes*, I, 6. I do not know exactly who are meant by "the usual war horses," but, from a check of the names most commonly appearing in the first three conferences, I would deduce the most likely men to be A. J. Gordon, A. T. Pierson, J. H. Brookes, and G. C. Needham.

46. *Ibid.* Those leading the S. V. M. conference were Moody, Gordon, Brookes, Nathaniel West, William G. Moorehead, W. W. Clark, A T. Pierson, and D. W. Whittle. Although I cannot do any more than refer to the subject, it is apparent that a thorough study of the origins of Fundamentalism ought to investigate the degree to which dispensationalism influenced foreign missions and the Bible institute movement. That dispensationalists were responsible for the foundation of most of the early Bible institutes, especially the Moody Bible Institute and the Bible Institute of Los Angeles, is already clear.

47. *The Fundamentals: A Testimony to the Truth* (Chicago, [1910-1915]).

and chief stockholders of the Union Oil Company of Los Angeles, has been common knowledge for many years. The Lyman Stewart correspondence reveals, however, that Lyman Stewart was the real sponsor of the series and Milton a silent partner,[48] that Lyman Stewart spent a great deal of energy attempting to bolster the Christian faith particularly through the printed word,[49] and, most significantly, that he was both a Presbyterian and a dispensationalist.[50] In the summer of 1909, Stewart met the Rev. Amzi C. Dixon, a dispensationalist Baptist minister at that time pastor of the Moody Memorial Church, and impressed by Dixon's militant defense of Christian truth, enlisted him as chairman of an editorial committee to supervise the publication of The Fundamentals.[51] This committee, selected by Dixon, was comprised of several laymen who were members of Dixon's church and three clergymen—Reuben A. Torrey, a dispensationalist who was Dean of the Bible Institute of Los Angeles (another Lyman Stewart project), Elmore Harris, possibly a dispensationalist, President of the Toronto Bible Institute and an editor of the Scofield Reference Bible, and Louis Meyer, a Jewish convert to Christianity then working among Jews under the auspices of the Presbyterian Board of Home Missions.[52]

What kind of editorial matter did Stewart and his committee produce? The tone of the volumes is quite calm. Though the articles are polemical, they are almost never vituperative. This approach seems to have been intentional. Stewart explained the rejection of an article by saying that its language was not ". . . the chaste and moderate language which causes even the opponent to stop and read."[53] In content the volumes seem broader than the make-up of the editorial committee might indicate. Altogether 64 authors furnished a total of 90 articles to The Fundamentals. In articles defending specific Christian doctrines, the subject of Biblical authority certainly

48. Milton did put up some of the money, but never showed much interest in the project other than that. (Lyman Stewart to J. M. Critchlow, April 14, 1911, Lyman Stewart Papers, Bible Institute of Los Angeles).
49. Lyman Stewart frequently expressed a desire to see his fortune turned into "living gospel truth" by which he apparently meant printed works of some kind. He contributed $1000.00 toward the publication of the Scofield Bible (L. S. to C. I. Scofield, July 21, 1908, Lyman Stewart Papers).
50. Stewart wrote to J. W. Baer, President of Occidental College, to which Stewart was contributing for the support of the Bible department, that he believed that ". . . a man who does not have a grasp of dispensational truth cannot possibly rightly divide 'the word of truth'" (L. S. to J. W. B., October 8, 1908, Lyman Stewart Papers). He was a member of the Immanuel Presbyterian Church in Los Angeles.
51. A. C. Dixon was born in Shelby, North Carolina, in 1854, educated at Wake Forrest College and ordained into the Baptist ministry in 1876. He served pastorates in Baltimore, Brooklyn and Boston before becoming pastor of Moody Memorial Church in 1906. He left the Fundamentals project in 1911 to become the pastor of Spurgeon's Tabernacle in London. His life is described in Helen C. A. Dixon, A. C. Dixon (New York, 1931), and his dispensationalism is evidenced in his papers (A. C. Dixon Collection, V-5, Southern Baptist Historical Commission Archives).
52. For Torrey, see W. G. McLoughlin, Jr., Modern Revivalism (New York, 1959), pp. 366 ff. For Harris, see W. S. Wallace, The Macmillan Dictionary of Canadian Biography (3rd ed., London, 1963). Louis Meyer is not known to me outside the Lyman Stewart Correspondence.
53. Lyman Stewart to A. C. Gaebelein, December 5, 1911, Lyman Stewart Papers.

predominated, 29 separate articles being devoted to it, including five specifically on Biblical inspiration. It is significant that all five were written by dispensationalists, and that the two articles by James M. Gray and L. W. Munhall, which most clearly attempt to structure a theological argument for verbal inspiration, depend upon and quote directly from the works of the Princeton theologians.[54] Dispensationalism as such was never made the subject of a separate article; when it did occur, it appeared only as the natural mode of expression of a dispensationalist author. Nineteen authors responsible for contributing 31 articles can be identified as dispensationalists.[55] Only three members of the Princeton Seminary faculty contributed articles (David James Burrell, Charles R. Erdman, and B. B. Warfield), but a great many Calvinist-oriented clergy were recruited to write for *The Fundamentals*.[56]

There apparently was no overall plan followed in the publication of *The Fundamentals*, although individual volumes do occasionally show a common theme—volume VII, for instance, was devoted almost entirely to biblical problems and volume XII to evangelism and missions. The volumes do tend toward a more popular level during the last half of the series, apparently intentionally, for Lyman Stewart wrote to his brother in 1911 when the series was about half published, ". . . thus far the articles have been more especially adapted to men of the highest culture, . . . and a series of articles adapted to the more ordinary preacher and teacher should follow."[57] Some his-

54. The five articles and their authors are: William G. Moorehead, "The Moral Glory of Jesus Christ as a Proof of Inspiration," Vol. III; James M. Gray, "The Inspiration of the Bible," Vol. III; George S. Bishop, "The Testimony of the Scriptures to Themselves," Vol. VII; L. W. Munhall, "Inspiration," Vol. VII; and A. T. Pierson, "The Testimony of the Organic Unity of the Bible to Inspiration," Vol. VII. I have already noted that Moorehead, Munhall and Pierson were leaders in the Niagara Group. George S. Bishop (1836-1914), a Princeton Seminary graduate, is identified as a dispensationalist by Kraus (*Dispensationalism*, p. 93). James M. Gray (1851-1935) was a Reformed Episcopalian clergyman who began his career working in A. J. Gordon's Bible Training School. He was named Dean of the Moody Bible Institute in 1904 and acted as one of the editors of the *Scofield Reference Bible*. He gives clear expression of dispensationalism in his writings (see, for instance, "God's Plan in This Dispensation," *Light on Prophecy* [New York, 1918], p. 129 ff.).

55. Among the nineteen authors, ten have already been discussed: A. C. Gaebelein, G. S. Bishop, A. C. Dixon, W. J. Erdman, James M. Gray, Wm. G. Moorehead, L. W. Munhall, A. T. Pierson, C. I. Scofield, and R. A. Torrey. Three others were members of the Plymouth Brethren in Britain or the U. S.: Robert Anderson, Philip Mauro and Algernon J. Pollock. Charles G. Trumbull was editor of the *Sunday School Times*, which became a dispensationalist paper under his leadership. Wm. Henry Griffith Thomas (1861-1924) was a Canadian Anglican who was a professor at Wycliffe College in Toronto in 1910. He was one of the founders of the dispensationalist Dallas Theological Seminary and would have served on its faculty if death had not prevented him. Henry W. Frost, a director of the China Inland Mission, was a disciple, according to his own testimony, of H. M. Parsons, one of the founders of the Niagara Conference. In his *Fundamentals* article, he claims relationships to James H. Brookes and other dispensationalists. The last two identifications are less certain, but there is some circumstantial evidence that Melvin G. Kyle, Professor of Biblical Archaeology at Xenia Seminary and one of the editors of the *Sunday School Times*, and George F. Pentecost, a close associate of Moody, were also dispensationalists.

56. Among American authors, fifteen were Presbyterian, eleven Baptist, three Dutch Reformed, three Congregationalist, four Methodist, two Episcopal, two Reformed Episcopal, and one Plymouth Brethren.

57. L. S. to Milton Stewart, March 3, 1911, Lyman Stewart Papers.

torians have speculated that *The Fundamentals* were part of an elaborate plan or the first shot in the subsequent controversy. There is no evidence that this series had any other intent than its title implied —the reaffirmation of fundamental truths.[58] That such a heavily dispensationally-dominated committee should produce such a balanced series would seem to demonstrate that these early Fundamentalists could still find some grounds for cooperation with other Christian leaders.

To this point in the paper I have refrained even from mentioning the "five points" of Fundamentalism. Historians and contemporary Fundamentalists have commonly talked as though there was a kind of Fundamentalist creed of five articles which all defenders of Christianity accepted and defended, and which all Modernists and Liberals attacked and rejected.[59] [The General Assembly of the Presbyterian Church in 1910 did adopt a five-point doctrinal deliverance —naming the inerrancy of the Scripture, virgin birth, substitutionary atonement, physical resurrection, and miracle-working power of Christ as essential Christian doctrines.[60] But this was the only occasion (relevant to early Fundamentalism) on which any denomination or group ever made a five-point statement. Stewart Cole was probably responsible for the confusion, for he carelessly stated that the Niagara Group had also adopted such a five-point declaration, but the only creedal statement ever produced by that group (in 1878) contained fourteen points.[61] What a career that simple mistake has made for itself. Through uncritical acceptance of Cole's mistake, generations of students have been taught to identify the Fundamentalist beast by its five points. That Fundamentalists defended many traditional doctrines is obvious. They did not define themselves in relation to any five particular points, however; and as has been pointed out, their innovations were more significant than their preservations.

[In the climate of the 1920's, the calm, determined spirit of *The Fundamentals* quickly gave way to the clangor and strife that has turned Fundamentalism into a term of reproach.] An analysis of the controversies of the 1920's is impossible in this paper, but it is essential to note the presence of dispensationalism and Princeton-Calvinism within the contending factions.

The Baptists did not split in the 1920's, but their denomination was as badly racked by controversy as any. Two organizations, The

---

58. Lyman Stewart himself said as much in a letter in which he also expressed some dissatisfaction with some articles published in the *Fundamentals* (Lyman Stewart to George S. Fisher, June 30, 1911, Lyman Stewart Papers).
59. See for example, Furniss, *The Fundamentalist Controversy*, pp. 13, 16, 50, 72, 119, 121, 122, 130, and *passim*.
60. Loetscher, *The Broadening Church*, p. 98.
61. Cole, *History of Fundamentalism*, p. 34. For Niagara, see note 23 above. The World Conference on Christian Fundamentals affirmed nine points in a doctrinal statement in 1919 (*Sunday School Times*, June 14, 1919).

National Federation of Fundamentalists and the Baptist Bible Union, led the attack upon Modernism in the Northern Baptist Convention. The Federation was determined in its program, but refused to carry the issue to the point of schism. The Baptist Bible Union, however carried on as if it had no real concern for the continuance of the denomination and did contribute eventually to several small schisms.[62] It has been common practice to ascribe only a difference in temperament to these two uncooperative Fundamentalist groups, whereas it is clear that the Baptist Bible Union was controlled largely by dispensationalists and the Federation by the other more Calvinistic wing of the movement. Many of the leaders of the militant Baptist Bible Union, such as A. C. Dixon, William B. Riley, T. T. Shields, and W. L. Pettingill, were dispensationalists.[63] Whatever the source of the militancy of Union members, it is clear that their dispensational doctrine of the church would have made it very much easier for them to countenance schism.

The other denomination badly shaken by the Fundamentalist controversy was, of course, the Presbyterian. The 1920's do not mark the beginning of the Fundamentalist controversy within Presbyterianism, but the disputes in the General Assembly, seminaries and mission boards of the church became much more acrimonious, leading to the reorganization of Princeton Seminary (and simultaneous establishment of the Fundamentalist Westminster Seminary in Philadelphia), the creation of a Fundamentalist Presbyterian mission board to rival the official denominational board, and, finally in 1936, the withdrawal of about 100 ministers and many congregations to form the Orthodox Presbyterian Church.[64] Breathing a rhetorical sigh of relief, J. Gresham Machen, the leader of the new denomination, wrote, "On Thursday, June 11, 1936, the hopes of many long years were realized. We became members, at last, of a true Presbytrian Church. . . ."[65] But within a matter of months the members of the new Presbyterian denomination were fighting with each other more bitterly than they had with the Modernists. Another schism quickly ensued in which the two groups, which we have traced in the origins of the Fundamentalist movement, appeared as opponents. Machen's group, representing the Princeton element, separated from a group

62. Furniss, op. cit., pp. 103 ff.
63. Both Riley and Pettingill were conveners of the dispensationally dominated 1918 Conference on Prophecy held in Philadelphia and gave several addresses there which reveal their dispensational theology (Light on Prophecy). For a dispensational address by T. T. Shields see The Fundamentalist, July-August, 1923, p. 4. Although I cannot substantiate his assertion, William Carver, Professor of Missions at Southern Baptist Seminary in Louisville, asserted that the other two barons of the Union, John R. Straton and J. Frank Norris, were dispensationalists and worked throughout the South spreading dispensationalism in close collaboration with C. G. Trumbull and the Sunday School Times (William O. Carver, Out of His Treasure [Nashville, 1956], pp. 76 ff.).
64. Loetscher, op. cit., especially chapter XV.
65. Presbyterian Guardian, June 22, 1936.

calling itself the Bible Presbyterian Synod which was heavily influenced by dispensationalists.[66] The Princeton faction, in an editorial in their periodical, apologized for not recognizing the dangers of dispensationalism earlier. "We cannot," they wrote, "offer a very good reason for a failure to raise the issue at an earlier time. Evidently the only reason is that we were absorbed in fighting that great enemy, Modernism."[67]

* * * * * *

This kind of analysis should be pursued more widely in denominational studies and nearer to our own day. The skeleton of the thesis has been made plain, however—Fundamentalism of the late nineteenth and early twentieth centuries was comprised of an alliance between dispensationalists and Princeton-oriented Calvinists, who were not wholly compatible, but managed to maintain a united front against Modernism until about 1918. This is a working hypothesis, subject to criticism and correction. The implications of the thesis have been examined for only a few major denominations. Perhaps other historians will attempt to test the thesis within their own areas of special competence.

Whether or not this thesis can provide a new context for the historiography of Fundamentalism, historians ought to be able to agree that the old explanations cannot be defended. A critique of traditional Fundamentalist historiography can be phrased in four parts.

First, this study has demonstrated that the Fundamentalist considered himself a champion of certain religious truths and worked within the scope of definable beliefs. Each Fundamentalist spokesman ought to be examined in the light of his theological position. The sociologist, or historian equipped with tools of sociological analysis, is welcome to work in this field but let us hope that he will approach his subject with enough sophistication in theology to recognize the factors of religious belief that played a discriminating role in the controversy. Most previous studies of the sociology of Fundamentalism have proved nothing because they could never produce an adequate definition of the subject. Even those historians concerned only with a narrative explanation of the events of the 1920's ought to recognize the inadequacy of a purely denominational-political explanation of the struggle. Power blocks do confront each other within the denominations, but to explain the origin of these power blocks without treating theological questions has reduced much of this agonizing struggle to buffoonery and monkey shines.

66. Carl McIntire, leader of the new schism, clearly aligned himself with dispensationalists in the *Presbyterian Guardian*, November 14, 1936.
67. *Presbyterian Guardian*, March 13, 1937, p. 217.

Second, this study has shown that the Fundamentalist's assertion of his own orthodoxy and conservatism cannot be accepted uncritically. Both dispensationalism and the Princeton Theology were marked by doctrinal innovations and emphases which it is mistaken to confuse with apostolic belief, Reformation theology or nineteenth-century evangelism. It is almost incredible that a dispute over the nature of orthodox Christianity could be discussed by a generation of historians without any of them analyzing the validity of Fundamentalist claims. This is especially significant in view of the fact that the Fundamentalist arguments—before, during, and after the controversy of the 1920's—rested entirely upon their claim to be defending the truths of an historic faith. Some Fundamentalists were only attempting to conserve their traditional faith, that is true. But the assumption that only the Modernists reconstructed their theological position during the intellectual crisis of the late nineteenth century cannot be maintained.

Third, though my own attempt at the formulation of a definition of Fundamentalism will not satisfy everyone, the task of formulating a historical definition is not insuperable. Fundamentalism was a movement which people joined, not an amorphous entity or an abstract category. By discovering what Fundamentalists believed and taught, who they counted among their friends and confidents, and what they said about themselves, a representative, honest portrait of Fundamentalism can be drawn. Our earlier historiographic surrender to semantic confusion has been unnecessary and unfortunate.

Fourth, we ought to stop referring to Fundamentalism as an agrarian protest movement centered in the South. Only by uncritically accepting the setting and conduct of the Scopes trial as the model of all other Fundamentalist activity can such a parody of history be sustained. If one turns to Fundamentalist periodicals and conference platforms, he does not find them dominated by ill-taught stump preachers or demagogues. In the nineteenth century, especially, the proto-Fundamentalists were frequently men in high esteem in their own denominations and communities. Only in the later twentieth century (if then) did Fundamentalism became particularly a phenomenon of the South. Fundamentalism was not a sectional controversy but a national one, and most of its champions came from the same states as their modernist opponents. Fundamentalism originated in the northeastern part of this continent in metropolitan areas and should not be explained as a part of the populist movement, agrarian protest or the Southern mentality.

Fundamentalism as an American Phenomenon, A Comparison with English Evangelicalism

GEORGE MARSDEN

"Fundamentalism" is used in so many ways that a definition is the only place to begin. As I here use the term, "fundamentalism" refers to a twentieth-century movement closely tied to the revivalist tradition of mainstream evangelical Protestantism that militantly opposed modernist theology and the cultural change associated with it. Fundamentalism shares traits with many other movements to which it has been related (such as pietism, evangelicalism, revivalism, conservatism, confessionalism, millenarianism, and the holiness and pentecostal movements), but it has been distinguished most clearly from these by its militancy in opposition to modernism. This militancy has typically been expressed in terms of certain characteristic theological or intellectual emphases: whereas modernism or liberal theology tended to explain life and much of religion in terms of natural developments, fundamentalists stressed the supernatural. Accordingly their most distinctive doctrines (although not all have been held by everyone in the movement)[1] were the divinely guaranteed verbal inerrancy of Scripture, divine creation as opposed to biological evolution, and a dispensational-premillennial scheme that explained historical change in terms of divine control. In America, where fundamentalism originated, adherence to the first of these teachings became a test for the purity of denominations, the second a symbol for efforts to preserve the Christian character of the culture, and the third a basis for fellowship among fundamentalists themselves.

During the 1920s, fundamentalists in America engaged in furious and sensational battles to control the denominations and the wider culture.

*Earlier versions of this paper were read at a history colloquium of the Harvard University Divinity School, to the Reformed Fellowship of New England, and at Calvin College. I am grateful for a number of helpful critical comments on each occasion. I am also very much indebted for fine advice from William Hutchison, George Selement, Harry Stout, Barbara Thompson, Peter Toon, Edwin Van Kley, Grant Wacker, and especially Ian Rennie (see below).

1. Fundamentalism especially in the 1920s was a coalition of rather diverse co-belligerents. For helpful accounts of some varieties within the leadership see the essays of C. Allyn Russell collected in *Voices of American Fundamentalism: Seven Biographical Studies* (Philadelphia, 1976).

Mr. Marsden is professor of history in Calvin College, Grand Rapids, Michigan, and visiting professor of church history in Trinity Evangelical Divinity School, Deerfield, Illinois for 1976-1977.

When these efforts failed they became increasingly separatist, often leaving major denominations and flourishing in independent churches and agencies.[2] They continued however to have an impact on large areas of American Protestantism and most of the pietistic or conservative movements with which they had contact took on some fundamentalist traits.

The phenomenon that I have defined as "fundamentalism" was overwhelmingly American in the sense that almost nowhere else did this type of Protestant response to modernity have such a conspicuous and pervasive role both in the churches and in the national culture.[3] An examination of fundamentalism should reveal some significant traits of American culture and, conversely, the American context will provide a key for understanding fundamentalism.

The crucial variables in the American environment can best be identified by comparing the American development of fundamentalism with its closest counterpart, English evangelicalism. The approach is particularly revealing since from the time of the Puritans down through the awakenings to the end of the nineteenth century British and American evangelicalism had been in many respects parts of a single transatlantic movement. Ernest R. Sandeen has even argued (although too simplistically) that on the basis of one of the many connections— millennarianism—the origins of fundamentalism were essentially British.[4] In any case, British-American ties were taken for granted even as late as the beginnings of the organized fundamentalist crusade; in *The Fundamentals*, published from 1910 to 1915, one-fourth of the authors were British.[5] Yet, strikingly, by the 1920s when the American fun-

---

2. In "From Fundamentalism to Evangelicalism: An Historical Analysis," *The Evangelicals,* ed. David Wells and John Woodbridge, (Nashville, 1975) I have discussed the changes in the character of fundamentalism since the 1920s. Among those close to the movement the meaning of the term, "fundamentalist," has narrowed in recent decades to include almost solely doctrinally-militant premillennialist revivalists. Cf. George W. Dollar, *A History of Fundamentalism in America* (Greenville, S.C., 1973).

3. Ulster appears to be an exception—one that would offer another illustration of the relationship of fundamentalism to relatively unique cultural experiences. Canada has some fundamentalism, although I have the United States primarily in mind in the "American" comparison. In many nations, confessionalists and churchly conservatives survived and in some, such as the Netherlands, they had considerable influence; but these lacked the revivalist ties and some of the intellectual emphases characteristic of fundamentalists. Evangelical or pietist revivalism, sometimes with genuinely fundamentalist traits, could be found throughout the world in the twentieth century, but even if vigorous, as scattered minorities often operating with an aspect of a religious underground.

4. *The Roots of Fundamentalism: British and American Millenarianism 1800-1930* (Chicago, 1970). My criticisms are found in a review article, "Defining Fundamentalism," *Christian Scholar's Review* I:2 (Winter, 1971): 141-151; see Sandeen's reply, 1, 3 (Spring, 1971): 227-233. See also LeRoy Moore, Jr., "Another Look at Fundamentalism: A Response to Ernest R. Sandeen," *Church History* 37 (June, 1968): 195-202.

5. A number of the British authors, however, were no longer living.

damentalists were engaged in intense spiritual warfare, there were few
on the English front willing to sound the battle cry.

As will be seen, a number of English evangelicals during the 1920s
firmly resisted the almost overwhelming trend to accept liberal theology;
yet, despite their similarity to American fundamentalism, most of their
efforts lacked its aggressiveness and militancy and certainly had no
comparable role in the culture and the churches. For Englishmen the
Scopes trial, for instance, was totally foreign to their own experience and
almost inconceivable. "Perhaps no recent event in America stands more
in need of explanation . . . " wrote one British observer in 1925.[6] Even
those who closely followed English church life saw no counterpart to
militant fundamentalism. "Perhaps it was [his] greatest service," observed the *Times* of London in 1929 concerning A. S. Peake, a moderate
British evangelical who had done much to introduce the public to biblical
criticism, " . . . that he helped to save us from a fundamentalist
controversy such as that which has devastated large sections of the church
in America."[7]

There had been, of course, considerable controversy when the new
evolutionary and higher critical views were first publicized in Great
Britain, but it never grew to the proportions of the American reaction.
In fact, one of the striking differences between the patterns of reactions
is that, while in America the controversies intensified from the 1860s to
the 1920s, in England the peak of popular furor had been reached
already by the 1860s. Initial reactions in English churches to *Origin of
Species* (1859), *Essays and Reviews* (1860), and the first volume of Bishop
John Colenso's *The Pentateuch and Book of Joshua Critically Examined*
(1862-1879) were largely negative. Yet the sensational and emotional
aspects of the controversies had already largely passed by the end of the
1869s.[8] After that, biological evolution never became a divisive issue of
nearly the proportions reached in America.[9] The question of the nature

6. S. K. Ratcliffe, "America and Fundamentalism," *Contemporary Review* 128 (September,
   1925); now in *Controversy in the Twenties: Fundamentalism, Modernism, and Evolution,* ed.
   Willard B. Gatewood, Jr., (Nashville, 1969), p. 414. Other British commentators seem
   to have agreed that fundamentalism was peculiarly American; see Gatewood in ibid.,
   pp. 409-412.
7. August 20, 1929, obituary of Dr. Arthur Samuel Peake, quoted in David G. Fountain,
   E. J. Poole-Connor (1872-1962): "Contender for the Faith," (London, 1966), p. 91.
8. This interpretation and that immediately below follow that of Willis B. Glover,
   Evangelical Nonconformists and Higher Criticism in the Nineteenth Century (London, 1954).
   Owen Chadwick, *The Victorian Church: Part II,* 2d ed. (London, 1972) provides a
   similar account of Anglican reactions to *Essays and Reviews* (pp. 75-90) and Colenso
   (pp. 90-97). A recent general account of British reaction to Darwinism in the 1860s is
   M. J. S. Hodge, "England," in *The Comparative Reception of Darwinism,* ed. Thomas F.
   Glick, (Austin, Texas, 1972). See Edward J. Pfeifer's interesting essay, "United States,"
   in ibid.
9. Chadwick, *op. cit.,* p. 23, says that evolution was fully accepted and respectable among
   clergymen by 1896. G. Stephen Spinks, "Victorian Background," in Spinks *et al.,
   Religion in Britain since 1900* (London, 1952) remarks that it was easier for the British to
   come to terms with the'new biology than with Biblical criticism, p. 20.

of Scripture was more difficult to resolve; but in general, once moderate historical-critical ideas were advanced by evangelicals known as reverent defenders of the faith, the new attitudes were accepted with remarkable swiftness. By the 1890s most of the clergy had abandoned traditional assumptions concerning the full historical accuracy of Scripture for some form of higher criticism.[10] Considerable numbers of church members still did not accept the newer ideas;[11] but most were at least familiar with the major issues so that there was little potential for an outbreak of public alarm after that time. In all, this rather peaceful development suggests that nineteenth-century British evangelical religion, like British politics, was closer in style to Edmund Burke than to Oliver Cromwell.

What accounts for the relatively smooth and rapid acceptance in England of the same views that caused so much turmoil in America? Both a strong tradition of theological latitude dating back to the Elizabethan settlement and a policy of toleration since at least the Act of Toleration of 1689 were major factors. These policies, however, were at least officially parts of the American religious heritage, and in fact Americans since the Revolution had been proud of their country's unusual degree of religious liberty and tolerance. The fact that often in American religious life there was not the degree of toleration that the popular mythology proclaimed is in part the phenomenon that needs to be explained.

Given the generally greater tolerance among evangelicals in nineteenth-century England, other factors are still needed to explain why the revolutionary new views concerning higher criticism and evolution did not foment a long and major controversy in England. Clearly the English were prepared in some way for the new ideas, but the initially strong opposition of the 1860s indicates that this preparation was

10. Glover, *op. cit.*, pp. 71-90, 109-110. On Anglican parallel see p. 9 and Chadwick, *op. cit.*, pp. 1-111. Cf. H. D. McDonald, *Theories of Revelation: An Historical Study 1860-1960* (London, 1963), pp. 101-118. By 1900, English champions of higher criticism thought "the battle was won" and that higher criticism had already "penetrated to the country clergymen," p. 116. McDonald, however, shows that at least the former of these statements was an overestimate since a few conservative attacks continued, pp. 118-136, 203-217. Already by the time of *Lux Mundi* (1889) which helped promote higher criticism among Anglicans, all the bishops except J. C. Ryle of Liverpool (a rather tolerant conservative) reportedly accepted the new ideas. See Marcus L. Loune, *John Charles Ryle 1816-1900: A Short Biography* (London, 1953), pp. 47-48, 56-57. Anti-evolution does not seem to have been a major issue even for the most conservative twentieth-century English evangelicals cited below.

11. Chadwick, *op. cit.*, p. 24, says that "for a decade or two after 1896 some members of the Church of England, especially among the evangelicals . . . and most of the simple worshippers among the chapels of the poor, continued to know nothing of evolution or to refuse to accept it on religious grounds. . . ." This estimate would still place the general popular acceptance by World War I. Cf. Glover, *op. cit.*, p. 217 for a similar observation regarding acceptance of higher criticism.

not one of direct familiarity. English theologians seem not to have had, for instance, a great deal more of sympathetic contact with the earlier German higher criticism than did their American counterparts. More basic than any specific preparation seems to have been a general intellectual climate—that is, the concepts of natural and historical development on which both Darwinism and higher criticism were based were closely akin to trends that had been developing in British thought for some time.[12] The whole English constitutional system (in contrast to America, where newness demanded written and rational definition) reflects a sense of gradually developing tradition that appears characteristic of English thought generally. Regarding the acceptance of higher criticism in nineteenth-century England, Willis Glover in his careful study of the Nonconformists correctly makes much of this point. He says:

> But the most essential presuppositions of criticism, such as the unity and continuity of history, were a part of the general climate of opinion shared by traditionalists and critics alike. The historical sense of the century was so strong that the defenders of tradition found it extremely difficult to deny higher criticism in principle. In the last quarter of the century even those who upheld tradition against the critics on every count were often ready and even anxious to make it clear that they did not oppose the critical and historical study of the Bible but merely the conclusions of "rationalistic" critics.[13]

This estimate of Glover applies well, for instance, to the work of James Orr of Scotland, the leading British theological critic of liberalism around the turn of the century. Because of his reputation as a defender of the faith, Orr had close and cordial relations with the American revivalists who organized *The Fundamentals* and was a major contributor to that series. Yet unlike the American leaders of the emerging fundamentalist movement, Orr not only was amenable to limited forms of biological evolution but also accepted historical criticism of Scripture in principle, even while vigorously attacking most of its usual applications. The attempt to defend the faith on the basis of "inerrancy," said Orr, was simply "suicidal."[14]

12. Darwinism quite evidently reflected tendencies developing in the British intellectual climate for some time. The sense of history as "a natural and organic development" was commonplace by the mid-nineteenth century. See, for example, the account in Walter E. Houghton, *The Victorian Frame of Mind: 1830-1870* (New Haven, 1957), pp. 29-31, and *passim*. Romanticism, which had been a major force in England since before 1800, also encouraged emphasis on process rather than on fixed or static truth. (Cf. note 43, below).

13. Glover, *op cit.*, p. 25. Cf. Chadwick, *op. cit.*, p. 59, who attributes the widespread agreement on new views of the Old Testament " . . . not only to German criticism and to English scholarship but to the general growth of historical consciousness. . . ." Cf. p. 462.

14. Orr, *Revelation and Inspiration* (New York, 1910), p. 198; cf. 209-210, 214-215. Orr was critical of the emphasis of the Princeton theologians on inerrancy although he thought

While most British evangelicals in the twentieth century were moving much further than James Orr,[15] and few were doctrinally militant, some conserved traditional views chiefly through vigorous piety. Outstanding in this respect was the Keswick Convention, founded in 1875 in the wake of the Moody revivals. The Keswick summer conferences became the informal meetingplace for British conservative evangelicals and its emphases on Bible study, evangelism, missions, personal piety, and "victory over sin" had wide influence. Like Moody, Keswick teaching took for granted a conservative view of Scripture, yet explicitly avoided any controversy.

This non-controversialist stance gave Keswick and much of the British conservative evangelicalism that it nourished an emphasis rather different from twentieth-century American fundamentalism. While many American fundamentalists adhered to Keswick teaching concerning the "victorious life," its irenic emphases were overshadowed in the early decades of the twentieth century by anti-liberal militancy. By the 1920s Keswick was becoming suspect even to fundamentalist leaders. After a visit in 1928, William B. Riley, president of the World Christian Fundamentals Association, criticized Keswick for "carelessness" in tolerating doctrinal error and noted, no doubt in reference to himself, that "'a controversialist' could never be on its platform."[16]

While in England enthusiasm for controversy was hardly considered evidence of true faith, some British conservative evangelicals more or

---

he had much in common with them regarding Scripture. On evolution see Orr, "Science and the Christian Faith," in *The Fundamentals; A Testimony to the Truth* (Chicago, 1910-1915), 4:91-104. The inclusion of Orr's moderate statement on evolution in *The Fundamentals* indicates that the lines had not yet firmly hardened on this point among the American revivalists.

15. Even some evangelicals who protested against the more liberal trends were rather progressive themselves. Charles H. Vine, ed., *The Old Faith and the New Theology: A Series of Sermons and Essays on Some Truths Held by Evangelical Christians* (New York, 1907), being protests by British Congregationalists against the "new theology," and B. Herklots, *The Future of the Evangelical Party in the Church of England* (London, 1913) both parallel fundamentalist concerns over questions such as miracles yet assume a tolerance toward higher criticism not found among American fundamentalists. See, e.g., Vine, *op. cit.*, pp. 225 and 227; Herklots *op. cit.*, pp. v, 57-68, 107, 113. Cf. general accounts by E. L. Allen, "The Acids of Modernity," in Spinks, *Religion in Britain*, pp. 49-64 and John Kenneth Mozley, *Some Tendencies in British Theology from the Publication of* Lux Mundi *to the Present Day* (London, 1951), p. 24-46.

16. *The Christian Fundamentalist*, 2 (1928): 7, 17. General accounts of Keswick are found in Bruce Shelley, "Sources in Pietistic Fundamentalism," *Fides et Historia* 5 (1973): 68-78 and Steven Barabas, *So Great Salvation: The History and Message of the Keswick Convention* (Westwood, N.J., 1952). At this same time Riley was engaged in an all-out attack on the well-known conservative-evangelical British preacher, G. Campbell Morgan, sometimes also associated with Keswick, and a contributor to *The Fundamentals*. Morgan, concluding a brief and stormy stay at the Bible Institute of Los Angeles, described as "frankly impossible" the attitude of fundamentalists. "They separate themselves, not only from those who accept evolutionary theory, but from those who deny the literal inerrancy of Scripture." Quoted from *The British Weekly* in *The Christian Fundamentalist* 2 (1928): 14.

less in the Keswick tradition responded to the threats of liberalism by maintaining doctrinal purity and a degree of separateness.[17] Prominent among such efforts was the Inter-Varsity Fellowship organized in 1928 among university student groups that had been steering a course separate from the more liberal Student Christian Movement. The constitution of IVF affirmed "the fundamental truths of Christianity," including the infallibility of Scripture, and decreed its continued non-cooperation with liberals. Despite these fundamentalist resemblances IVF placed far more emphasis on the personal piety and evangelism reminiscent of the Moody-Keswick era (when its progenitor the Cambridge Inter-Collegiate Christian Union had originated) than on the doctrinal militancy of the fundamentalist era.[18] Its ties to America were confined largely to the moderate variety of fundamentalism eventually known as "neo-evangelicalism."[19]

A similar development was the split in 1922 of the Church Missionary Society, the missions agency of the evangelical party in the Church of England. A rather distinguished group of conservatives, who made the historical trustworthiness of Scripture a doctrinal test, withdrew in protest over inclusivist tendencies and formed the Bible Churchmen's Missionary Society. This move closely paralleled American controversies concerning missions following World War I, yet it was effected without prolonged dispute;[20] questions of separation and independent action

17. I am very greatly indebted to Ian S. Rennie of Regent College, Vancouver, for pointing me toward much of the information used in the following sections on English evangelicalism. In two very extensive critiques of an earlier version of this essay he argues that there was "an identifiable movement known as English Fundamentalism" and that "its controversies were only different in the fact that England provided a somewhat different context." While I am impressed by the evidence used to support this conclusion, and hope that Professor Rennie will publish his own account of it, I nevertheless remain convinced that the English movement differed significantly from American fundamentalism. However, I do not object strongly to calling the British movement (as Rennie does) "fundamentalist," which I think is consistent with British parlance. With such a broader definition, my thesis would be that there is a qualitative difference between British and American fundamentalism as well as a difference in impact on the churches and the culture.

18. J. C. Pollock, *A Cambridge Movement* (London, 1953) gives a very complete account of the background and origins of I.V.F. His work may be supplemented by broader accounts in Frederick Donald Coggan (ed.), *Christ and the Colleges: A History of the Inter-Varsity Fellowship of Evangelical Unions* (London, 1934), which contains the constitution, and Douglas Johnson (ed.), *A Brief History of the International Fellowship of Evangelical Students* (Lausanne, 1964).

19. George W. Dollar in his militantly fundamentalist *A History of Fundamentalism in America* (Greenville, S.C., 1973) includes I.V.F. in "An Enemy Within: New Evangelicalism," p. 205; cf. p. 258. On the other hand more liberal critics in England called I.V.F. "fundamentalist" in the 1950s, e.g., Gabriel Hebert, *Fundamentalism and the Church of God* (London, 1957); cf. J. I. Packer's defense, *"Fundamentalism" and the Word of God* (London, 1958). The "fundamentalism" in these debates is more sophisticated, scholarly, and flexible than all but a very small portion of American fundamentalism.

20. Complementary accounts of this dispute are found in Gordon Hewitt, *The Problems of Success: A History of the Church Missionary Society 1910-1942* (London, 1971) and in G. W. Bromiley's sympathetic biography of the leader of the conservatives, *Daniel Henry Charles Bartlett: A Memoir* (Burnham-on-Sea, Somerset, Eng., 1959).

could readily be resolved within the wider spirit of Anglican comprehension. Within the Bible Churchmen's Missionary Society and in numerous older agencies[21] evangelicals felt free to operate without purging established ecclesiastical structures.

Although the foregoing examples illustrate that uncompromising conservative evangelicalism survived in England, the contrast to America becomes apparent when we consider the fragmentary scope of English attempts to organize something like a militant anti-liberal crusade. The prototype of such efforts was the separation of Charles Haddon Spurgeon from the Baptist Union in the "Downgrade controversy" of 1887. This action near the end of the career of this illustrious London preacher seems to have had little wider impact.[22] It was not that no effort was made to carry on his controversialist work, for A. C. Dixon, one of the editors of *The Fundamentals*, came from America in 1911 to serve as pastor of Spurgeon's Metropolitan Temple but departed again in 1919 without leaving any substantial fundamentalist organization behind him.[23] Slightly more successful in organizing a full-fledged controversialist fundamentalist movement in England[24] was E. J. Poole-Connor,

21. Ian Rennie points out that there was a substantial infra-structure of such agencies. Among those he mentions are: the Church Pastoral-Aid Society, the South American Missionary Society, the (now-named) Commonwealth and Continental Missionary Society, and the Church's Ministry to the Jews (all Anglican), the Scripture Union and Children's Special Service Mission, the Christian Alliance of Women and Girls (a secession in 1919 from the YWCA), the Bible League, and the Victoria Institute (a center for anti-evolution thought); among theological schools, Tyndale, Clifton, Oak Hill, and to some extent St. John's Highbury, a few Bible colleges, but no colleges at the universities; publications, *The Christian*, *The Life of Faith* and the annual *Keswick Week*. Correspondence with author.

22. Ian Murray, *The Forgotten Spurgeon* (London, 1966), whose title is revealing, documents a sympathetic account of Spurgeon's role as a controversialist in Downgrade, pp. 145-206. Willis B. Glover, "English Baptists at the Time of the Downgrade Controversy," *Foundations* 1 (1958): 46, goes so far as to conclude of Downgrade, "its chief interest is the fact that it has so few long-range effects."

23. Among the Baptists, however, a "Baptist Bible Union" was organized by the 1920s. See a report on its fundamentalist activities by its founder, John W. Thomas, "Modernism and Fundamentalism in Great Britain," *The King's Business* 14 (1923): 817-821. The impact, however, appears to have been slight. E. J. Poole-Connor, *Evangelicalism in England*, rev. ed. (London 1965 [1951]), p. 249, laments that modernism had triumphed almost completely among British Baptists by 1925. Arthur H. Carter, in "Modernism: the Outlook in Great Britain," *The King's Business* 15 (1924): 691, remarks: "But the saddest aspect of the situation lies in the fact that the entire body of English Nonconformity accepts their theological position, and, save in a few isolated cases, the whole body of the Free Churches has gone *holus bolus* over to the ranks of Modernism." No doubt more non-militant Biblicism survived among Nonconformists than these estimates would allow. Yet contrast American fundamentalists at this time who often claimed to represent the majority of American church members, e.g. "A Divided House," *The King's Business* 15 (1925): 347.

24. Another Englishman who qualified as a full-fledged fundamentalist was W. H. Griffith Thomas (1861-1924). He was associated with *The Fundamentals*, Keswick, dispensationalism, and the founding of Dallas Theological Seminary. However, he left a position as principal at Wycliffe Hall, Oxford, in 1910 to become professor of Old Testament at Wycliffe College, Toronto, hence reducing his influence in England. C. G. Thorne, Jr., "William Henry Griffith Thomas," *The New International Dictionary of The Christian Church*, ed. J. D. Douglas (Grand Rapids, 1974), p. 972.

who also thought he was carrying on Spurgeon's cause. Poole-Connor opposed any cooperation with or tolerance for theological liberalism, and in1922 he founded the Fellowship of Independent Churches which he described in 1925 as having "a strongly fundamentalist credal basis."[25] The organization remained quite small (perhaps 100 to 150 congregations and six to seven thousand members in its first twenty years)[26] and Poole-Connor himself compared the non-militant stance of most conservative evangelicalism in England unfavorably to American fundamentalism.[27] The general extent of Poole-Connor's influence in English church life is revealed by the remark of his admiring biographer, "Truly he was a prophet 'without honour.'"[28]

Paradoxically one factor contributing to this notable lack of success of such separatist fundamentalist efforts was the significant presence in England of the Plymouth Brethren. The Brethren had many of the same traits as American fundamentalists, and no doubt attracted some persons who in the American context might have become involved in wider denominational struggles. Between 1910 and 1960 the principal (Open) Brethren group increased in adherents by roughly half, reaching a total of perhaps ninety thousand.[29] However, by the nature of the case, Brethren separatism left them with little ecclesiastical influence outside their own circles. Even among other conservative evangelicals they were viewed with some suspicion and regarded as operating too much like a secret society.[30] In all, Brethren influence in England was much like a religious underground and did not gain the role in the churches and the culture that fundamentalism had in America.

In conclusion it appears that the English conservative evangelicals differed from their American counterparts in two major respects: (1) a lack of widespread militancy, but instead Keswick-type emphases on

25. Quoted in Fountain, *E. J. Poole-Connor*, p. 126.
26. The statistics are approximations, Fountain, *Poole-Connor*, p. 18. Nearly three hundred congregations were claimed by the time of Poole-Connor's death in 1962, p. 211.
27. Fountain, *Poole-Connor*, pp. 34, 44, 131-134. Fountain, p. 119 observes, "In the United States the conflict was sharper than in this country for two reasons. The Liberals were more extreme and the Evangelicals more faithful and more able."
28. Fountain, *Poole-Connor*, p. 134. Conservative evangelical scholarship also seems to have been at its nadir during the period between the wars. H. D. McDonald, *Theories of Revelation*, in a very sympathetic account remarks nevertheless on this era that "There was, on the whole, however, no serious conflict, because, not only were other interests uppermost, but evangelicals were in the backwood as far as convincing Biblical scholarship was concerned," p. 208; cf. pp. 280-282. The founding of the *Evangelical Quarterly* in 1929 signaled the reversal of this trend.
29. F. Roy Coad, *A History of the Brethren Movement* (London, 1968), p. 185. The figures are for the entire British Isles.
30. Coad, ibid., p. 284. They included, nonetheless, notable elements from higher economic and social standing. Rennie correspondence with author.
    With respect to the possibility of a more general social factor, Rennie, noting a variety of evidences, observes, "Thus a significant difference does appear—English Fundamentalism often seems upper middle class while its American counterpart is usually much more plebian."

non-controversialist piety; (2) a lack of general impact on the churches and the culture.[31]

AMERICA

Compared to the English, what in the American situation fostered militant fundamentalism as a major and sometimes influential religious force? The answers to this intriguing question inevitably will be rather speculative. The most significant factors on the American scene can be broken down conveniently into three interrelated categories, the social, the religious, and the intellectual.

SOCIAL FACTORS

Although a number of social factors might be explored, the most apparent involves the communication of ideas. Every observer has noticed, for instance, that fundamentalism sometimes flourished in isolated rural areas. Such cases suggest that in a very large, recently settled, and rapidly changing country, cultural pockets developed that were effectively insulated from the central intellectual life. The importance of this phenomenon can be seen more clearly by comparison with England. English intellectual and cultural life is relatively centralized. Ancient and well-established channels of communication made it difficult for an issue to be discussed in the universities, for instance, without soon being well-known throughout the parishes. Although there might have been a few "backward" areas,[32] the dissemination of new trends seemed to proceed at a relatively even pace.

In America there were great lags in communication. These resulted primarily from sociological, ethnic, and geographical factors, but were also reinforced by denominational differences. Congregationalists, Presbyterians, and Baptists, for instances, became familiar with the novel ideas at differing times. Within these groups, Northerners and Southerners, or Easterners and Westerners, might seriously encounter the ideas as much as generations apart.[33] Theological discussion could proceed in one section of the country, in one denomination, or among the educated

31. These conclusions may be compared to those of William R. Hutchison, "The Americanness of the Social Gospel; An Inquiry in Comparative History," *Church History* 44 (1975): 367-381, who stresses the essential similarities between British and American social gospel. Hutchison does find American liberals to have been more optimistic in their humanism and affirmations of the present age than were their British counterparts. It might be added that by the 1920s liberalism appeared to have triumphed far more completely in British churches (cf. note 23 above).

32. The fact that the industrial revolution was earlier in England than America and hence the transitions from rural to urban cultural patterns more nearly completed by the late nineteenth century helped also to reduce such cultural pockets. Cf. Harold Perkin, *The Origins of Modern English Society 1780-1880* (London, 1969).

33. Albert H. Newman, "Recent Changes in Theology of Baptists," *The American Journal of Theology* 10 (1906): 600-609, made essentially this point at the time.

elite while many people in other areas were virtually oblivious. In a period of rapid intellectual change, the potential for theological warfare once these diverse groups discovered each other was immense. In fact, the principal moment of discovery came just following World War I, when a general sense of cultural alarm heightened the intensity of fundamentalist reactions.

Although during the controversies of the 1920s fundamentalism appeared to many as primarily a social phenomenon, especially related to rural-urban tensions, such factors, while very important, only partially explain its development.[34] First of all (as Ernest Sandeen has pointed out), fundamentalism was not necessarily rural; its principal centers were initially urban and Northern. Furthermore (as Sandeen has also argued), if fundamentalism were to be adequately explained by social tensions, rural-urban themes, problems of communication and the like, then fundamentalism should have generally disappeared, as many in the 1920s predicted it would, once the crises of social transitions were past. Since in fact fundamentalism survived the 1920s and continued to flourish, its roots must have been considerably deeper. Sandeen finds these deep roots particularly in the millenarian movement.[35] It remains to be explained why they took their strongest hold in America, and not in England where the fundamentalist forms of millenarianism in fact originated. The lasting appeal of fundamentalism must be explained by elements deep in the American religious and cultural traditions themselves.

RELIGIOUS TRADITIONS

The primary force in the American religious experience that prepared the way for fundamentalism is what can be called "the dynamics of unopposed revivalism."[36] Although revivalism has flourished in many other countries since the eighteenth century, in America it came to be almost unchallenged by other formidable traditions and institutions. The comparison with England is again instructive. While revivalism was

---

34. The classic statement of this interpretation is Stewart G. Cole, *History of Fundamentalism* (New York, 1931). In "From Fundamentalism to Evangelicalism," *The Evangelicals*, I have attempted to explain some social factors by suggesting that fundamentalism might involve a White Anglo-Saxon Protestant experience analogous to that of elements in immigrant groups.
35. Sandeen, *Roots*, esp. pp. ix-xix, and his "Fundamentalism and American Identity," *The Annals of the American Academy of Political and Social Science* 387 (January, 1970): 56-65. Paul A. Carter, "The Fundamentalist Defense of the Faith," in *Change and Continuity in Twentieth Century America: The 1920s*, ed. John Braeman *et al.*, also offers an effective ciritique of primarily social and social-economic or political interpretations of fundamentalism, which he himself had endorsed in his earlier work.
36. This phrase is borrowed in part from Stanley Elkins, *Slavery: A Problem in American Institutional and Intellectual Life* (Chicago, 1959) who refers to "the dynamics of unopposed capitalism."

long a transatlantic phenomenon, in England the universities, the established church, and the pre-revivalist traditions of most of the Nonconformist groups were among the venerable forces promoting moderation and restraint. Tradition in general was much stronger in England than in America. The strength of resurgent evangelicalism in early nineteenth-century England, for instance, as a force for theological conservatism was substantially offset toward the middle of the century by the High Church movement which made even stronger claims upon traditionalist sentiments.[37]

In America such forces either were absent or had little effect, thus leaving revivalism an almost open field for determining the distinctive characteristics of American religious life.[38] Many of these traits (such as individualism, Biblicism, and primitivism) are conspicuous both in the mainstream of nineteenth-century American Protestantism and in twentieth-century fundamentalism. Such continuities suggest that fundamentalism can best be understood not primarily as an outgrowth of the movements promoting millenarianism and inerrancy (as Sandeen suggests), but rather to a large extent as a sub-species of revivalism in which certain types of new emphases became popular as part of the anti-modernist reaction.

Unopposed revivalism often fostered anti-intellectualism, as Richard Hofstadter has described,[39] yet perhaps even more important for the development of fundamentalism was the revivalists' tendency to promote and reinforce a particular type of intellectual emphasis—that is, a tendency to think in terms of simple dichotomies. The universe was divided between the realm of God and the realm of Satan; the supernatural was sharply separated from the natural; righteousness could have nothing to do with sin. The central impulse of revivalism was to rescue the saved from among the lost, and its whole way of conceiving reality was built around this central antithesis. In such a dichotomized view of things, ambiguities were rare. Like the conversion experience itself, transitions were not gradual, but were radical transformations from one state to its opposite.

Such intellectual categories left almost no room for the motifs of

37. Cf. Poole-Connor, *Evangelicalism*, p. 220. Well into the twentieth century Anglican conservative evangelicals had to deal with two fronts—the liberals and the High Church party.
38. Cf. Donald G. Mathews, "The Second Great Awakening as an Organizing Process 1780-1830: An Hypothesis," *American Quarterly* 21 (1969): 23-43. See also William G. McLoughlin, "Revivalism," in *The Rise of Adventism: Religion and Society in Mid-Nineteenth-Century America*, ed. Edwin S. Gaustad, pp. 119-154, who goes so far as to suggest that revivalism is the key to understanding American life generally.
39. *Anti-Intellectualism in American Life* (New York, 1962). Cf. *The Paranoid Style of American Politics and Other Essays* (New York, 1965). Hofstadter is correct in seeing anti-intellectualism as an important component of fundamentalism, although this single emphasis obscures many other aspects.

thought that were characteristic of liberal theology and scientific naturalism in the later nineteenth century. Both Darwinism and higher criticism emphasized gradual natural development, and the new theology saw God working through such means, emphasizing the synthesis of the natural and the supernatural rather than the antithesis. Wherever revivalism had been relatively unopposed in American religious life, there was virtually no preparation for the acceptance of the new categories—indeed there was hardly a way to discuss them. The reaction of many American Protestants, then, was not only to reject them outright as antithetical to the faith, but to assert the antitheses even more decisively. In reaction to naturalism, the supernatural aspects of the faith, such as the Virgin Birth, were emphasized in lists of fundamental doctrine. The three most distinctive doctrines of fundamentalism itself, inerrancy, opposition to evolution, and the premillenial return of Christ, all uncompromisingly accentuated the supernatural in the way God works, drawing the sharpest lines against any naturalistic or developmental explanations.

Although the dynamics of revivalism appear central to understanding the popularity of militant defenses of such doctrines in America, much of American revivalism and more broadly, pietism, had developed in the context of one other major religious tradition—that of Calvinism. From the beginning, Calvinism in America supported a tendency to demand, among other things, intellectual assent to precisely formulated statements of religious truth in opposition to all error.[40] Revivalists often modified and simplified the doctrines involved, yet many of them preserved both the emphasis on antitheses and the general point that assent to rightly-stated doctrine could be of eternal significance.

This tradition helps explain the paradox between Americans' reputation for religious tolerance and the actual intolerance in most of their ecclesiastical life. Both Calvinists and their revivalist heirs accepted and even endorsed civil tolerance of religious diversity by the eighteenth century, but civil tolerance was quite different from intellectual tolerance. One might allow Quakers or Roman Catholics full political equality and yet consider semi-Pelagianism to be legitimate grounds for fierce theological debate and separation. For Calvinists, separation of church and state often meant, among other things, that toleration did not have to extend to the churches.

Such tendencies, initiated in Calvinism, were preserved to some extent in American revivalism and hence continued into twentieth-century fundamentalism. This point is confirmed by the fact that fundamen-

---

40. Seventeenth-century Puritans, for instance, were fascinated by dichotomies and antitheses as the popularity of the Ramist method and their concerns over precisely distinguishing between the regenerate and the unregenerate indicate.

talism appeared primarily among groups with Reformed origins, such as Baptists and Presbyterians, but was rather rare on the side of American revivalism with Methodist origins where ethical rather than intellectual aspects of Christianity tended to be emphasized.[41] Furthermore, in the late nineteenth and early twentieth centuries, the most natural allies of the revivalist fundamentalists were the Princeton theologians who for generations had been firing heavy theological artillery at every idea that moved and who were almost indecently astute at distinguishing Biblical and Reformed truth from all error.

Intellectual Factors

The wider fundamentalist battle against the new ideas was fought with materials drawn from both the Bible and the common stockpile of American assumptions and concepts. The relative popularity of the fundamentalist account of things reveals something, therefore, about the character of American intellectual life generally. Continuing the comparison with England, it appears that the historical sensitivities of the mid-nineteenth century had inclined Englishmen toward a rather ready acceptance of new and sometimes startling ideas concerning biological evolution and the historical development of the Bible. Much the same might be said of some portions of America (such as New England) where many people were well-prepared to accept the new ideas; yet there were important countervailing forces as well.

Perhaps the best way to describe the difference on this point between America and Europe is to say that in America the romantic era was truncated. America came of age during the Enlightenment and remained generally content with mid-eighteenth-century modes of thought long after these had gone out of style in Europe. The American intellectual community remained rather isolated during the early national period; well into the second half of the nineteenth century the type of philosophy taught in almost all American colleges was the "common sense realism" of the Scottish Enlightenment. Although this philosophy was susceptible to a romantic interpretation in which persons could intuit truth, the truths involved were basically fixed aspects of reality from which could be derived rather definite law, so that there was little concept of development. Although it is difficult to document, a version of this common sense approach to reality appears to have been

41. Pentecostalism is the movement of this tradition that parallels fundamentalism. Pentecostals also rejected modern culture but more in terms of intense personal piety that separated individuals from the world, rather than in terms of doctrinal warfare. The two movements should be kept distinct, I think, even though they sometimes overlapped and had some common origins in American revivalism and hence many common traits. W. J. Hollenweger, *The Pentecostals* (Minneapolis, 1972), comments on fundamentalist traits in Pentecostalism, p. 9 and elsewhere. Various holiness teachings are likewise found among fundamentalists.

strong in shaping the popular philosophy of nineteenth-century America as well.

By the mid-century, of course, there were many manifestations of romanticism on the American philosophical, religious, and artistic scenes, yet even at that time these had to contend with strong counter-forces that still embodied Enlightenment categories. Among evangelicals, who controlled most of American higher education, the tension between these two tendencies was far from resolved at mid-century. Theologians such as Horace Bushnell, Henry B. Smith, and John Nevin were just emerging, and those closer to the eighteenth-century tradition such as Charles Hodge or Nathaniel William Taylor were still strong influences.[42] Revivalism did provide a popular romanticism emphasizing personal sentiment and piety, but this had little if anything to do with the sort of romanticism that since before the beginning of the nineteenth century had been fostering among Europeans a sensitivity to the dynamics of change and a suspicion of rational and fixed definitions of experience.[43]

The result was that in the second half of the nineteenth century many Americans were only just beginning romantic explorations when the second scientific revolution, associated with Darwinism, demanded that the new historical and developmental views be placed on the theological agenda. Even the intellectual community, then, was not always thoroughly prepared for the post-romantic modes of thought. As a result some rather well-educated Americans were among those who met naturalist challenges with pre-romantic rationalistic defenses. Emerging fundamentalism at the popular level accordingly did not entirely lack intellectual leadership, especially from a number of older theological institutions. Its modes of thought were not simply shaped by revivalist eccentricities, but by the substantial pre-romantic and rather rationalistic intellectual trends that survived in the American academic and theological communities.

The fundamentalist response to Darwinism, for instance, generally was not an anti-intellectualistic one framed in terms of the incompatibility of science and religion. It was an objection rather to a *type* of

42. The above generalizations about nineteenth-century American evangelicals are illustrated (among other places) in Marsden, *The Evangelical Mind and the New School Presbyterian Experience* (New Haven, 1970).
43. It is common practice to set the peak of European (including English) romanticism in the period from 1780 to 1830; e.g. Arthur O. Lovejoy, "The Meaning of Romanticism for the Historian of Ideas," *Journal of the History of Ideas*, 2 (June, 1941): 260-261 and Jacques Barzun, *Romanticism and the Modern Age* (Boston, 1943), pp. 134-139. The tiny Mercersburg movement in America compared to the Anglo-Catholic movement in England during the same era suggests something of the contrast in the strength of romanticism in the religious life of the two countries; cf. James Hastings Nichols, *Romanticism in American Theology: Nevin and Schaff at Mercersburg* (Chicago, 1961).

science—a developmental type—which they almost always branded as "unscientific."[44] Seldom did they denounce science in principle. Fundamentalist theology likewise reflected high esteem for being "scientific" in the sense of organizing, classifying, and rationally ordering data.[45] Similarly their view of Scripture tended to be positivistic: the Bible contained only firm evidence and no error.[46]

There is in fact little reason to suppose that many Americans would be inclined to reject science outright as an authority. The opposite would be more likely in a highly technological society with strong Enlightenment roots. The fundamentalist view of science was thus not wholly incompatible with the American intellectual climate.

Perhaps even more striking in suggesting American cultural traits is the attitude toward history found in fundamentalism. Here the contrast to English evangelicalism is particularly instructive. The characteristic view of history among fundamentalists has been dispensational-premillennialism (although not every fundamentalist, especially in the 1920s, held this position). Dispensational-premillennialism originated in England in the early nineteenth century. Yet in the twentieth century it apparently has had relatively few adherents in England except among Plymouth Brethren while in America it remains tremendously popular.[47] What, then, accounts for the remarkable popularity in America of this imported British view?

Dispensationalism is essentially an anti-developmental and anti-naturalistic way of explaining historical change. History is divided into seven "dispensations," each representing "some change in God's method of dealing with mankind," and each involving "a new test of the natural man." Man fails these tests, so that each dispensation ends in judgement and catastrophe.[48]

44. For examples, John Horsh, "The Failure of Modernism," (Chicago, 1925) (pamphlet), pp. 22-23 says "The science with which Scripture conflicts is unproved theory; it is science falsely so-called." William B. Riley defines science as "knowledge gained and verified by exact observation and correct thinking; especially as methodologically arranged in a rational system," which he takes to exclude "theory," "hypothesis," and "assumptions," p. 5, "Are the Scriptures Scientific?" (Minneapolis, n. d.) (pamphlet).
45. A good example is Reuben A. Torrey, *What the Bible Teaches*, 17th ed., (New York, 1933 [1898]), which he describes as " . . . simply an attempt at a careful, unbiased, systematic, thorough-going, *inductive* study and statement of Bible truth. . . . the methods of modern science are applied to Bible study—through analysis followed by careful synthesis," p. 1.
46. Cf. John Opie, "The Modernity of Fundamentalism," *Christian Century*, May 12, 1965, pp. 608-611.
47. One example is the popularity of Hal Lindsey, *The Late Great Planet Earth* (Grand Rapids, 1970). As of the July 1974 printing the publisher claimed 4,300,000 copies in print.
48. These dispensations are 1) "Innocence," ending with the Fall; 2) "Conscience," ending with the Flood; 3) "Human Government," ending with Babel; 4) "Promise," ending in the bondage in Egypt; 5) "Law," ending with the death of Christ; 6) "Grace," which will end with a period of great tribulation, immediately followed by Christ's return to earth, victory at Armageddon; and 7) the millennium or personal reign of Christ, ending with Satan "loosed a little season" but quickly defeated. After the

Two general tendencies found in fundamentalist thought are particularly evident in this scheme. First is a fascination with dividing and classifying.[49] The second is a heightened supernaturalism. Human efforts and natural forces have almost nothing to do with historical change. Instead, God periodically intervenes with a series of spectacular supernatural events that suddenly transform one age into another. In fact, God and Satan are virtually the only significant historical forces, and they are armed forces at that. In contrast to any romantic and developmental interpretations, dispensationalism explains history as a series of supernatural impositions of highly abstract, logical, and almost legalistic principles that humans might either accept or reject.[50]

The popularity of such views in twentieth-century America reveal an important American thought pattern—that in comparison with other Western countries many Americans lacked certain typically modern concepts of history. In the views of history popular in America the elements that were missing were precisely the assumptions central to most modern historical scholarship. These were the assumption that history is a natural evolutionary development and the corollary that the present can be understood best as a product of developing natural forces from the past. American historiography had long been dominated by supernatural, or at least providential, interpretations.[51] Furthermore, even the secular histories frequently emphasized the newness of America, dwelling on the past only to accentuate progress and the future. To Americans, who had relatively little history of their own, their national experience often seemed like a new dispensation, discontinuous with the past.

This widespread absence of a sense of gradual or natural historical change had been reinforced and partly created by the revivalist tradition, and in many places preserved from encounters with developmental ideas by social and geographical factors. These influences combined to dispose many persons to declare every aspect of the new views to be anathema and to oppose them with various non-negotiable logical antitheses. The greater the claims and the greater the influence of the naturalistic developmental views, the more firmly the fundamentalists stressed the opposing paradigms.[52]

millennium is the "new heavens and new earth" of eternity. C. I. Scofield, *"Rightly Dividing the World of Truth"* (Revell paper edition, New York, n.d. [1896], pp. 12-16.
49. Scofield says, "The Word of Truth . . . has right divisions . . . *so any study* of that Word which ignores these divisions must be in large measure profitless and confusion." Ibid., p. 3.
50. C. Norman Kraus, *Dispensationalism in America* (Richmond, 1958), pp. 66-67 and 125-126 comments perceptively on this point.
51. Ernest Lee Tuveson, *Redeemer Nation: The Idea of America's Millennial Role* (Chicago, 1968) gives many examples of this point.
52. The total lack of communication between fundamentalists and modernists concerning both history and science fits well the now-familiar patterns of paradigm conflict described in Thomas S. Kuhn, *The Structure of Scientific Revolutions* (Chicago, 1962).

This observation should not be interpreted to mean that because modern developmental paradigms were newer they were for that reason necessarily superior to fundamentalist emphases on antitheses. Furthermore, it seems to me incorrect, at least as far as Christian thought is concerned, to regard antithesis and natural historical development as incompatible categories, since central to Christianity itself is the wholly-other God revealing himself and acting in history. In America in the early twentieth century both fundamentalists and liberals tended to oversimplify the issue on this point. Fundamentalists, seeing clearly that the Bible spoke of antitheses, would hear almost nothing of natural development; liberals, enamored of historical and developmental explanation, proclaimed that the old antitheses must be abandoned. Perhaps in part because of the novelty in America of the modes of thought associated with modernism, both sides oversimplified the issues and each overestimated the degree to which recognition of historical development necessitated the abandonment of traditional Christian teaching.

Fundamentalist Institutions and the Rise of Evangelical Protestantism, 1929-1942

JOEL A. CARPENTER

In April of 1952 an article in *Christian Life* magazine proclaimed Chicago "the evangelical capital of the U.S.A."[1] To back this claim, editor Russell T. Hitt cited a host of evangelical agencies in greater Chicago: mission boards, denominational offices, colleges, Bible institutes, seminaries, publishing concerns (including *Christian Life* itself) and youth organizations. In total, the author mentioned over one hundred different agencies such as Youth For Christ International, the Slavic Gospel Association, Scripture Press and the Swedish Covenant Hospital.[2] At first glance, the article appears to present a confusing list of unrelated organizations, but closer inspection reveals a coherent pattern. The agencies in the Chicago area represented the swiftly growing evangelical movement which observers have labelled the third force of American Christianity.[3] Most institutions listed did not belong to the older, more prestigious denominations. The mission boards, such as Wycliffe Bible Translators, the Worldwide Evangelization Crusade and the International Hebrew Christian Alliance were independents. The denominational headquarters, including those of the Conservative Baptist Association, the Evangelical Mission Covenant Church, the North American Baptist General Conference and the General Association of Regular Baptist Churches, represented fundamentalists and other evangelicals. The schools—the Moody Bible Institute, North Park College, Trinity Seminary and Bible College, Wheaton College, the Mennonite Biblical Seminary, the Salvation Army Training College and Emmaus Bible Institute—came from the same source.[4]

Whether or not Chicago was the capital of evangelicalism is not as

1. Russell T. Hitt, "Capital of Evangelicalism," *Christian Life* 5 (April 1952):16.
2. Ibid., pp. 16–18, 46–48. *Christian Life* itself was an interesting symbol of a growing evangelical wing of Protestantism. *Christian Life* was formed in 1948 by enterprising young evangelical publishers who wanted a market for a breezy, "Christian" version of *Life* magazine. *Christian Life* 1 (July 1948):3.
3. "The Third Force in Christendom," *Life* (June 9, 1958):113–121; and Henry P. Van Dusen, "The Third Force's Lesson for Others," *Life* (June 9, 1958):122, 125. See also William G. McLoughlin, "Is There a Third Force in Christendom?" *Daedalus* 96 (Winter 1967):43–68; Winthrop S. Hudson, *American Protestantism* (Chicago, 1961), pp. 153–176.
4. Hitt, "Capital of Evangelicalism," pp. 16, 18, 46, 48.

Mr. Carpenter is assistant professor of history in Trinity College, Deerfield, Illinois.

important as the image the article revealed. Chicago was a regional evangelical stronghold in the 1950s when the evangelicals were leading a revival of popular religious interest. This revival developed largely from the institutional base which evangelicals had established in the previous decades. The fundamentalists were especially prominent in the postwar evangelical revival. This fact might seem surprising to one who supposed that their movement had been crushed twenty years earlier. That was scarcely the case, as we shall see. Fundamentalism was not a defeated party in denominational politics, but a popular religious movement which in the 1930s developed a separate existence from the older denominations as it strengthened its own institutions. By the 1950s, this building phase had paid off and Billy Graham, a fundamentalist favorite son, became the symbol of evangelicalism's new prominence.

As Hitt's article suggests, evangelicalism was not a monolithic fundamentalism but rather a broad mosaic comprised of clusters of denominations and institutions with different ethnic and doctrinal heritages. One of this mosaic's most visible segments is rightly called fundamentalism, a movement of conservative, millenarian evangelicals who came mostly from Presbyterian, Baptist and independent denominations, such as the Evangelical Free Church. Other segments include the Holiness Wesleyans, such as the Church of the Nazarene; the pentecostals, including the Assemblies of God; the immigrant confessional churches, such as the Lutheran Church, Missouri Synod and the Christian Reformed Church; southern-based conservatives, notably the Southern Baptists and the Churches of Christ; peace churches of Anabaptist, Quaker or pietist backgrounds; and black evangelicals of Methodist, Baptist, Holiness and pentecostal denominations.[5] As the twentieth century progressed, the evangelicals cut a progressively wider swath through the ranks of the American churches. By 1960 they comprised an estimated half of the nation's sixty million Protestants.[6]

When the term fundamentalist is used to designate any or all of these churches, it becomes an ambiguous and derogatory term. But by precise and

5. Hudson, *American Protestantism*, pp. 155-165. Here I am especially indebted to my colleagues in a research project funded by the National Endowment for the Humanities, "The American Evangelical Mosaic." For information on evangelical groups, see Ernest R. Sandeen, *The Roots of Fundamentalism, British and American Millenarianism, 1800-1930* (Chicago, 1970); Timothy L. Smith, *Called Unto Holiness, The Story of the Nazarenes: The Formative Years* (Kansas City, Mo., 1962); Klaude Kendrick, *The Promise Fulfilled: A History of the Modern Pentecostal Movement* (Springfield, Mo., 1961); James DeForest Murch, *Christians Only: A History of the Restoration Movement* (Cincinnati, 1962); William Wright Barnes, *History of the Southern Baptist Convention, 1845-1953* (Nashville, 1954); Milton L. Rudnick, *Fundamentalism and the Missouri Synod* (St. Louis, 1966); John Henry Kromminga, *The Christian Reformed Church: A Study in Orthodoxy* (Grand Rapids, 1949); Cornelius J. Dyck, ed., *An Introduction to Mennonite History* (Scottdale, Pa., 1967); E. Franklin Frazier, *The Negro Church in America* (New York, 1964).

6. Hudson, *American Protestantism*, pp. 155, 162.

historical definition, fundamentalism is a distinct religious movement which arose in the early twentieth century to defend traditional evangelical orthodoxy and to extend its evangelistic thrust. The movement combined a biblicist, generally Calvinist orthodoxy, an evangelistic spirit, an emphasis on the higher Christian (Holy Spirit directed) life and a millenarian eschatology. Because the urban centers were strongholds of Protestant liberalism and the most challenging home fields for evangelism, they became the principal centers of early fundamentalist activity. The movement drew its name from *The Fundamentals*, a twelve-volume series of articles published by conservative leaders between 1910 and 1915 to affirm and defend those doctrines which they considered essential to the Christian faith, such as the verbal inspiration and infallibility of the Bible and salvation only by faith in the atoning death of Jesus Christ. Fundamentalism was a popular movement, not merely a mentality; it had leaders, institutions and a particular identity. Fundamentalists recognized each other as party members as it were, and distinguished themselves from the other evangelicals listed above.[7]

As a complex aggregate entity, evangelical Protestantism in the twentieth century demands closer attention than it has received. Studies of the 1930s and early 1940s in particular have yielded little understanding of its development. The prevailing opinion among historians is that Protestantism suffered a depression during at least the first half of the 1930s which was relieved only when neo-orthodox theology renewed the vision and vitality of the old-line denominations.[8] Evangelical Protestants fit into this scheme only tangentially. Sydney Ahlstrom noted that "something like a revival took place" among the holiness, fundamentalist and pentecostal churches; and William McLoughlin credited evangelicals with keeping alive the tradition of revivalism during the depression. Other historians, however, viewed the activity of this third force as a symptom of Protestantism's depressed condition rather than a sign of grassroots vitality.[9] The institutional growth in the 1930s of the most vocal and visible evangelicals, the fundamentalists,

7. The problem of defining fundamentalism is discussed in Ernest R. Sandeen, "Toward a Historical Interpretation of the Origins of Fundamentalism," *Church History* 36 (March 1967):66-83; LeRoy Moore, Jr., "Another Look at Fundamentalism: A Response to Ernest R. Sandeen," *Church History* 37 (June 1968):195-202; George M. Marsden, "Defining Fundamentalism," *Christian Scholar's Review* 1 (Winter 1971):141-151; Ernest R. Sandeen, "Defining Fundamentalism: A Reply to Professor Marsden," *Christian Scholar's Review* 1 (Spring 1971):227-233; Ernest R. Sandeen, "*The Fundamentals:* The Last Flowering of the Millenarian-Conservative Alliance," *Journal of Presbyterian History* 47 (March 1969): 55-73.
8. Martin E. Marty, *Righteous Empire: The Protestant Experience in America* (New York, 1970), pp. 233-243; Robert T. Handy, *A Christian America, Protestant Hopes and Historical Realities* (New York, 1971), pp. 217-219. See also Paul A. Carter, *The Decline and Revival of the Social Gospel* (Ithaca, N.Y., 1954).
9. Sydney E. Ahlstrom, *A Religious History of the American People* (New Haven, 1972), p. 920; William G. McLoughlin, *Modern Revivalism: Charles Grandison Finney to Billy Graham* (New York, 1959) pp. 462-468; Handy, *A Christian America,* p. 203; Marty, *Righteous Empire,* p. 237.

challenges the widespread notion that popular Protestantism experienced a major decline during that decade. What really transpired was the beginning of a shift of the Protestant mainstream from the older denominations toward the evangelicals.

The older denominations did experience what Robert T. Handy called a "religious depression," beginning in the middle of the 1920s until the late thirties, when their fortunes revived somewhat. For example, membership in the northern Presbyterian and the Protestant Episcopal denominations declined 5.0 and 6.7 percent respectively between 1926 and 1936. The foreign missionary enterprise lost momentum as budgets tightened and many missionaries returned home at mid-career for lack of funds.[10] Social programs also suffered from the loss of contributions as the churches had to cut off the lower end of their priority lists.[11] At the onset of the great depression of the 1930s, many Christians wondered if a revival would descend, bringing with it the return of prosperity. But Samuel C. Kincheloe reported to the Social Science Research Council in 1937 that "the trend over the past thirty years" had been "away from emotional revival services"; and that the depression did "not seem to have produced much variation in this major trend."[12] When Robert and Helen Lynd revisited "Middletown" in 1935, they saw little evidence of a religious awakening. "If the number of revivals is any index of religious interest in the depression," they concluded, "there has been a marked recession."[13] McLoughlin and Ahlstrom recognized, however, that the slight overall growth in Protestant membership in the 1930s stemmed largely from what the Lynds had called "working-class churches."[14]

In singular contrast to the plight of the major denominations, fundamentalists and other evangelicals prospered. During the 1920s, fundamentalists had grown more vocal and apparently more numerous, but the leaders had been publicly defeated in denominational battles, and had made themselves look foolish in the anti-evolution crusade.[15] Adverse publicity from public controversy had discredited fundamentalists and established the Mencken-

10. Robert T. Handy, "The American Religious Depression, 1926-1935," Church History 29 (March 1960):4-5; percentages computed from membership statistics in U.S. Department of Commerce, Bureau of the Census, Religious Bodies, 1936, 2 vols., vol. 2 Denominations (Washington, D. C., 1941):1386, 1478.
11. Handy, "The American Religious Depression," pp. 5-9.
12. Samuel C. Kincheloe, Research Memorandum on Religion in the Depression, Social Science Research Council #17 (New York, 1937), p. 93. See also, "Why No Revival?" The Christian Century 52 (Sept. 18, 1935):1168-1170; "Billy Sunday, the Last of His Line," The Christian Century 52 (Nov. 20, 1935):1476.
13. Robert S. and Helen M. Lynd, Middletown in Transition (New York, 1937), p. 303.
14. Ahlstrom, A Religious History of the American People, p. 920; McLoughlin, Modern Revivalism, p. 464.
15. Norman F. Furniss, The Fundamentalist Controversy, 1918-1931 (New Haven, 1954), pp. 103-176; Stewart G. Cole, The History of Fundamentalism (New York, 1931), pp. 65-225; Sandeen, The Roots of Fundamentalism, pp. 250-264.

esque image which has dogged them ever since.[16] Yet these defeats by no means destroyed the movement. Fundamentalism cannot be understood by studying only its role in headline-making conflicts. Rather, we must examine the growing network of institutions upon which fundamentalists increasingly relied as they became alienated from the old-line denominations.

One of the most important focal points of fundamentalist activity in the thirties was the Bible institute, a relatively new type of institutional structure. The two pioneers of Bible institute education were A. B. Simpson, founder of the Christian and Missionary Alliance, who in 1882 established the Missionary Training Institute in New York City, and Dwight L. Moody, who founded in 1886 what became the Moody Bible Institute of Chicago.[17] The idea of a teaching center for lay Christian workers caught on quickly, and other schools sprang up across the country. By 1930 the fundamentalist weekly *Sunday School Times* endorsed over fifty Bible schools, most of which were in major cities.[18]

The Bible institutes became the major coordinating agencies of the movement by the 1930s, as popular fundamentalist alienation toward the old-line denominations reached new heights. True, most fundamentalists had not left the older denominations, but after the controversies over evolutionary theory and theological liberalism in the 1920s, they were more aware than before of the intellectual attitudes engendered by church-related colleges and seminaries. While the nondenominational Bible institutes had been founded to train lay and paraministerial workers such as Sunday school superintendents and foreign missionaries, now they faced demands for educating pastors and for other services that denominations formerly provided.[19]

Since the Bible institutes had already branched out into activities not directly connected with in-residence instruction, they were well equipped to meet such demands. Some of the schools had extension departments, such as those of the Philadelphia School of the Bible, or the Moody Bible Institute of Chicago. These agencies organized week-long summer and other shorter Bible conferences, supplied staff evangelists for revival meetings and provided churches with guest preachers.[20] Many schools ran publishing

16. Furniss, *The Fundamentalist Controversy*, pp. 76–100; Cole, *The History of Fundamentalism*, pp. 259–280.
17. S. A. Witmer, *The Bible College Story: Education With Dimension* (Manhasset, N. Y., 1962), pp. 34–37.
18. "Bible Schools That are True to the Faith," *Sunday School Times* 72 (February 1, 1930): 63 (hereafter cited as *SST*).
19. Ernest R. Sandeen suggests this development, pointing out in *The Roots of Fundamentalism* (pp. 241–243) that the scope of Bible institute activity was such that the schools functioned as denominational surrogates.
20. Renald E. Showers, "A History of Philadelphia College of Bible," (M.Th. Thesis, Dallas Theological Seminary, 1962), pp. 69, 81, 86; *Brief Facts About the Moody Bible Institute of Chicago* (Chicago, 1928); *Moody Bible Institute Bulletin* 12 (November 1932):14; 16 (November 1936):15.

and/or distributing ventures, including the Bible Institute of Los Angeles' BIOLA Bookroom, Approved Books of the Philadelphia School of the Bible (PSOB) and the mammoth Bible Institute Colportage Association at Moody.[21] In addition many magazines provided their schools with publicity and the readers with fundamentalist literature and opinion: *The Moody Monthly, The King's Business* of BIOLA, *Serving and Waiting* of PSOB, Northwestern (Minneapolis) Bible and Missionary Training School's *The Pilot* and Denver Bible Institute's *Grace and Truth.*[22] As centers of religious enterprise, the Bible institutes soon saw the potential impact of radio broadcasting both as a religious service opportunity and a way to increase their constituency. BIOLA led the way with its own station, KJS, in 1922. Moody installed WMBI three years later, and, although they did not own stations during the 1930s, Providence (R. I.) Bible Institute, Columbia Bible College in South Carolina and Denver Bible Institute all sponsored radio programs.[23]

With so many services to provide to fundamentalist individuals and small Bible classes and congregations, the Bible schools became regional and national coordinating centers for the movement. Moody Bible Institute (MBI) became the national giant of institutional fundamentalism. The MBI Extension Department held weekend Bible conferences in nearly 500 churches during 1936, more than doubling its exposure of six years earlier. By 1942, WMBI was releasing transcribed programs to 187 different stations, and the radio staff had visited nearly 300 different churches since the station's inception, returning to many churches several times. The Institute had over 15,000 contributors in 1937 and about the same number enrolled in Correspondence School, while the *Moody Monthly* showed a net increase of 13,000 subscribers over the decade to total 40,000 by 1940.[24] Other schools could not match MBI in scale, but they carried strong regional influence. By the mid-1930s, for instance, Gordon College of Theology and Missions had supplied 100 pastors in greater Boston, and 48 out of the total 96 Baptist pastors in New Hampshire. At one time in the 1930s, every

21. "Institute Items," *The King's Business* 3 (November 1912):295–296; Showers, "Philadelphia College," pp. 69, 89; *A Brief Story of the Bible Institute Colportage Association of Chicago: Forty-five Years of Printed Page Ministry* (Chicago, 1939).
22. "Interdenominational Christian Magazines," *SST* 73 (February 7, 1931):72.
23. Daniel P. Fuller, *Give The Winds A Mighty Voice: The Story of Charles E. Fuller* (Waco, Texas, 1972), pp. 75–77; "WMBI," *Moody Monthly* 30 (January 1930):270; "Radio Station WMBI," *Moody Monthly* 31 (May 1931):480; "The Sunday School Times Radio Directory," *SST* 73 (May 30, 1931):313. Hereafter, *Moody Monthly* is cited as *MM*.
24. *Moody Bible Institute Bulletin* 12 (November 1932):14; 16 (November 1936):15; "Miracles and Melodies," *MM* 42 (April 1942):487. Figures on radio staff itineraries compiled from *Annual Report of the Radio Department of the Moody Bible Institute of Chicago* for the years 1929–1941; "President's Report," *Moody Bible Institute Bulletin* 17 (October 1937):3; typescript table taken from file six, "Enrollment," The Moodyana Collection, Moody Bible Institute; "And Now For 50,000," *MM* 41 (September 1940):4; *N. W. Ayer and Son's Dictionary of Newspapers and Periodicals,* 65th anniversary edition (Philadelphia, 1933).

Baptist pastor in Boston proper was either a Gordon alumnus, professor or trustee.[25] In Minnesota, William Bell Riley, the pastor of Minneapolis First Baptist Church, held virtually a fundamentalist bishopric by virtue of the 75 pastors statewide who had attended his Northwestern Bible and Missionary Training School.[26] BIOLA had 180 alumni Christian workers in California by 1939.[27] Considering all the activity Bible institutes engaged in, the influence they wielded through direct contact and alumni, and the support they received, it is no wonder that one confused reader of the *Moody Monthly* asked, "Why don't you publish something on the other denominations once in a while?"[28]

Fundamentalists who desired a Christian liberal arts education for their children in the 1930s sought it for the most part outside the movement proper. The fundamentalists themselves operated only a few such schools, notably Wheaton College near Chicago and Bob Jones College then located in Cleveland, Tennessee, while Gordon College of Missions and Theology in Boston was developing an arts and sciences division. Advertisements in fundamentalist periodicals show, however, that colleges sponsored by other evangelicals, including Taylor University in Upland, Indiana and Grove City College in Pennsylvania, attracted students from fundamentalist congregations.[29] These evangelical colleges prospered during the thirties. A survey of evangelical higher education in 1948 found that the total enrollment of seventy such schools in the United States doubled between 1929 and 1940.[30]

Wheaton College, founded in 1857, provides perhaps the most striking example of the rapid growth of fundamentalist higher education. J. Oliver Buswell, Wheaton's president from 1926–1940, labored to improve its enrollment and academic standing. During his administration the college won a high accreditation rating, and for three years Wheaton led all liberal arts colleges in growth nationwide. By 1941 Wheaton's enrollment of 1100, up from about 400 in 1926, led all liberal arts colleges in Illinois.[31] The

---

25. Nathan R. Wood, *A School of Christ* (Boston, 1953), pp. 165–166.
26. "The Sweep of Northwestern Schools," *The Pilot* 17 (January 1937):108.
27. "BIOLA's Workers in the Homelands," *The King's Business* 30 (July 1939):268–270.
28. "A Magazine For All," *MM* 40 (February 1942):249.
29. Several such colleges were advertised in *MM* issues from September 1930 to August 1931.
30. Harry J. Albus, "Christian Education Today," *Christian Life* 1 (September 1948):26, 46; quoted in Louis Gasper, *The Fundamentalist Movement* (The Hague, 1963), p. 104.
31. "The World Is Wondering About Wheaton," *Baptist Bulletin* 3 (March 1938):14; "Wheaton College, 'For Christ and His Kingdom,'" *Baptist Bulletin* 1 (November 1935):5, 11–12; "Dr. Buswell to be President of the National Bible Institute," *SST* 83 (May 24, 1941):434; "Wheaton Annuities," *SST* 79 (September 4, 1937):61; "Wheaton College," *Watchman-Examiner* 29 (October 16, 1941):1067. The survey of higher education in the October 11, 1936 issue of *The New York Times*, section 2, p. 5, confirms the claims for that year.

school had become the "Harvard of the Bible Belt," a producer of such evangelical leaders as theologian Carl F. H. Henry and Billy Graham.[32]

As millions of Americans motored each summer to popular resorts, a growing number of summer Bible conferences competed with tourist camps and resort hotels for the patronage of vacationing fundamentalists.[33] From the Boardwalk Bible Conference in Atlantic City, and the Montrose Summer Gatherings in the Pennsylvania hills, at Winona Lake in Indiana, at Redfeather Lakes in the Colorado Rockies, and at Mount Hermon, California,[34] Bible conferences offered a unique vacation: a blend of resort style recreation, the old-fashioned camp meeting and biblical teaching from leading fundamentalist pulpiteers.[35] Enrollees might hear Harry A. Ironside of Chicago's Moody Memorial Church, Paul Rood of BIOLA or Martin R. DeHaan of the "Radio Bible Class." The conferences offered different programs so one could choose among sessions featuring missions, young people, the pastorate, Bible study, "Victorious Living," prophecy, sacred music, business men, business women, or Sunday school.[36] The lists of forthcoming conferences published each summer by the *Moody Monthly* grew steadily larger during the thirties, from twenty-seven sites and 88 conference sessions in 1930, to over 200 sessions at more than fifty different locations in 1941.[37]

A report in the Baptist *Watchman-Examiner* of the Bible Conference at Winona Lake, Indiana in 1941 portrays the character of such meetings. Each summer, the whole Winona Lake community became a religious resort with thousands of fundamentalists renting cottages and streaming to the conference grounds. The meetings that capped off the 1941 summer schedule at Winona Lake attracted more than 2,000 enrollees, including some 400 ministers. They were joined by perhaps 2,000 more daily visitors. Participants listened to as many as six sermons a day out of the thirteen to fourteen total sessions scheduled between seven in the morning and ten in the evening. The men on the platform included several fundamentalist celebrities: William Bell Riley of First Baptist Church, Minneapolis; Harold T.

32. "Wheaton College, Harvard of the Bible Belt," *Change* 6 (March 1974):17–20; McLoughlin, *Modern Revivalism*, p. 486; Carl F. H. Henry, "Twenty Years a Baptist," *Foundations* 1 (1958):46–47.
33. Dixon Wecter reported that as many as thirty-five million Americans went on vacation trips each summer in the thirties. Wecter, *The Age of the Great Depression, 1924–1941* (New York, 1948), p. 225.
34. "Forthcoming Conferences," *MM* 35 (August 1935):589.
35. "Shall I Go to a Summer Bible Conference?" *SST* 77 (May 18, 1935):337; C. H. Heaton, "The Winona Lake Bible Conference," *Watchman-Examiner* 29 (September 4, 1941):826.
36. "Forthcoming Conferences," *MM* 34 (July 1934):528; "Forthcoming Conferences," *MM* 35 (August 1935):589.
37. "Forthcoming Conferences," *MM* 30 (June 1930):517; "Forthcoming Conferences," *MM* 41 (June 1941):614.

Commons, executive director of the Association of Baptists for the Evangelization of the Orient; and evangelists J. C. Massee, Ralph E. Neighbour and J. Hoffman Cohn. The master of ceremonies was Billy Sunday's former partner Homer Rodeheaver, "the leading song director of America."[38] The reporter sensed a brotherhood at the conference which knew no denominational bounds when he saw that

> a Methodist Bishop, a Baptist evangelist, a Presbyterian professor, a Lutheran pastor, a Christian layman and a Rescue Mission superintendent could stand on the same platform and preach the common tenets of the Christian faith while multitudes of believers wept and rejoiced together as if some glorious news had for the first time burst upon their ears.[39]

Such events were a powerful force for cementing the bonds of commitment within the movement. In 1937 the *Sunday School Times* reported a poll taken at a small Bible college which showed that all but fifteen of the 150 students had attended a summer conference. Sixty-five first accepted Christ or made a recommitment to the Christian life there, and sixty-two claimed that they were in Bible college because of a summer conference. According to the editor, the Bible conference had become "one of the most powerful factors in spiritual life of the church."[40]

In the 1930s the rapidly rising commercial radio industry provided the fundamentalists with a new medium through which to send out their "old gospel" to the rest of the nation. The number of radio sets had doubled between 1930 and 1935 to over eighteen million. By 1938 a *Fortune* survey named radio listening the first preference for leisure time entertainment in America. Fundamentalist preachers quickly took to the airwaves. A casual, reader-contributed directory in the January 23, 1932 *Sunday School Times* lists over 400 evangelical programs on eighty different stations nationwide.[41]

Interest in religious broadcasting was not limited to the fundamentalist movement. For a time they and other evangelicals feared restrictive network policies would force them off the airwaves. The Federal Council of Churches, the United Synagogues of America and the National Catholic Welfare Conference cooperated with the Columbia Broadcasting System and the National Broadcasting Corporation to produce nonsectarian programs on free network time. The CBS "Church of the Air" featured such prominent preachers as Harry Emerson Fosdick, Bishop Fulton J. Sheen and Rabbi Stephen S. Wise. This venture reflected the intention of CBS and NBC to

38. C. H. Heaton, "Winona Lake Bible Conference," p. 826.
39. Ibid.
40. "Why Attend A Summer Conference?" *SST* 79 (May 15, 1937):348.
41. Herman S. Hettinger, "Broadcasting in the United States" and Spencer J. Miller, "Radio and Religion," *Annals of the American Academy of Political and Social Sciences* 177 (January 1935):6, 140; "Fortune Survey: Radio Favorites," *Fortune* (January 1938): 88; "A Directory of Evangelical Radio Broadcasts," *SST* 74 (January 23, 1932):44-45.

limit religious broadcasting to a few hours a week and to "representative" national religious bodies.[42] Father Charles E. Coughlin's controversial radio blasts over the CBS network had led that network to adopt a policy that would insure bland, "safe" religious programming.[43] This change directly affected the evangelical broadcasters, especially the fundamentalists, many of whom had paid for network time. These preachers were often controversial or sectarian in tone and received no place in ecumenical broadcasting schemes. Yet the fundamentalists and friends were by no means driven from the air. The religious programs were too attractive a market for commercial stations to turn down, and hundreds of local stations sold them time, as did the new Mutual Broadcasting System until 1944.[44] Indeed, it became clear by the late thirties that paid programs drew the greater share of popular support. Charles E. Fuller's weekly "Old-Fashioned Revival Hour" became the most popular religious program in the country.[45]

The "Old-Fashioned Revival Hour" climbed rapidly to national prominence. From modest beginnings in 1925, Charles E. Fuller, fundamentalist pastor of Calvary Church in the Los Angeles suburb of Placentia, expanded the work in its early years to include three weekday broadcasts, two Sunday broadcasts from Calvary Church and a Sunday broadcast sponsored by BIOLA over the CBS Pacific Coast network. Fuller left his church for full-time radio ministry in 1933 and soon was heard each Sunday on the Mutual Network. Six years later the "Old-Fashioned Revival Hour" was broadcast weekly coast to coast and overseas to an estimated fifteen to twenty million listeners. Fuller's coverage consisted of 152 stations in 1939 and 456 three years later, the largest single release of any prime time radio broadcast in America.[46]

As a whole fundamentalist forays into national broadcasting were immensely successful. Other programs captured regional and national audiences, most notably Martin R. DeHaan's "Radio Bible Class," Philadelphia Presbyterian Donald Gray Barnhouse's "Bible Study Hour", and the "Miracles and Melodies" series transcribed by Moody Bible Institute's studio. Programs with smaller coverage supplemented them to fill the

---

42. Miller, "Radio and Religion," pp. 136–139; Fuller, *Give the Winds a Mighty Voice*, pp. 101–103.

43. "Directory of Evangelical Radio Broadcasts," p. 44; "Another Year of Miracle Gospel Broadcast," *SST* 81 (October 21, 1939):720–722.

44. Fuller, *Give the Winds a Mighty Voice*, pp. 151–157.

45. "Another Year of Miracle Radio Broadcast," p. 720; Fuller, *Give the Winds a Mighty Voice*, pp. 113–122.

46. Untitled listing of radio programs, *SST* 73 (April 12, 1931):184; "*The Sunday School Times* Radio Directory," *SST* 73 (May 30, 1931):313; Charles E. Fuller File, Baptist Ministers and Missionaries Benefit Board Registry, the Samuel Colgate Library of the American Baptist Historical Society, Rochester, New York; Gasper, *The Fundamentalist Movement*, p. 77; "Another Year of Miracle Gospel Broadcast," p. 720; Fuller, *Give the Winds a Mighty Voice*, p. 140.

airwaves with the old-time gospel. More than any other medium, radio kept revivalistic religion before the American public.[47]

Of all the activities pursued by both fundamentalists and major Protestant denominations during the 1930s, their foreign missionary work portrayed most starkly their contrasting fortunes. The great missionary enterprise of the Protestant churches had entered the twentieth century with unbounded hope and zeal; but liberal disillusion with evangelism, inflation and constituents' dislike of liberal programs depleted the denominational mission budgets and stifled enthusiastic young volunteers. For instance, the Northern Baptist Convention experienced an extremely heavy decline in its mission program. Its staff dwindled from 845 in 1930 to 508 in 1940. The year 1936 was particularly disastrous as NBC contributions for missions totaled $2.26 million, down 45 percent from 1920. That year no new missionaries went out, and many in the field came home for lack of money.[48]

Fundamentalists wanted missionaries who preached the old gospel of individual repentance and redemption. They recoiled from the denominational boards because of alleged theological liberalism, social gospel programs and high overhead costs.[49] But fundamentalist interest in missions did not flag. Fundamentalists supported independent, "faith" missions which were not denominationally connected and did not solicit funds directly. They also founded new denominational agencies. While the Laymen's Foreign Missions Inquiry reported in 1932 that evangelism in missions was passé, the fundamentalist-backed missions grew stronger, better financed, more evangelistically aggressive and more successful in recruiting volunteers than ever before.[50] The China Inland Mission (CIM), a giant among the independents, experienced the greatest growth of its history during the thirties. Even though China was then involved in conflict with Japan and suffering internal strife, CIM sent out 629 new missionaries in 1930-1936, for a total force of almost 1400.[51] CIM was but one of a growing group of

47. Gasper, *The Fundamentalist Movement*, pp. 19-20, 76-78; George A. Dollar, *A History of Fundamentalism* (Greenville, S. C., 1973), pp. 255-257; "Hear WMBI Favorites On Your Station," *MM* 41 (September 1940):31.

48. Curtis Lee Laws, "Shall Baptists Go Out of Business?" *Watchman-Examiner* 24 (January 2, 1936):13; "The Tragedy of the Northern Baptist Convention," *Watchman-Examiner* 24 (June 11, 1936):699; "The Tragedy of It All," *Baptist Bulletin* 6 (July 1940):1; *Annual of the Northern Baptist Convention, 1937* (Philadelphia, 1937), p. 28.

49. A classic fundamentalist exposé of the missions situation is in Robert T. Ketcham, *Facts for Baptists To Face* (Chicago, 1937), pp. 5-15. Lewis A. Brown, "A Missionary Speaks Plainly," *Watchman-Examiner* 25 (March 18, 1937):300, and Carey S. Thomas, "Is Non-Cooperation Justifiable?" *Watchman-Examiner* 25 (February 18, 1937):179-181, are lamentations of the nonsupport of conservatives.

50. Handy, *A Christian America*, pp. 190-196. William E. Hocking, ed., *Rethinking Missions* (New York, 1932) is the major report of the findings of the Laymen's Foreign Missions Inquiry.

51. Robert Hall Glover, "What Is a Faith Mission?" *Missionary Review of the World* 58 (September 1935):409-411; "Suggestions for Your Christmas Giving," *SST* 73 (December 26, 1931):737; Ernest Gordon, "A Survey of Religious Life and Thought," *SST* 81 (June 24, 1939):430; Robert Hall Glover, "Decrease in Missions Giving—Its Real Cause and Cure," *Revelation* 7 (June 1937):241.

independent fundamentalist missions. Each year, the *Sunday School Times* published a list of fundamentalist mission agencies, which showed forty-nine in 1931 and seventy-six by 1941. These missions ranged in size from the tiny Layyah Barakat (Syria) Home for Orphan Girls to the Sudan Interior Mission, which received $250,000 in 1937 and doubled its army of missionaries during the decade.[52]

These missions worked in close association with Bible institutes which trained missionaries, housed mission offices and helped raise funds. From the Moody Bible Institute alone came over 550 new missionaries from 1930 to 1941, while BIOLA housed both the Orinoco River Mission and the United Aborigines Mission offices.[53] Unfortunately, we know little about this wave of missionary recruits from the Bible institutes and evangelical colleges. Likewise, we know next to nothing of the collective impact of the independent boards and those of conservative denominations since 1900. Yet this brief glimpse at activities during the 1930s shows that a movement of great proportions was underway. Evangelical fervor for missions generated by the Student Volunteer Movement had not died but rather had changed its institutional base. As a traditional indicator of religious vitality, missionary activity demonstrated the vigor of fundamentalism no less than the movement's other enterprises.

In these four areas of fundamentalist activity—education, summer Bible conferences, radio broadcasting and foreign missions—the evidence shows a growing, dynamic movement. Other activities thrived also: publishing houses such as Fleming H. Revell, Loizeaux Brothers and Moody Press; and seminaries, notably Evangelical Theological College in Dallas, Texas, and Westminster Theological Seminary of Philadelphia, Pennsylvania. Even this brief survey, however, demonstrates that the fundamentalist movement did not decline during the thirties. Rather, there was a shift of emphasis within the movement. Fundamentalist efforts to cleanse the denominations of liberal trends had seemed to fail. Rather than persisting along the 1920s lines of conflict, fundamentalists during the 1930s were developing their own institutional base from which to carry on their major purpose: the proclamation of the evangelical gospel.

Was there an "American Religious Depression" among Protestants during the 1930s? Not among fundamentalists and apparently not among other evangelicals either. Fundamentalist activities mentioned here had parallels. The other evangelical groups grew during the 1930s, some very

---

52. "Suggestions for Your Christmas Giving," *SST* 73 (December 26, 1931):737; "A New Missionary Board for the Old Faith," *SST* 76 (May 5, 1934):287; "The Presbyterian Controversy," *MM* 35 (May 1935):411; "Suggestions for Your Christmas Giving," *SST* 83 (December 6, 1941):1010–1111; Gordon, "A Survey of Religious Life and Thought," p. 430.

53. President's yearly reports, *MBI Bulletin* 12 (November 1932):5; 13 (April 30, 1933):3; 14 (November 1934):4; 16 (November 1936):6; 17 (February 1938):3; 18 (February 1939):3; 19 (February 1940):8; 20 (February 1941):6; 21 (February 1942):5; *The Appeal of the Century* (Chicago, ca. 1937), p. 5.

rapidly indeed. The Assemblies of God increased fully fourfold from 47,950 in 1926 to 198,834 members in 1940. The Church of the Nazarene more than doubled its membership from 63,558 to 165,532. The Southern Baptists gained almost 1.5 million members over the same period to total 4,949,174, while the Christian Reformed Church counted 121,755 members in 1940, an increase of 25 percent.[54] Perhaps old-line Protestantism was depressed but popular evangelicalism flourished.

Did the evangelicals provide the impetus for the post–World War II revivals? The fundamentalist community played a leading role. Billy Graham's crusades and other agents of revivalism such as Youth For Christ were not merely throwbacks to the Billy Sunday era. They were the postwar descendants of a continuing revival tradition preserved and transformed by the fundamentalist movement. For instance, Youth For Christ held its first nationwide convention at the Winona Lake Conference in the summer of 1944. Its first president, Torrey Johnson, was a Wheaton College graduate. Of course, Graham was a Wheaton graduate also. His evangelistic team included George Beverly Shea, a former soloist at WMBI, and song leader Cliff Barrows, a Bob Jones College graduate.[55] Revivalism had not died during the depression. Rather, the fundamentalist movement nurtured that tradition, introduced innovations and produced a new generation of revivalists.

The evidence is compelling, therefore: we need a reassessment of the nature and influence of fundamentalism. The revivalistic, millenarian movement that flourished in the urban centers of North America in the late nineteenth and early twentieth centuries continued under the banner of fundamentalism and left no break in the line of succession from Dwight L. Moody to Billy Graham. Fundamentalism bears all the marks of a popular religious movement which drew only part of its identity from opposition to liberal trends in the denominations. The movement had its own ideology and program to pursue. As Ernest R. Sandeen has shown, millenarian eschatology was an important ideological component.[56] Yet fundamentalism's commitment to urban evangelism and foreign missions suggests that the movement was primarily concerned with preaching the evangelical gospel in the twentieth century, both at home and abroad. The evidence shows that it pursued this goal with increasing success during the 1930s.

Once again, as had happened so many times in the past, part of Christianity had taken the form of a vigorous popular movement. Fundamentalists surged out of the bonds of older denominational structures to

54. United States Department of Commerce, Bureau of the Census, *Denominations;* Benson Y. Landis, ed., *Yearbook of American Churches, 1941* (New York, 1941); compare Ahlstrom, *A Religious History of the American People,* p. 920.
55. McLoughlin, *Modern Revivalism,* pp. 480–487.
56. Sandeen, *Roots of Fundamentalism,* pp. xviii–xxiii.

create flexible, dynamic institutions, such as independent mission agencies, radio programs and Bible schools. Despite or perhaps in part because of opposition, the movement grew. According to anthropologists Luther P. Gerlach and Virginia H. Hine, movements arise to implement changes, to pursue goals that people think the established order is unsuccessful in attaining. Thus, a movement often grows in opposition to the established order from which it came. Because movements are decentralized and based on popular support, they are virtually irrepressible.[57] So it has been with fundamentalism. This widely dispersed network of conservative evangelicals became increasingly at odds with the old-line Protestant establishment. Defeats in the denominational conflicts of the 1920s forced fundamentalists to strengthen their own institutional structures outside of old-line denominations. They responded creatively to the trends in contemporary popular culture and made a lasting place for themselves in American Protestantism. Fundamentalists and other evangelicals prospered. The outlines of a changed Protestant order began to emerge by 1950. However, the task of studying the growth of popular evangelical movements in the context of American cultural history remains. How these movements were involved in the larger process of cultural change has yet to be seen.

57. Luther P. Gerlach and Virginia H. Hine, *People, Power, Change: Movements of Social Transformation* (Indianapolis, 1970), pp. xvi–xix.

C. Allyn Russell

A previous contributor to the JOURNAL, Dr. Russell is
Professor of Religion, Boston University,
The College of Liberal Arts.

William Jennings Bryan:
Statesman—Fundamentalist

WILLIAM JENNINGS BRYAN, LAWYER, professional politician, unequaled American orator, editor, social-political reformer, self-proclaimed and widely accepted "champion of the people," and popular lay-preacher, became fundamentalism's best-known spokesman. National in his influence in contrast to the regional impact of other representatives of ultraconservative theology, the colorful Bryan in his twilight years brought fundamentalism to the attention of the masses through his relentless opposition to the theory of Darwinian evolution.

Bryan's fundamentalism was not an appendage of later years. The ingredients of that theological tendency and life-style had been with him from his earliest days. Yet this charismatic leader, to whom the fundamentalists looked for deliverance from the threat of religious liberalism as once farmers and blue-collar workers had looked to him for rescue from the monied interests of the East, did not fit many of the traditional fundamentalist stereotypes. He was as unique in person and in emphasis as he was in the quality and persuasiveness of his legendary voice. Following a cursory review of Bryan's life and thought, this article will trace the development of his religious posture, with particular emphasis upon parental training; thereafter it will seek to indicate the indigenous nature of Bryan's fundamentalism.[1] Such a study is important because it further illustrates the considerable diversity among fundamentalist leaders; it gives us a truer picture of the "great Commoner" (as he came to be called) than that of the aged, outmaneuvered, humiliated anti-evolutionist at the Scopes Trial.

Moreover, in our own day when many are calling for renewed integrity in government, it shows us the difficulties in the application of religious principles to national and international political problems.

I

William Jennings Bryan was born of rugged Irish-English-Scotch ancestry in Salem, Illinois, 19 March 1860.[2] He gained his education at Illinois College, where he majored in classical studies, and at the Union College of Law in Chicago. Bryan initially practiced law in Jacksonville, Illinois, and later in Lincoln, Nebraska. In 1884, the fledgling, impecunious attorney married Mary Baird, the educated and refined daughter of a prosperous merchant. To be of assistance to her husband, she also studied law and was admitted to the bar just four years later. She was the only woman in her class at the Union College of Law, and, scholastically, ranked third among seventeen candidates.

Bryan entered upon his political career in earnest in 1891 after failing to distinguish himself as a lawyer. Twice the people of the First Congressional District of Nebraska elected him to serve as their congressman in the House of Representatives of the United States. A campaign for the Senate followed. Although Bryan, a Democrat, captured a substantial majority of the votes, he lost the race in the midst of a Republican landslide because at that time senators were chosen by members of the state legislature rather than by popular election. Following this defeat, Bryan became editor-in-chief of the *Omaha World-Herald,* and continued as a popular lecturer on bimetallism. Then came the dramatic campaign of 1896, which was the first of three occasions when he was the Democratic Party's nominee for President of the United States.

At the youthful and unprecedented age of thirty-six Bryan won the nomination "by a perfect blend of oratorical brilliance, political finesse, and sheer luck."[3] The key issue which he espoused to handle the financial difficulties of the day was the free and unlimited coinage of silver without waiting for the aid or consent of any other nation. The emotional turning point of the nominating convention at Chicago was reached when Bryan concluded an eloquent and moving address with classical words based upon biblical metaphors:

> Having behind us the producing masses of this nation and the world, supported by the commercial interests, the laboring interests, and the toilers everywhere, we will answer their demand for a gold standard by saying to them: "you shall not press down upon the brow of labor this crown of thorns, you shall not crucify mankind upon a cross of gold."[4]

Despite receiving a total popular vote larger than that of any previously elected president, Bryan lost his bid for the nation's highest office to William McKinley who received 271 electoral votes to Bryan's 176. The popular vote was much closer with McKinley drawing 7,107,822 or 50.88 per cent to Bryan's 6,511,073, or 46.77 per cent.[5] While the apex of Bryan's political career had already been reached in 1907, he continued as the standardbearer of his party for nearly twenty years. In 1900 he ran once more against McKinley on a platform which advocated "free silver" and opposed "imperialism," which he applied to his country's annexation of the Philippines. In 1908, Bryan's opponent was William Howard Taft, supported by Theodore Roosevelt. The issue was the nature and role of financial trusts, although personal charges and counter-charges beclouded the campaign.[6] In each instance, Bryan lost more decisively than in 1896, capturing 155 electoral votes of a possible 477 in 1900 and winning 162 of a possible 482 in 1908.

In 1898, at the time of the Spanish-American War, Bryan, the erstwhile proponent of world peace, organized and commanded a regiment of soldiers from Nebraska, holding the rank of Colonel. On the day the war ended, he resigned his position in the Army as an expression of opposition to the developing imperialism of his country. In 1901, after the second of his unsuccessful efforts to gain the presidency, Bryan founded and edited a weekly journal called *The Commoner* which became an important channel for the dissemination of his political, social, and religious views.

Frustrated three times in his own attempt to become the leader of his country, the dauntless Bryan persisted in state and national politics until destiny and political opportunism provided him the occasion to become a "kingmaker" for someone else. When the Democrats seemed hopelessly locked in their endeavor to choose a presidential candidate at Baltimore in 1912, Bryan strategically switched his loyalties from Champ Clark of Missouri, Speaker of the House of Representatives, to Woodrow Wilson of New Jersey.[7] This unexpected support eventually led to Wilson's nomination on the forty-sixth ballot, and ultimately to his election as President. The reward for Bryan was an appointment as Secretary of State in Wilson's cabinet, a position for which he was eminently unqualified. Bryan, however, held the office with some distinction from 1913 to 1915. The achievement in which he took the greatest pride was the successful negotiation of peace treaties with thirty foreign nations whereby these powers agreed to submit all disputes to an impartial inquiry and to delay military hostilities a full year while arbitration was taking place. Ironically, at the time

of the first World War, President Wilson and his cabinet were unwilling to apply this principle to Germany; whereupon, in protest,
the courageous Bryan resigned as Secretary of State.[8]

During the last decade of his life, while continuing to exercise
leadership in the councils of his party, Bryan gave himself increasingly to religious interests. He continued to lecture on the Chautauqua circuit, which he had been doing since 1896, and where he
was widely hailed as the most inspirational of speakers. He fought
tenaciously against the teaching of evolution in public schools. He
taught an immensely popular Bible class in Miami, Florida, where
he had moved in 1921. In time, he arose to the defense of the fundamentalist movement both within and without his own Presbyterian denomination. The weary warrior of many a political and
religious campaign died at Dayton, Tennessee, 26 July 1925, five
short days after the conclusion of the renowned Scopes Trial.
Paradoxically, the intermittent pacifist was buried, at his own re-
quest, in the national military cemetery at Arlington, Virginia.

II

For many, the Scopes Trial has greatly distorted the image of
William Jennings Bryan. If he had died a year earlier, the public
undoubtedly would have remembered him as the political-social
reformer which he was—for good or ill depending upon one's evaluation of individual reforms—rather than as the zealous defender
of a literalist interpretation of the Scriptures. As a reformer, Bryan
had championed the cause of the underprivileged, especially the
farmer and other laborers, endeavoring to assist them in securing
equal rights with those whom he thought were the overprivileged.
While living in an era that was dominated by a passion for social
progress, including reform in politics, business, and morals, Bryan
had been frequently in the forefront of the struggle, both in time
and in program. Actually, his political-social views had been an
unusual combination of progressive and conservative elements
invariably intertwined with his religious beliefs.

Among the reforms which Bryan championed and which eventually became legalized in his own day were the popular election
of United States senators (supported in part as the result of his
own experience), a federal income tax, women's suffrage, national
prohibition, the publication of the names of the owners of newspapers, the creation of a Department of Labor in the national government with its executive officer holding cabinet rank, the outlawing of contributions by corporations to national campaign
funds, the publication of individual campaign contributions *be-*

fore elections, the enlargement of the powers of the Interstate Commerce Commission, tighter governmental control of such diverse groups as the meat-packing industry and the railroads, and the introduction of the initiative and referendum into contemporary voting practices.

The Great Commoner fought for other causes before their time had come, reflecting his forward-looking and crusading spirit. These included the limitation of the presidency to a single six-year term (advocated by him *prior* to each of his own three candidacies); a national primary; the abolition of the electoral college with its replacement by direct, or popular, voting for the presidential office; a bipartisan national bulletin for the purposes of keeping the electorate informed on political issues; compulsory signing of all newspaper editorials; a referendum on war except in case of actual invasion; a national minimum living wage law; and, a change in the Constitution to make it more easily amendable.[9] Furthermore, Bryan favored a single standard of sexual morality and, in conjunction with his support of women's suffrage, believed that the vote of women, when it became a reality, would outlaw both war and the legalized saloon.

In the realm of international politics, Bryan hailed the League of Nations as "the greatest step toward peace in 1,000 years" stressing the fact that the substitution of reason for force (as he had advocated when Secretary of State) was in itself an epoch-making advance.[10] But not even Bryan's attempt to secure world peace received more attention or reflected greater conviction than did his struggle against alcoholic beverages.

Throughout his lifetime Bryan saw intoxicants as the nation's worst enemy and the liquor traffic its greatest evil.[11] He viewed the drinking of alcohol as a sin against the individual, against society, and against God.[12] Drink, he reasoned, brought no advantage, decreased a man's efficiency, imparted a constitutional weakness to one's offspring, was a waste of money, formed a dangerous habit, caused poverty and crime, and, from a Christian standpoint, provided a poor example.[13] Bryan did more than "preach" against the practice of drinking. He was a total abstainer who signed the temperance pledge even before he knew its meaning. As a Bible teacher he asserted that the wine mentioned in the account of Jesus' first miracle at Cana of Galilee was, in reality, unfermented grape juice, and used the New Testament record to attack the liquor interests.[14] When Secretary of State, with President Wilson's approval, Bryan and his wife declined to serve alcoholic beverages at social occasions, substituting instead White Rock water and grape juice. At one state dinner the Russian ambassador told his dinner partner that it was the first time that he had tasted water in years.[15]

As a politician, Bryan initially favored local option. Later, in 1908, he advocated county option, state prohibition in 1914—and, finally, the cause of national prohibition. No one was a more ardent spokesman for outlawing liquor than the tireless Nebraskan who criss-crossed the country speaking with the zeal of an evangelist on behalf of the Eighteenth Amendment. To legalize alcohol, Bryan told the crowds, was insanity. "How absurd it is to license a man to make men drunk and then fine men for getting drunk . . . it [is] like licensing a person to spread the itch and then fining the people for scratching."[16]

Bryan's personal presence at the State Department when the Eighteenth Amendment was signed into law was high recognition of his leadership in making America "dry." Furthermore, the National Dry Federation (of which he was then president) presented Bryan with a Silver Loving Cup for his gallantry in the cause of prohibition. Shortly thereafter, in his sixty-first year, Bryan optimistically predicted that within his lifetime he expected to see the outlawing of the saloon "in every civilized nation."[17]

There was no doubt that Bryan, along with Billy Sunday, had been America's foremost spokesman for the cause of temperance. Yet there was something deeply ironical about the fact that the portly Bryan with a gargantuan appetite could be so abstemious about drink, and so undisciplined about food. It was a lifelong vice, for as a boy Bryan had carried extra pieces of bread in his pockets to satisfy his appetite between meals. As an adult, in the last years of his life, he carried a pith helmet filled with radishes also to satisfy his hunger. On one occasion, the constantly hungry but never "thirsty" Bryan said with deep conviction, ". . . I am fond of radishes; my good wife knows it and keeps me supplied with them when she can. I eat radishes in the morning; I eat radishes at noon; I eat radishes at night; I like radishes. . . ."[18]

Frequently progressive in an era known for its progressiveness, Bryan was not without his conservative counterpoints, the most notable of which was his attitude on race. For Bryan, the voice of the people was the voice of God (vox populi; vox Dei) but this hardly applied when it came to the voice of the Negro. The Negro's voice had been muted by centuries of slavery and, despite theoretical emancipation in Bryan's time, the black person in the South was still imprisoned by an educational qualification for suffrage which virtually left him without expression. Because Bryan's strength came in large measure from the South, he could not afford to lose that support by pronouncements of a too-liberal nature on racial issues. His political dilemma and personal roots compounded the situation. With the white southern vote already in his column, realism dictated that the black vote, at least in the South, was not crucially needed. Consequently, in the words of Louis W. Koenig, "Bryan faced the political task of reassuring his Southern supporters, while subtly encouraging a Negro breakaway from the Republicans."[19]

Bryan's specific statements appeared to reflect more his own southern cultural heritage than his religious idealism of brotherhood and understanding. He declared that the white man had a more advanced civilization and higher ideals than the black man; that the white people of the South, if in control, would be more apt to deal justly with the blacks than vice versa; and, that the white people already realized that it was in their interest to raise the standard and elevate the condition of the black man.[20] In Bryan's judgment, the presence of an educational qualification for suffrage did not deny the natural and the inalienable rights of the black man. If he could not vote today, then he could look forward to voting tomorrow. Besides, the suffrage amendments in the South which were introduced in self-defense and for self-preservation, were not nearly so severe as the Republican colonial policy in the Philippines![21] Bryan added that the Republican leaders were stirring up racial antagonism in the country to keep the "colored" vote solid for the Republican party. He concluded that the amalgamation of the races, which would be the result of social equality was not the solution to the racial question. The solution was for the "colored" man to establish a reputation for virtue, sobriety, and good sense. Then he could devote himself to the building up of the society which would satisfy his needs.[22]

III

William Jennings Bryan was not a trained theologian, but he liked to speak on religion even more than on politics, and the world knew that he enjoyed doing the latter! As Bryan was progressive in his political-social outlook with few important exceptions, he was conservative in his theological beliefs, also with a few important exceptions. Since he was not a theologian, he never set forth a systematic presentation of his religious ideas. A review of his books and speeches, however, indicates his major Christian emphases.

Bryan was an avowed supernaturalist whose religious convictions were rooted in the Bible. The Scriptures for him were so divinely inspired as to be free from error and to be an infallible authority concerning what God has said and done.[23] Bryan stated flatly that these Scriptures have never yet been incorrect on any issue. The proofs of such inspiration for him were internal ones— fulfilled prophecies, the harmony of the biblical documents, and the unity of their structure despite the fact that they were written by many individuals whose lives spanned many centuries.[24] The Bible, for Bryan, had pragmatic advantages. It gave meaning to existence, supplied each individual with a working plan for life, and answered the longing of the heart with a satisfying knowledge of God.[25] The God revealed by these Scriptures, and in whom Bryan believed, was all-wise, all-powerful, and all-loving. Bryan preferred to begin with such a God and "reason down" from that high premise rather than begin with dull, inanimate matter and "reason up" to God. He argued that belief in this God was not an optional matter. It was the necessary basis of every moral code as well as the required foundation for the establishment of justice and brotherhood among nations and men.[26]

Like most religious conservatives, Bryan gave Jesus a central place in his theology. He stated that the life of Jesus, his teachings, and his death, as well as the general impression that these have made upon the human race, led him to conclude that Jesus was divine. A particularly interesting argument for Jesus' divinity was Bryan's conviction that one cannot contemplate the fact of Christ's life without feeling that in some way that life is related to those now living. "Somehow," he mused, "there is a cord that stretches from the life of Jesus to modern man."[27] Acceptance of Jesus' divinity by Bryan included a belief in the virgin birth, the vicarious atonement, and the bodily resurrection of Christ. Such a supernatural Christ, Bryan declared, is the only Christ of whom the Bible speaks, and to believe his divinity is the easiest way to explain his personality and the miracles which he performed.[28]

Despite his acceptance of a supernatural Christ, Bryan's theology was strongly "this-worldly." He did emphasize immortality, but even in this respect he stressed that a belief in the future life restrains the individual from evil deeds in this life.[29] Unlike many other fundamentalists, he was not a pronounced millenarian.[30] On one occasion he commented that there were too many people who didn't believe in the first coming of Christ to worry about those who didn't believe in the second.[31] Bryan's almost complete silence on the second coming and consequently his silence on premillennialism tied in with his understanding of the social application of the Gospel and the nature of man. Rather than expecting the world to grow progressively worse until the second coming of

Jesus, as the premillennialists believed, Bryan stressed the possibility of a better society in his own day through Christians who applied the teachings of Jesus to every human situation.

Bryan's specific suggestions as to where such applications should be carried out were far-ranging in their scope. He proposed the establishment of church loan societies to assist the "temporarily embarrassed;" he begged for sympathy and fellowship between capital and labor; he condemned the rich tax-dodger and the "corrupter of government" and pled for a new sense of fairness, duty, and patriotism.[32] He also suggested that love, forgiveness, friendship and cooperation be the foundation of a new world order instead of hatred, revenge, and war. In this latter context Bryan was applying to international relations the teaching of Jesus to "love your enemies" and the words of Paul: ". . . if thine enemy hunger, feed him; if he thirst, give him drink; for in so doing thou shalt heap coals of fire on his head" (Romans 12:20). In the bitter days which followed World War I, Bryan declared that there never was a time in the world's history when the returning of good for evil was more imperatively needed.[33]

Such idealism was linked with Bryan's understanding of the nature of man. He did not believe that man was perfect, but neither did he accept man's total depravity as many Calvinists have done. Between the extremes of perfection and total depravity Bryan leaned in the direction of stressing what man might become through "pure purpose and persevering diligence."[34] Surprisingly for one who was so orthodox in other beliefs, Bryan encouraged men to have faith both in themselves and in mankind. He cautioned against confidence in oneself which would lead to egotism, but also indicated that egotism was not the worst of possible faults. Bryan quoted with approval his father who had said on one occasion that if a man had the big head, it could be whittled down, but that it he had the little head, there was no hope for him.[35] Bryan's exhortation was to have faith in one's self, but a faith which was conditioned by moral, intellectual, and physical preparation. Bryan was equally realistic about faith in mankind. He reasoned that it was better to trust others and to be occasionally deceived by them than to be distrustful and live alone.

> Mankind deserves to be trusted. There is something good in every one, and that good responds to sympathy. . . . The heart of mankind is sound; the sense of justice is universal. Trust it, appeal to it, do not violate it. . . . Link yourselves in sympathy with your fellowmen; mingle with them; know them and you will trust them and they will trust you. . . .[36]

This high regard for humanity was at the heart of Bryan's political-social philosophy as well as being important to his theology. As early as 1888 he declared that history teaches that, as a general rule, truth is found among the masses and emanates from them. He stated: "earth has no grander sight than the people moving forward in one compact mass, destroying evil, suppressing wrong, advancing morality. Divinity itself might look with admiration upon such a picture."[37] Those were strong words, indeed, for one who eventually enlisted in the fundamentalist camp.

As one would expect, Bryan placed a high premium upon service. He believed that a wonderful transformation would occur in society when all were animated by a desire to contribute to the public good rather than by an ambition to absorb as much as possible from others.[38] Such service, Bryan stated, was the measure of greatness and happiness for both the individual and the nation. Bryan, the patriot as well as the Christian, especially hoped that the goodness associated with service would be a trait of his own beloved country. He wanted America to "destroy every throne on earth, not by force of violence, but by showing the world something better than a throne—a government resting upon the consent of the governed—strong because it is loved, and loved because it is good."[39]

A knowledge of Bryan's religious tenets makes it easier to understand his deep-rooted opposition to evolution. For him the acceptance of Darwinism raised serious questions about biblical authority because its conception of the origin of man differed from the account in Genesis where man was made by the direct fiat of God. Evolution also undercut social reform since, by this theory, life was governed by the slow, gradual "survival of the fittest," not by the deliberate application of Christian principle. Bryan also argued that this unproven hypothesis weakened faith in God, threatened the orthodox interpretation of Jesus, made a mockery of prayer, lessened the sense of brotherhood, contributed to class struggle, encouraged "brutishness" and undermined belief in immortality.[40] His "answer" to this major threat was to be found in clergymen who taught the biblical account of creation, in trustees who permitted only Christian teachers (meaning those who rejected evolution) to teach in Christian schools, and in taxpayers who prevented the teaching in public schools of atheism, agnosticism, Darwinism, or any other theory that "linked man to the brute."[41] Finally, the hope of the world, in Bryan's judgment lay not in the heartless speculation of the survival of the fittest, but in the conversion of society through the silent influence of a noble Christian example which returned good for evil.[42]

A last belief held by Bryan which is important to an understanding of his life, and one which many biographers have neglected, was his firm conviction in retribution. He believed deeply in a God of justice who in this life stood behind the truth and was able to bring victory to his side. He also proclaimed the existence of a future life in which the righteous would be rewarded and the wicked would be punished.[43] Such convictions led Bryan to say: "One can afford to be in a minority, but he cannot afford to be wrong; if he is in a minority and right, he will someday be in the majority."[44] This belief in the righteousness of his causes and in that elusive "some day" when they—and he—would be vindicated, provided the Great Commoner with much of his motivation. It also made his life—especially the concluding years—all the more melancholy.

IV

The clue to understanding Bryan's religious and political convictions lies in the nature of his training as a child and youth. His parents were deeply religious people who transmitted their faith to Bryan. Consequently, by the time he left home for college their beliefs had become his and these convictions were not to change appreciably in the years which were to follow. Just as he never varied the style of his clothes from youth to old age (he wore a black frock coat) so Bryan did not deviate noticeably from the religious persuasions of his own home.[45]

Bryan's religious conditioning began with his paternal great-grandfather. William Bryan was such a prominent Christian in the Blue Ridge Mountains of Virginia, near Sperryville, that the Baptist Church in his neighborhood became known as the "Bryan Meeting House." Silas Willard Bryan, the father of William Jennings Bryan, inherited such a religious tradition, gave it to his children, and, in the process, experienced religion himself. One event in his life was especially formative. Gravely ill with pneumonia, he prayed earnestly for healing, promising God that if his health were restored he would pray three times daily the rest of his life. When he did regain his health, he kept the promise despite a demanding schedule as lawyer, politician, circuit judge, and farmer.

Bryan's father was a devout Baptist; his mother, Mariah Jennings Bryan, was a conscientious Methodist who eventually transferred her church membership to her husband's congregation in 1877 at the age of forty-three. The parents conducted family devotions regularly as well as faithfully attending church, Sunday School, and prayer meeting. The piety of the family was evident in the practices of Bryan's father who had once used tobacco and snuff but eventually became convinced that these were harmful and gave them up just before the birth of his son. One biographer wrote of the Bryan family: "The Protestant [evangelical] religion—the gospel service, the revival, the emphasis on missions, and the appeal to the heart rather than to the mind in soul-winning—amply sufficed for them, their community, and the Middle West."[46]

While loyal to the Baptist and Methodist churches, Bryan's parents were not rigid denominationalists. Annually they invited the ministers of the churches in Salem, including a Catholic priest, to their home for a social occasion and annually a load of hay was sent at harvest time to each of the same clergymen from the Bryan farm. Of these various churches and their representatives, Bryan indicated that there was agreement upon the "fundamentals" and charity upon all the nonessentials.[47]

Reflecting upon the richness of his own heritage, Bryan commented in his *Memoirs:*

> . . . I was born in the greatest of all ages. No golden ages of the past offered any such opportunity for large service, and therefore, for the enjoyment that comes from [the] consciousness that one has been helpful.
>
> I was born a member of the greatest of all the races—the Caucasian Race, and had mingled in my veins the blood of English, Irish, and Scotch . . .
>
> I was born a citizen of the greatest of all lands. So far as my power to prevent was concerned, I might have been born in the darkest of the continents and among the most backward of earth's peoples. It was a gift of priceless value to see the light in beloved America, and to live under the greatest of the republics of history.
>
> And I was equally fortunate in my family environment. I cannot trace my ancestry beyond the fourth generation and there is not among them . . . one of great wealth or great political or social prominence, but . . . they were honest, industrious, Christian, moral, religious people—not a black sheep in the flock, not a drunkard, not one for whose life I would have to utter an apology. The environment in which my youth was spent was as ideal as any that I know.[48]

With such a high evaluation of his heredity and environment, it is no surprise that Bryan's beliefs did not change during his lifetime. In his opinion he had received the best—he was beginning at the highest level—therefore, there was little, if any, room for growth. The challenge was not to grow; the challenge was to share the legacy of one's good fortune.

The shaping of Bryan's own religious fortune, in addition to the influence received at home, centered in the churches of Salem, Illinois. He attended the Methodist Sunday School in the morning with his mother and the Baptist Sunday School in the afternoon with his father. Bryan was converted at the age of fourteen at a revival service in the Presbyterian Church after which he joined that congregation with seventy other young people. His father had secretly hoped that his son would become a Baptist, but he declined to interfere with his decision, stating: "I am thankful that my son's convictions are sufficiently deep that he has a preference."[49] Although his conversion necessitated no change in patterns of thought or personal habits because of the prior training in his own home, Bryan declared that it had more influence for good upon his life than any other experience.[50] Following his conversion, young Bryan continued to live an exemplary pietistic life. He did not smoke, swear, drink, dance, play cards, or gamble. Furthermore, at the conclusion of his second year in high school, he already believed in the superiority of faith over reason.

Following two years of high school, with his head full of politics, fortified with a protecting insulation of conservative religion and armed with his church letter of transfer, Bryan prepared to meet the "mind worshippers" at Whipple Academy, the preparatory school of Illinois College.[51] At Whipple and at Illinois, he increased his knowledge, majored in classical studies, but clung to earlier convictions. The Bible remained his primary source of authority. Other books which he came to value, in the order of their importance, included the works of Jefferson; Tolstoy's essays; Fairchild's *Moral Philosophy;* Carnegie Simpson's *The Fact of Christ;* the poems of William Cullen Bryant; Plutarch's *Lives;* the writings of Shakespeare; Homer's *Iliad* and *Odyssey,* and the novels of Charles Dickens.[52]

While at college, Bryan joined the Presbyterian Church at Jacksonville and remained a member there until 1902, when he transferred his membership to the First Presbyterian Church of Lincoln, Nebraska. Bryan faced an important vocational decision while at college. As a boy he had given some thought to entering the ministry, but when he learned that to become a Baptist minister (this was before his Presbyterian days) one had to be immersed in water, he quickly gave up the idea of that profession![53] Later, as a lad whose family was about to move to a 500 acre farm, he considered the possibility of becoming a farmer himself and raising pumpkins.[54] Eventually, he settled upon a career as a lawyer, influenced by having heard his father successfully argue cases in court. Finally, he chose to become a politician, motivated largely by the desire to please his mother and to compensate for a narrow political defeat suffered by his father. Bryan's daughter, who may be supposed to have written from intimate knowledge, declared: "I have no doubt that the desire to please his mother, to give her the satisfaction of seeing her son in the position that her husband failed to reach [he had lost a campaign for Congress by a scant 240 votes] greatly influenced my father in his decision to accept his first nomination to the House of Representatives."[55]

One may conclude that early in his career Bryan had adopted a way of life consistent with what fundamentalism later came to be. The Bible, literally interpreted, had become his central religious authority; there was agreement in his family upon the basic doctrines of Christianity; faith was already recognized as superior to reason; and the pietistic life was being followed. Furthermore, Bryan was highly motivated by love for his mother and respect for his father to win an influential Congressional seat. Undoubtedly he also felt that the God of retribution was on his side.

Bryan continued his devout religious stance during the years of his professional political career. In so doing he combined the socially progressive and the personally conservative in a unique way. As a public servant, Bryan's endeavor to bring international peace and his specific condemnation of war were based upon loyalty to Jesus, the Prince of Peace, and a literalist interpretation of Jesus' teaching found in the Bible. For the same reasons, Bryan opposed what he believed was American "imperialism" in the Philippines.

> If true Christianity consists in carrying out in our daily lives the teachings of Christ, who will say that we are commanded to civilize with dynamite and proselyte with the sword?
>
> Imperialism finds no warrant in the Bible. The command, "Go ye into all the world and preach the gospel to every creature," has no gatling gun attachment. Compare . . . the swaggering, bullying brutal doctrine of imperialism with the Golden Rule and the commandment "Thou shalt love thy neighbor as thyself."
>
> Love, not force, was the weapon of the Nazarene; sacrifice for others, not the exploitation of them, was His method of reaching the human heart.[56]

Bryan also found support for numerous progressive causes in the Scriptures, even populism itself; in fact, he was fond of quoting the text that said of Jesus: "the common people heard him gladly."[57] Invariably, whatever his audience, Bryan's speeches and addresses were saturated with biblical quotations, expressions, and metaphors. Probably no politician in American history quoted so copiously from the Scriptures as did Bryan. For this reason, some taunted him for being a dreamer and a visionary, but even for these Bryan had a biblical rejoinder—as well as a touch of humor. He replied that it was not so bad being a dreamer so long as, like Joseph in the Old Testament, one had the corn![58] For others who could not be handled so easily, Bryan manifested love, forbearance, and the turning of the other cheek.

Throughout his public career, Bryan also showed a great deal of moral courage, no more so than at the national convention of his party in San Francisco in 1920. To delegates who were "wet" in their attitudes toward prohibition and altogether hostile to his principles, Bryan pled with deep conviction:

> Are you afraid that we shall lose some votes? O, my countrymen, have more faith in the virtue of the people! If there be any here who would seek the support of those who desire to carry us back into bondage to alcohol, let them remember that it is better to have the gratitude of one soul saved from drink than the applause of a drunken world.[59]

There were other occasions when Bryan's impassioned opposition to the "money men" of the East drew such belligerent responses that his life was threatened—yet Bryan, the politician of moral stamina, declined to retreat from his position and, at the same time, refused physical protection.[60]

In his personal life during the years of his political leadership, Bryan, who believed that religion and politics were entirely consistent, kept the simple, pietistic faith of his fathers. Until his children grew up and went their own separate ways, Bryan led his family in daily devotions, comprised, as they had been in his own home, of Scripture, prayers, and the singing of hymns. To find time to do this was not easy since Bryan worked a twelve to thirteen-hour day—but it was normally done. The Bible was also at Bryan's bedside at night and instinctively, in the time of trial and indecision, he turned to it and to personal prayer for guidance.

Another important facet of Bryan's religious expression was his strict Sabbatarianism, both in theory and in practice. Theoretically, Bryan reasoned that it was natural and proper that the day which is observed religiously by the general public should be selected as the day of rest also, with respect still being shown to those who conscientiously observe another day. He made allowance for variation of practice in different communities on Sundays, but declared that his own experience led him to two basic propositions: (1) every citizen should be guaranteed *time* for rest and for worship, and (2) every citizen should be guaranteed the *peace* and *quiet* necessary for both rest and worship.[61] In practice, Bryan endeavored whenever possible to spend his Sundays at home. He declined to accept speaking engagements of a political nature on Sundays; what few addresses he did deliver were religious talks or "sermons" without compensation.[62] Normally, Bryan was at home on Sundays—whether in Lincoln, Washington, or Miami.

While at home, the members of the Bryan household regularly attended public services of worship. When Bryan moved in 1923 from Lincoln, Nebraska, to his country home in nearby Fairview, he transferred his membership to, and attended, the Westminster Presbyterian Church, where he was elected a ruling elder.[63] Bryan remained a member of this congregation until he took his letter to the First Presbyterian Church of Miami in 1921, although he frequently attended the neighboring Methodist Church as well.[64] When in Washington, the Bryans worshipped at the New York Avenue Presbyterian Church, the church where his funeral services were to be conducted in 1925. While traveling, Bryan visited churches of several faiths—reflecting the denominational-ecumenical pattern established by his parents.

V

Loyal to his religious beliefs both in public and in private, it was a natural move for Bryan, the Christian statesman, to join the fundamentalists and for them initially to welcome his leadership. The issue which drew them together was a strong mutual dislike of religious liberalism including the espousal by many of the liberals of the theory of evolution. Psychologically, this union between Bryan and the fundamentalists was beneficial to both sides. Bryan needed the fundamentalist-liberal controversy as the subjective matter for another crusade after his seemingly successful espousal of women's suffrage and prohibition (as well as after his earlier failures to gain the presidential prize). The fundamentalists were pleased to have a spokesman of such obvious prominence as Bryan who would support their views and give their cause national visibility.

With typical flourish and zeal, Bryan attacked theological liberalism through his books, speeches, and popular "Bryan Bible Talks."[65] The latter were delivered on Sunday mornings at the First Presbyterian Church of Miami, Florida, the city to which the Bryans had moved in 1921. These talks reflected Bryan's fundamentalist theology and eventually drew such large crowds (2,000 to 6,000 people) that it became necessary to hold the meetings outdoors on the edge of Biscayne Bay. The talks were also syndicated in many newspapers throughout the country reaching thereby an estimated fifteen million people.

Bryan accused the religious liberals of disturbing the harmony of the Church and robbing Christian theology of its true meaning. In a sermon with the intriguing title, "They Have Taken Away My Lord," Bryan affirmed that the modernists "have robbed our Saviour of the glory of a virgin birth, of the majesty of His Deity, and of the triumph of His resurrection . . . and are attempting to put in His place a spurious personage, unknown to the Scriptures, and as impotent to satisfy the affections of Christians as a painted doll would be to assuage the sorrow of a mother mourning for her first born."[66]

Bryan touched base on the issues to which the fundamentalists were giving attention. The great question of the day, he affirmed, surpassing all national and international questions, was whether the Bible was true or false. If the Bible is not true, he reasoned, then Christ ceases to be a divine character, and his words are no longer binding upon man. Bryan declared his own belief in the verbal theory of inspiration in the original biblical manuscripts.[67] The proofs of such inspiration, he declared, were not only the words themselves but the influence which these words had exerted upon the hearts and lives of millions of people.[68] Of the virgin birth of Jesus, Bryan argued that no writer in the Scriptures denied it and the only ones who mentioned his birth [Matthew and Luke] mentioned his virgin birth. Bryan stated that there was nothing more mysterious about the birth of Christ than the birth of anybody else. "The birth of every person is a mystery. Christ's birth was simply different."[69] Defending the vicarious atonement of Jesus, Bryan said that winning hearts through love expressed in sacrifice is a natural way of redemption.[70] On the subject of the resurrection, Bryan, in a typically graphic presentation, asserted that Christ, by his resurrection, had made immortality sure. "He has transformed death into a narrow, star-lit strip between the companionship of yesterday and the reunion of tomorrow."[71] Bryan believed in the power and willingness of God to perform miracles and in the actual performance of miracles in biblical and modern times. Of modern miracles, he declared that the feeding of the five thousand with a few loaves and fishes was not nearly so great a mystery as the cleansing of a heart and the changing of a life.[72] He was particularly incensed that the liberals sought to allegorize the biblical miracles. "Give the modernist the words, 'allegorical,' 'poetical,' and 'symbolical' and he can suck the meaning out of every vital doctrine of the Christian Church and every passage of the Bible to which he objects."[73]

Within the context of the liberal-fundamentalist controversy, Bryan had harsh words to say about the biblical scholars known as the "higher critics."[74] He referred to the average higher critics as "men without spiritual vision, without zeal for souls, and without any deep interest in the coming of God's Kingdom." Their opinions, Bryan was convinced, were formed before their investigations. Like many other liberals, in their handling of the Scriptures, they were "tampering with the main spring" and mutilating the inspired biblical books. In Bryan's judgment, they lacked the "spiritual fluids" to digest the miraculous and the supernatural in the Bible. Furthermore, Bryan pointed out, they were opposed to revivals! Such a position prompted Bryan to defend revivals, in general, and the methods of Billy Sunday, in particular.[75]

Politically, Bryan's fortunes with the fundamentalists were about as frustrating as they had been in national politics. Most of his endeavors were within his own denomination although in 1922 he spoke to the Southern Baptist Convention and the Fundamentalist Federation of the Northern Baptist Convention. The latter appearance brought the accusation that he was an "outsider" meddling in a serious family quarrel.[76] Among the Presbyterians, Bryan attended several General Assemblies. In 1923 he ran for moderator of that body in hopes of turning the attention of its delegates to his particular concern, evolution. He was defeated on the third ballot by Charles F. Wishart, probably because "his policy of coyly concealing his desire for office placed his supporters in a disadvantageous position."[77] This was Bryan's last major bid for elected office. A year later, in 1924, Bryan nominated and supported Clarence E. Macartney in a close and successful bid for the moderatorship of the General Assembly. In appreciation, Macartney appointed Bryan Vice Moderator, primarily an honorary position, in the hope that "in a year or two" he might become moderator apart from the conflict and tensions of recent elections.[78] In the Assembly of 1925, Bryan broke with the extreme fundamentalists when he supported the candidacy of Dr. W. O. Thompson, President of Ohio State University, who advocated the settling of denominational problems engendered by the fundamentalist-liberal controversy through peaceful means. Thompson humiliated Bryan by rejecting his support.[79] Thompson's defeat by Charles R. Erdman, a theological conservative, did little to salve Bryan's wounded feelings. By this time, Bryan's political and religious-political fortunes were at an alltime low and the God of retribution seemed embarrassingly far off. Bryan continued to battle, although his testiness, unusual for him, was beginning to show. At a meeting of the Men's Fellowship of the General Assembly in 1925 he criticized both major political parties "for their failure in their platforms to say 'a single word about the greatest need of the world—religion.' His criticism was unsparing. 'One candidate in particular,' he added, 'I should have expected to recognize this need, but he did not. That was my brother.'"[80] Bryan's shrewd brother, Charles, had assisted him in editing *The Commoner*, and had held office first as mayor of Lincoln, Nebraska, and later, with Bryan's active support, as governor of the state.

At this low ebb in his career, Bryan needed a victory badly. He sought it through his sincere and continuing opposition to evolution. Like many of his other convictions, Bryan's opposition to the theory of evolution had come early in life. While in college he had experienced a low-keyed struggle about the origin of man but quickly resolved the difficulty in an act of faith in a personal God who was the creator of all. This subject matter of evolution and creation lay relatively dormant as far as Bryan was concerned until the last five years of his life (1920–1925) when he campaigned ardently against Darwinism with much of his old-time fervor and wit. The catalyst which brought about the change was not only Bryan's need of a cause but the belief that the teaching of evolution as fact rather than theory was destroying or chilling the faith of many college students. It also tied in naturally with Bryan's opposition to liberalism, because he firmly believed that evolution was the basis of liberalism.[81] Evolution, carried to its logical conclusion, he argued, would annihilate revealed religion. In colleges and universities, in state legislatures and religious assemblies, Bryan's theme was much the same:

> . . . let the atheist think what he pleases and say what he thinks to those who are willing to listen to him, but he cannot rightly demand pay from the taxpayers for teaching their children what they do not want taught. The hand that writes the paycheck rules the school. As long as Christians must build Christian colleges in which to teach Christianity, atheists should be required to build their own colleges if they desire to teach atheism . . . modernist teachers . . . by endorsing unproven guesses, undermine confidence in the Bible as a divine authority. . . . With from one to three millions of distinct species in the animal and vegetable world, not a single species has been traced to another. . . . Why should we assume without proof that man is a blood relative of any lower form of life?[82]

The Bryan anti-evolutionist crusade, with fundamentalist support all along the way, led in 1925 to Dayton, Tennessee, the village which proved to be the Waterloo of the religious politician from Nebraska. The facts about the Scopes Trial are so well known that they shall not be repeated here.[83] A few observations, however, are in order. Initially, one letter to Bryan indicates that at least in an indirect way he contributed to the events which brought about the Butler Law with its banning of the teaching of evolution in Tennessee. A Nashville lawyer, W. B. Mann, informed Bryan that a lecture of his entitled, "Is the Bible True?" delivered earlier in the state, had been published and distributed generally. Five hundred copies were sent to the members of the state legislature, including John Washington Butler, the sponsor of the aforementioned bill. Mann theorized that "evidently this caused Mr. Butler to read and think deeply on this subject and prompted him to introduce his bill. '[84] After the bill was introduced, the pamphlets containing Bryan's talk were sent once again to the members of the state legislature. Admittedly, this link between Bryan and Butler was a tenuous one, but it shows at least that Bryan did contribute to the general atmosphere out of which the Butler Law arose.

In perspective, there appeared to be two highlights to the trial itself. One was the persuasive speech of Dudley Field Malone, a member of the team of defense lawyers, and, ironically, the third Undersecretary of State under Bryan during the years 1913–1915. Malone, in what was generally considered the most eloquent speech of the trial, declared that the actual effect of the Butler Bill was the declaration by the Legislature that the truth must not be taught in the schools of the state. For his part, Malone reasoned that the denial of the truth of nature is atheism disguised as religion. He pointed out that truth is of the most imperishable order. "It may inconvenience us, it may disturb us, it may completely upset many of our scientific ideas, it may run counter to our religious views; our duty is not to avoid the consequences of the truth but to face them and overcome them."[85] Malone concluded by declaring that truth, imperishable as it was, needed neither governmental support nor the support of Mr. Bryan. "We feel we stand with fundamental freedom in America. We are not afraid. Where is the fear? We deny it!"[86] When Malone finished his oration, women shrieked their approval and men cheered. After the courtroom had cleared, Bryan said to his former colleague in the state department: "Dudley, that was the greatest speech I ever heard." "Thank you, Mr. Bryan," Malone replied, gathering up his papers, "I am terribly sorry that I was the one who had to do it."[87]

The second highlight of the Scopes Trial was Bryan's unfortunate decision to permit himself to appear on the witness stand to be cross-examined as an authority on the Bible. It was surprising that Bryan, untrained in theology, would subject himself to the merciless and persistent questioning of Clarence Darrow because just three years earlier he [Bryan] had written:

> I know of no reason why the Christian should take upon himself the diffi-cult task of answering all questions and give to the atheist the easy task of asking them. Anyone can ask questions, but not every question can be answered. If I am to discuss creation with an atheist it will be on condition that we ask questions. . . . He may ask the first one if he wishes, but he shall not ask a second one until he answers my first.[88]

Going contrary to his own advice, Bryan permitted Darrow (an agnostic, not an atheist) to crowd him into an embarrassing corner where, among other things, he admitted that the biblical days of creation mentioned in Genesis were not necessarily six days of twenty-four hours, but lengthy periods of time—possibly six mil-lion to six hundred million years. Such an opinion ran directly con-trary to a literalist interpretation of the Scriptures held by the fun-damentalists and, therefore, did much to discredit Bryan in their eyes. This was ironical because on at least one previous occasion, Bryan had expressed the same judgment in a letter to a friend, al-though, it had not, of course, received such widespread publicity as his expression at Dayton.[89] More ironical was the fact that Bryan, who had supported many a fundamentalist cause, and who had been used widely by the fundamentalists to champion their programs, was practically boycotted by the fundamentalist leaders at the Scopes Trial. J. Frank Norris wrote Bryan shortly before the trial telling him that he (Bryan) was engaged in the greatest work of his life, "and [you] are rendering ten thousand times more service to the cause of righteousness than a dozen presidents."[90] Norris also volunteered to provide a court stenog-rapher at his expense to cover the trial. But, when the court opened at Dayton, Norris, T. T. Shields, and W. B. Riley, the President of the World's Christian Fundamentals Association (the organization which had asked Bryan to prosecute John Scopes) were in Seattle, Washington, for the battle royal between the fundamentalists and the liberals at the annual meeting of the Northern Baptist Conven-tion. John Roach Straton originally had agreed to be present at Dayton, but eventually found his vacation retreat in the Adiron-dacks too attractive to leave. J. C. Massee, in the midst of his own defection from the fundamentalists, attended neither Seattle nor Dayton. J. G. Machen wrote Bryan that he did not consider himself a specialist on the subject of evolution. James J. Gray declined Bryan's invitation because of "summer conference work which will keep me on the go until September."[91] P. H. Welshimer, min-ister of the First Christian Church of Canton, Ohio, asserted that he would be unable to attend because his congregation was in the midst of constructing a new building.[92] Alfred W. McCann, a Roman Catholic author whose books dealt with evolution, refused to become a witness because he disapproved of the entire procedure ("men will go on thinking their thoughts regardless of any inhibition or dictum to the contrary").[93] Billy Sunday wrote Bryan indicating that he could not be present, but sent a few "ideas" that he thought might be used (e.g. "if man evolved from a monkey why are there any monkeys left? Why don't they all evolve into humans?") So Bryan, who came to Dayton in failing health, with an invalid wife, and who had not been in a courtroom in twenty-eight years, was left to dangle alone on Darrow's cruel and inescapable hook. While technically the decision of the Bible-reading mountaineers who constituted the jury favored Bryan's side, the verdict of the public went strongly against him. The Great Commoner had lost his final battle.

VI

William Jennings Bryan was a strange and fascinating combination of the conservative and the progressive, a union of what Sydney Ahlstrom has called "the nostalgic and the forward looking."[94] Bryan's conservatism was rooted in his rural, Protestant, evangelical background including a firm grounding in a literalist interpretation of the Scriptures. The seeds of Bryan's progressivism were found in Jeffersonian thought; in the democracy of the American frontier where his ancestors became prominent self-made men and women; and in the application of the same Scriptures, again literally interpreted, to new and broad-ranging social problems. The influence of Bryan's parents, as we have seen, was crucial in shaping his religious beliefs and helping him to become America's foremost layman. Furthermore, Bryan's conviction in the superiority of his inheritance, his religion, his country, and the era in which he was born, gave him considerable motivation but also prevented noticeable growth and led beyond patriotism and pietism to a dangerous chauvinism and a marked expression of self-righteousness. A sense of melancholy surrounded Bryan during his concluding years. After 1896 the trail led downward as the Great Commoner seemed finally deserted not only by politicians and fundamentalist leaders, but by the God of retribution as well. As people viewed Bryan in perspective, few were neutral about his person or contributions. Most agreed that he was America's foremost orator, but after that opinions varied widely.[95] His friends saw him as an idealistic, sincere, honest, courageous, champion of the masses, the real leader of progressivism in his generation. Some termed him the John Bright and the William Gladstone of American Politics; others, in the field of religion, went so far as to call Bryan the greatest lay preacher in the history of America and the St. Paul of the twentieth century.[96]

Bryan's enemies emphasized his lack of originality, called him variously a buffoon, a demagogue, and a deceiver, and stressed his dogmatism, opportunism, self-centeredness, and superficiality. H. L. Mencken wrote that Bryan was "a charlatan, a mountebank, a zany without sense or dignity . . . a peasant come home to the barnyard."[97] Another person wrote: "When I reflect upon what this Nation and perhaps the World escaped through his [Bryan's] successive defeats for the Presidency, I am almost persuaded that there is a Providence which looks after fools, drunkards, and the United States."[98] The truth probably lies somewhere between such polarized extremes.

> . . . no one wants to be remembered for his old age, while the rest of his life is ignored. A fairer label would be based upon all the years of his life rather than those at the end when the flame flickered out of a core of hardened dogma, and that label would have to be neither martyr nor buffoon. Bryan was, like most of us, an individual who had contributed both good and bad, and a fair man would pity him for the bad he brought to the world, and love him for the good.[99]

That interpretation seems even more impressive when one realizes it came from John Scopes! In the category of the "good" mentioned by Scopes, it seems more fitting to remember Bryan for his attempts to gain international peace through arbitration and his opposition to nationalistic imperialism long before others (including many liberals) sought such worthy goals or made such needed protests, than for his excessive claims for the Scriptures in Dayton.

As a fundamentalist, Bryan provides one further illustration that there is no single stereotype into which one or all of the ultra-conservatives may be placed. William Jennings Bryan was one of a kind, primarily because the sphere of national politics in which he worked was different from that in which other fundamentalist leaders found themselves. He was national in influence; they were mostly regional in impact. The company he kept was much more cosmopolitan than that known by other fundamentalists, as he associated with both the humble and the great, the conservative and the liberal. The champion of the common man crossed paths with presidents and counts; the defender of the faith who was linked with the World's Christian Fundamentals Association spoke at the famous Edinburgh Missionary Conference in 1910 and once called the Federal Council of Churches "the greatest religious organization in our nation."[100] Bryan, as we have seen, also gave more attention to political and social reform than did other ultraconservative leaders. Despite believing in original sin, Bryan possessed more faith in human nature to bring about such reform than did most fundamentalists. It was confidence in the average person which led Bryan to oppose what he felt to be elitism in politics, in religion, and in education. Bryan was also more of a pacifist in his leanings than his contemporary conservative stalwarts. One would hardly expect the latter to declare, as Bryan did, that "war is not necessary. It is the philosophy of Nietzsche, not the doctrine of the Nazarene."[101] Furthermore, despite being a fighter for the faith as he understood it, Bryan seemed less hostile to his enemies than did his fellow fundamentalist leaders, many of whom were widely known for their bitter invectives.[102] Lastly, Bryan was neither a millenarian nor a dispensationalist. He preferred to emphasize proper conduct in this life and a general belief in immortality based upon a conviction in retribution—that elusive retribution which seemingly had failed to reward Bryan during his own earthly career. One, however, should be careful not to blame the God of retribution for Bryan's mundane defeats. More likely, notwithstanding his cosmopolitan associations, Bryan was too closely identified with the farmer and the farmer's religion in an age when America was becoming increasingly urbanized and industrialized. More likely Bryan was ahead of his time while Secretary of State but behind his time at Dayton. More likely his ideals were stronger than his organizational abilities to realize them.

While Bryan was this unique fundamentalist, let it be said clearly that he remained a fundamentalist to the end, in belief and in attitude. The last words of his last prepared speech were these: "Faith of our fathers! living still . . . , we will be true to thee till death."[103] Quite properly, therefore, they carved on Bryan's gravestone at Arlington Cemetery the words: "He kept the faith."

NOTES

[1]The primary sources for this article have included Bryan's books plus his personal papers which are located in the Manuscript Division of the Library of Congress, Washington, D. C. Secondary sources dealing with Bryan's life include Paolo E. Coletta, *William Jennings Bryan* (Lincoln: University of Nebraska Press, 1964-1969), 3 vols.; Paul W. Glad, editor, *William Jennings Bryan, A Profile* (New

York: Hill and Wang, 1968); Paxton Hibben, *The Peerless Leader, William Jennings Bryan* (New York: Farrar and Rinehart, Inc., 1929); Louis W. Koenig, *Bryan, A Political Biography of William Jennings Bryan* (New York: G. K. Putnam's Sons, 1971); Lawrence W. Levine, *Defender of the Faith William Jennings Bryan: The Last Decade, 1915–1925* (New York: Oxford University Press, 1965); and Charles M. Wilson, *The Commoner, William Jennings Bryan* (Garden City, New York: Doubleday Co., Inc., 1970).

²The first known Bryan to settle in the United States was William Smith Bryan who came to Gloucester County, Virginia, in 1650. "William Jennings Bryan: Biographical Notes, His Speeches, Letters and Other Writings," enlarged and edited by Grace Dexter Bryan, 10 December 1941, 2. Typescript, BP, Box 64. Hereafter referred to as Grace D. Bryan, "Biographical Notes."

³Koenig, *Bryan*, 178.

⁴William J. Bryan, *The First Battle* (Chicago: W. B. Conkey Company, 1896), 206. Koenig points out that while Bryan's address was a silver speech, it was ingeniously unspecific. "Midway in the uproar which followed Bryan's speech, Altgeld [the governor of Illinois], turned to Clarence Darrow and asked . . . 'I have been thinking over Bryan's speech. What did he say, anyhow?'" Koenig, *Bryan*, 199.

⁵Koenig, *Bryan*, 251f. For other statistical analyses of the campaign of 1896, see: Bryan, *The First Battle*, 607, 611; William Jennings Bryan, *The Memoirs of William Jennings Bryan, by himself and his wife, Mary Baird Bryan* (Philadelphia: The John C. Winston Co., 1925), 267 (hereafter referred to as *Memoirs*); T. Harry Williams, Richard W. Current, and Frank Freidel, *A History of the United States [Since 1865]* (New York: Alfred A. Knopf, 1965), 230. Bryan pointed out that the election was so close that a properly distributed change of less than 25,000 votes in six key states would have given the Democrats a majority in the electoral college.

⁶In a letter written during the course of the campaign of 1908, President Theodore Roosevelt said: "Of course I do not dare in public to express my real opinion of Bryan. He is a kindly man and well-meaning in a weak way. . . . But he is the cheapest fakir we have ever proposed for President." S. E. Forman, *Our Republic* (New York: D. Appleton-Century Co., Inc., 1937), 724.

⁷*Memoirs*, 180–83.

⁸A further disagreement between Bryan and Wilson concerned the matter of neutral ships carrying ammunition to a belligerent nation. Bryan opposed such a policy believing it was contrary to the spirit of neutrality. For his views, Bryan was dubbed a pacifist, pro-German, and disloyal to his own country. Nevertheless, when the United States entered the war, Bryan offered his services, as a private in the Army.

⁹*The Washington Herald*, 7 Dec. 1916; Grace D. Bryan, "Biographical Notes," 305.

¹⁰Grace D. Bryan, "Biographical Notes," 388. Bryan's first recorded advocacy of peace through international arbitration appeared in an editorial in *The Commoner*, 17 Feb. 1905.

¹¹"Prohibition Address by Hon. William Jennings Bryan" (Washington: Government Printing Office, 1916), 11. Pamphlet, BP, Box 49.

¹²William Jennings Bryan, untitled, eleven page, handwritten speech on temperance, c. 1883–1885, 11, BP, Box 1.

¹³William Jennings Bryan, "Why Abstain?" Address delivered at Philadelphia, 15 March 1915, auspices the National Abstainers Union, 1–8. Pamphlet, BP, Box 49. In late nineteenth and early twentieth century America, "temperance" had come to mean total abstinence.

[14]Ferenc Szasz, "Three Fundamentalist Leaders: The Roles of William Bell Riley, John Roach Straton, and William Jennings Bryan in the Fundamentalist-Modernist Controversy," Ph.D. dissertation, the University of Rochester, 1969, 187.

[15]*Memoirs*, 351.

[16]"Prohibition Address," 11.

[17]William Jennings Bryan, *In His Image* (New York: Fleming H. Revell Company, 1922), 216. In this country the Eighteenth Amendment passed Congress in 1917; was ratified in 1919, and became law in 1920.

[18]*Ibid.*, 17.

[19]Koenig, *Bryan*, 449.

[20]William Jennings Bryan, "The Race Problem," *The Commoner*, 2 October 1903, 1.

[21]*Ibid.*

[22]*Ibid.*, 21 August 1903, 1. In 1924, Bryan led a successful movement at the Democratic National Convention against condemning the Ku Klux Klan by name. He believed that the Klan did not need the "advertisement" that such a motion would bring; the Roman Catholic Church did not need political aid with its rich spiritual heritage; and neither his party nor the Christian Church needed the division such a resolution would bring. William Jennings Bryan, "Religious Liberty," speech, Democratic National Convention, New York City, 28 June 1924. Pamphlet, BP, Box 49.

[23]William Jennings Bryan, *Seven Questions in Dispute* (New York: Fleming H. Revell Company, 1924), 16.

[24]*In His Image*, 39.

[25]*Ibid.*, 58.

[26]*Ibid.*, 30.

[27]William Jennings Bryan, *The Prince of Peace*, as found in Grace D. Bryan, "Biographical Notes," 269.

[28]William Jennings Bryan, "They Have Taken Away My Lord," sermon, Philadelphia, Pa., Bethany Presbyterian Church, 17 March 1925. Pamphlet, BP, Box 49.

[29]*In His Image*, 32.

[30]In the preface of his book *Seven Questions in Dispute* (1924), Bryan mentioned that he would give no attention to the question of premillennialism or postmillennialism. He added: "Both schools rely upon the Bible as their authority; it is not a question of inspiration but of interpretation. Both realize that Christ's Second Coming depends upon His first coming." Beyond that, Bryan had nothing to say on the Second Coming or the millennialist positions when discussing the crucial theological issues of his day.

[31]Szasz, "Three Fundamentalist Leaders," 187.

[32]*In His Image*, 144f; 190; 230.

[33]In response to the declaration by President Lowell of Harvard University that militarism must be fought with militarism, Bryan responded: "Those who advocate the policy of 'fighting the devil with fire' seem to overlook two important facts: (1) the devil is better acquainted with fire than his adversaries; (2) being at no expense for fuel he has an economic advantage which tells powerfully in any prolonged contest." *In His Image*, 158.

[34]William Jennings Bryan, "Perfection," handwritten essay, 23 October 1880, BP, Box 49.

[35]*In His Image*, 259.

[36]*Ibid.* Despite his high regard for man Bryan still believed in original sin, declaring that no one would doubt the doctrine if he studied [human] nature and then analyzed himself.

³⁷William Jennings Bryan, "Graduating Address," June 1888 [school not indicated] in Mary D. Bryan, "Biographical Notes," 21.

³⁸*In His Image*, 233.

³⁹Mary D. Bryan, "Biographical Notes," 382.

⁴⁰*In His Image*, 121–34. In his wár against evolution, Bryan relied upon humor as well as logic. On several occasions he said that he was not certain whether man was an improved monkey or whether the monkey was a degenerate man.

⁴¹*Ibid.*, 121f.

⁴²*Memoirs*, 513.

⁴³*In His Image*, 264.

⁴⁴*Ibid.*, 189.

⁴⁵Harry P. Harrison and Karl Detzer, *Culture Under Canvas: The Story of Tent Chautauqua* (New York, 1958), 156.

⁴⁶Coletta, *William Jennings Bryan*, I, 7.

⁴⁷*Memoirs*, 27.

⁴⁸*Ibid.*, 10f.

⁴⁹Alice Vaughan, "Memorials to William Jennings Bryan, at his birthplace, Salem" (DAR, Isaac Hull chapter, Salem, Illinois, February 1931), I, BP, Box 69.

⁵⁰*Memoirs*, 11.

⁵¹Coletta, *William Jennings Bryan*, I.

⁵²Grace D. Bryan, "Biographical Notes," 9.

⁵³*Memoirs*, 17. In older years Bryan justified his choice of a legal career above that of the ministry by a prooftext in Proverbs 21:3. "To do justice and judgment is more acceptable to the Lord than sacrifice."

⁵⁴At the age of forty-four, Bryan's father bought a 500-acre farm about a mile northwest of Salem and built "a mansion there which was the showplace of Marion County." Paola Coletta commented: "The Judge [Bryan's father] might call himself a commoner and often speak on the 'laboring and toiling masses,' yet by the time he retired from the bench in 1872 he was as close to being an aristocrat as rural conditions permitted. Few families in Salem owned 500 acres unencumbered, hired Negro servants, used silver at table, and had a piano in the parlor," Coletta, *William Jennings Bryan*, I, 3, 5.

⁵⁵Grace D. Bryan, "Biographical Notes," 7.

⁵⁶*Memoirs*, 501.

⁵⁷*Ibid.*, 204. Populism refers to a party or cause supported by the people. In America the Populists (People's) Party was formed in 1891. It advocated an increase of the currency, free coinage of silver, public ownership and operation of railroads and telegraphs, an income tax, and limitation in ownership of land.

⁵⁸Clarence E. Macartney, *Six Kings of the American Pulpit* (Philadelphia: The Westminster Press, 1942), 207.

⁵⁹Coletta, *William Jennings Bryan*, III, 188–91.

⁶⁰Bryan caused an uproar at the Democratic Convention of 1912 by proposing that his party declare itself opposed to the nomination of any candidate for president "who is the representative of or under obligation to J. Pierpont Morgan, Thomas R. Ryan, August Belmont, or any other member of the privilege-hunting and favor-seeking class." Mrs. Bryan claimed that the state of excitement was such that one delegate offered $25,000 to anyone who would kill her husband. *Memoirs*, 174–77.

⁶¹*In His Image*, 156.

⁶²*Memoirs*, 452.

⁶³Szasz, "Three Fundamentalist Leaders," 102.

⁶⁴*Ibid.*, 103.

[65]The increased time which Bryan gave to religious interests in older years is reflected in his books and the dates of their publication. *Heart to Heart Appeals* (1917); *The First Commandment* (c. 1919); *The Bible and Its Enemies* (1921); *In His Image* (1922); *Orthodox Christianity versus Modernism* (1923); *Famous Figures of the Old Testament* (c. 1923); *Seven Questions in Dispute* (1924); *Christ and His Companions* (c. 1925); *Memoirs* (1925).

[66]"They Have Taken Away My Lord."

[67]*Seven Questions*, 18.

[68]*Ibid.*, 25.

[69]William Jennings Bryan, "It Is Written," Pamphlet, address at the Southern Bible Conference, Miami, Florida, 17 February 1924, 7. BP, Box 49.

[70]*In His Image*, 80.

[71]*Ibid.*, 91.

[72]*Ibid.*, 120.

[73]*Ibid.*, 106.

[74]*Ibid.*, 40–6.

[75]William Jennings Bryan, "Billy Sunday and His Work," *The Commoner*, December, 1916, 1–2.

[76]Allyn K. Foster, "A Communication," *Christian Century*, 15 June 1922, 755.

[77]Norman F. Furniss, *The Fundamentalist Controversy, 1918–1931* (Hamden, Connecticut: Archon Books, 1963), 133.

[78]Koenig, *Bryan*, 630.

[79]*Ibid.*, Consistent with his opposition to the extreme fundamentalists, Bryan rejected in 1920 the document known as *The Protocols of the Learned Elders of Zion* which maliciously and falsely pictured the Jew as the arch troublemaker of the world. Bryan branded the work a spurious product of an enemy of the Jewish people and proceeded to indicate the many prominent Jews whom he counted as intimate friends. William Jennings Bryan, "The Protocols," *The Commoner*, December 1920.

[80]Koenig, *Bryan*, 631.

[81]"They Have Taken Away My Lord," 10.

[82]*Ibid.*, 13.

[83]For monographs on the Scopes Trial, see: John T. Scopes and James Presley, *Center of the Storm: Memoirs of John Scopes* (New York: Holt, Rinehart and Winston, 1967); Ray Ginger, *Six Days or Forever? Tennessee v. John Thomas Scopes* (Boston: Beacon Press, 1958); Leslie H. Allen, *Bryan and Darrow at Dayton* (New York, 1925); *D-Day at Dayton: Reflections on the Scopes Trial*, edited by Jerry Tompkins (Baton Rouge: Louisiana State University Press, 1965); Ferenc M. Szasz, "The Scopes Trial in Perspective," *Tennessee Historical Quarterly*, Fall 1971.

[84]W. B. Mann to Bryan, 6 July 1925, BP, Box 47.

[85]Dudley Field Malone, "Comments at the Scopes Trial," 9. Nine-page summary, BP, Box 47 (several pages missing).

[86]Lawrence W. Levine, *Defender of the Faith*, 346.

[87]Koenig, *Bryan*, 646.

[88]*In His Image*, 13.

[89]Bryan to Dr. Howard A. Kelly, 22 June 1925. BP, Box 47.

[90]J. Frank Norris to Bryan, 3 June 1925, BP, Box 47.

[91]James M. Gray to Bryan, 22 June 1925, BP, Box 47.

[92]P. H. Welshimer to Bryan, 23 June 1925, BP, Box 47.

[93]Alfred W. McCann to Bryan, 30 June 1925, BP, Box 47. It is possible that the fundamentalist leaders failed to see the importance of the trial or, more likely, felt unable to defend their position when confronted directly by evolutionists.

[94]Sydney E. Ahlstrom, *A Religious History of the American People* (New Haven: Yale University Press, 1972), 879.

[95]Bryan was taunted even about his oratorical skills. Joseph Foraker, a Republican, when asked if he thought Bryan's title, the "Boy Orator of the Platte," was an accurate phrase replied that it was, because the Platte River was six inches deep and six miles wide at the mouth. T. H. Williams, R. N. Current, and Frank Freidel, *A History of the United States [Since 1865]*, Second Edition, 226.

[96]Macartney, *Six Kings*, 200; *Christian Century*, 31 May 1923; *Memoirs*, 493.

[97]*The Vintage Mencken*, gathered by Alistair Cooke (New York: Vintage Books, 1955), 164.

[98]Charles S. Thomas to Professor J. T. Scopes, 26 June 1925; Charles S. Thomas to Bryan, 27 June 1925, BP, Box 47.

[99]Scopes and Presley, *Center of the Storm*, 217.

[100]The reference to the Federal Council of Churches is found in Szasz, "Three Fundamentalist Leaders," 182.

[101]William Jennings Bryan, "Peace and Prohibition," 14. Speech delivered at Democratic National Convention, San Francisco, 2 July 1920. Pamphlet, BP, Box 49.

[102]Szasz, "Three Fundamentalist Leaders," 182. Bryan is an exception to Furniss' judgment that the fundamentalist movement was characterized by violence in thought and language. Furniss, *The Fundamentalist Controversy*, 36.

[103]*Memoirs*, 556.

WILLIAM JENNINGS BRYAN, EVOLUTION,
AND THE
FUNDAMENTALIST-MODERNIST CONTROVERSY

By FERENC M. SZASZ

The Fundamentalist-Modernist controversy of the 1920's caused a social and religious upheaval of major proportions. It produced disruption in almost all the large Protestant churches, especially among the Baptists and the Presbyterians, and in all sections of the country, especially the South. The most important individual connected with the controversy was William Jennings Bryan, three-time candidate for the presidency of the United States. At the time of his death in 1925, five days after his clash with Clarence Darrow at the Scopes trial in Dayton, Tennessee, there were few who did not regard him as the main leader of the Fundamentalist position. The editor of the *Richmond Times-Dispatch* claimed that Bryan personified orthodox Christianity. The popular journalist Glenn Frank went even further. "Mr. Bryan is Fundamentalism," said Frank. "If we can understand him, we can understand Fundamentalism."[1]

Actually, however, these commentators were quite mistaken. Bryan was by no means the main leader of the Fundamentalist movement, and he certainly was not Fundamentalism itself. In fact, Bryan differed considerably from the other important Fundamentalist figures over such items as theology, tactics, and the nature of American society. He played no role at all in the late 19th and early 20th century formation of Fundamentalism, and when he did become involved, in the early years of the 1920's, he changed the nature of the controversy beyond recognition. When Bryan joined the conflict, he brought with him the issue of anti-evolution, which had been on the periphery until that time. But after his entry, the burden of the Fundamentalist message become the passage of anti-evolution laws in every state. Under the umbrella of the constant press coverage which was his wherever he went, Bryan allowed the various factions of Fundamentalism to work together for common goals. But on his death in 1925, each began to split off and go its separate way. The purpose of this article is to show how Bryan and the evolution issue changed the nature of the emerging Fundamentalist movement, and how, in so doing, he forced upon the nation his understanding of the condition of the world in the early 1920's.

First, however, it will be necessary to sketch a bit of background. In general, one may say that the period after the Civil War was not an easy time for American Protestantism. In those years the churches first faced a series of dilemmas which they still have not solved. On a social level they met the rapid growth of industrialism, the influx of Catholic and Jewish immigration, and the sudden rise of large, impersonal cities. On the intellectual level they met the European ideas of Darwinian evolution, comparative study of religion, and higher criticism of the Scriptures. The result of this was to produce an uneasy feeling of isolation in the average American Protestant. At the same time his faith in his fellow man was being shaken by the new immigration, so also was his faith in God being shaken by the new European intellectual currents.

These same years also witnessed the growth of two distinct groups within most of the Protestant denominations, and these, because there are no better terms, must be called "liberal" and "conservative." The liberal elements tended to absorb the idea of evolution (not, however, natural selection); to replace the traditional disputes over theology with a renewed concern over common ethics; and to accept the principles and findings of the Biblical critics. Moreover, by the first decade of the 20th century, the liberals had made considerable inroads into the various seminaries and denominational organizations. They also established their own organization in the Federal Council of Churches.

Conservatives, however, moved steadily in other directions. They harbored a suspicion of evolution, declared that correct theology was as important as shared ethics, rejected most of the ideas of higher critics, and spoke out against the menace of theological liberalism. They, too, began to form their own organizations in the various Bible schools and in permanent Bible conferences.[2]

Much of the early Fundamentalist concern seems to have been directed not toward evolution but toward the issue of higher criticism. The rise of critical Biblical scholarship distressed conservatives, for the goal of the critics — understanding the Bible — was theirs also. Who was to say, however, when the critics disagreed among themselves? As a reaction, conservatives declared that they believed in higher criticism only when it was "rightly used" or "correctly employed." In turn they placed their emphasis upon a literal reading of every Biblical passage. This emphasis was new in Protestantism, for the earlier conflicts over the Bible had generally involved the correct interpretation of passages, not the literal infallibility contained within.

It was not long before fear of critical scholarship began to seep down to a popular level. In 1904 an Anti-Higher Criticism League was formed in New York and that same year a woman threw one of the religious sessions at the St. Louis World's Fair into disruption when she arose and denounced the practice. "We of Saint Louis," she said, "[are] not going to be led astray by this dangerous trifling with the Bible."[3]

The gradual rise of the fundamentalist insistance on Biblical literalism can only be understood as part of the opposition to an equally gradual rise of higher criticism. Biblical criticism seemed to strike at the very heart of American Protestantism. If its basic principles were accepted, the average man could no longer read and understand the Bible for himself. He would be forced to turn to a new elite — the higher critics — to explain to him what God meant in the *Old* and *New Testaments*. Biblical literalism seemed to supply an answer to this dilemma: it alone could restore the Word of God to the ordinary citizen.

It would be a mistake, however, to equate orthodox 19th century evangelical Protestantism with Fundamentalism as it emerged in the 20th century. Ernest R. Sandeen has shown that the various conservative conferences introduced a whole new aspect into the traditional Protestant outlook — that of dispensationalism.[4] Dispensationalism is a complex theology which claims that all time is divided into separate ages or dispensations,

Secretary of State William Jennings Bryan (center, seated) on September 1, 1914, signed peace treaties with representatives of Spain, France, Great Britain, and China. Bryan resigned from the Cabinet in 1916.

that God demands special actions from man during each age, and that each age is brought to a violent close by His direct intervention. The role of the minister, therefore, is to draw out a select body of the faithful before the world is consumed. Numerous lists and charts of the "final days" appeared from 1875 to 1920, and while there might be disagreement as to specifics, all dispensationalists could agree that America was hurrying toward the end of time. It would not be long before Christ would return bodily, draw up the true church of believers into His bosom, and inaugurate the thousand years of peace.[5]

By the second decade of the 20th century, two distinct groups existed within many of the major Protestant denominations. Although close observers feared a split at any moment, the reform spirit of the Progressive Movement and the emotional challenge of World War I both served to mask these internal differences. Thus, an organized Fundamentalist movement did not really emerge until after the Armistice of 1918, when a period of great optimism swept over the nation. Flushed with victory over Kaiser Wilhelm of Germany and the success of prohibition, conservatives organized groups which moved out into the field to sway the various denominations and the nation to their own point of view. "I believe the hour has come," said E. A. Wollam in the magazine of Chicago's Moody Bible Institute, "when the evangelistic forces of this country, primarily the Bible Institutes, should not only rise up in defense of the faith, but should become a united and offensive power." Conservative Baptists began protesting against liberalism in their annual denominational meetings. Massive evangelical conferences were called in 1918 and 1919 and plans were laid for what some considered the "new Reformation." Waves of conservative evangelical speakers swept across the country catching the liberal elements completely off guard as they presented their brand of the gospel message to thousands. This wave of evangelism, one of the largest of the century, was primarily a call to the people to accept God's grace and be saved. Basically, it was couched in positive terms. It did not attack other sects, higher criticism, or evolution.[6] In fact, evolution was not mentioned in the first wave of organized Fundamentalism. That issue arose only with the entrance of William Jennings Bryan.

William Jennings Bryan had little to do with the late 19th and early 20th century origins of Fundamentalism. As far as can be determined, he was in no way connected with the interdenominational Fundamentalist conferences or the Bible schools. He was not even intimately involved with the emerging Fundamentalist wing of his own Presbyterian Church. Although he frequently attended the Presbyterian General Assembly as a delegate, he never went to lobby for any particular position, except, perhaps, prohibition.

From its earliest days Bryan's career was grounded in evangelical Protestantism, but he was never much concerned with theological liberalism or evolution. His chief interest lay in the development of character. For example, when the *Christian Herald* in 1909 asked him to comment on a liberal minister's claim that the 20th century might need a new religion, Bryan said: "The Christian life, is, after all, the unanswerable argument in support of the Christian religion, and the Christian life Dr. Eliott leads will probably aid Christianity more than his words can hurt it."[7] He didn't even mention the dangers of liberalism in a lengthy letter he wrote in 1918 expressing his ideas for improvement in the Presbyterian Church.[8] In fact, the collection of his correspondence at the Library of Congress shows that before 1921 his chief contact with the ministerial community was with theological liberals — men such as Washington Gladden and Charles Stelzle, who were both strong advocates of the social gospel. After the war he was elected to the general committee of the radical Interchurch World Movement of North America. In 1919 he stated that he felt the liberal Federal Council of Churches was the "greatest religious organization in our nation."[9]

Nor can one say that Bryan had a long-standing concern with evolution. Although his most famous lecture, *The Prince of Peace*, which he began delivering shortly after his 1900 defeat, contains a passing reference to evolution, too much can easily be made of this. He never gave any speeches specifically against evolution, and in *The Prince of Peace* he cautioned his listeners that he was not attacking those who did believe in Darwinism. He simply said that he felt more proof was needed. His chief objection to evolution was teleological, for he felt that acceptance of the theory would cause man to lose the consciousness of God's presence in his daily life. Surely there is a difference between a passing comment against evolution and the decision to devote one's whole life to stopping it.[10]

The outbreak of World War I totally changed Bryan's outlook. In 1915 Baptist minister A. C. Dixon convinced him that the rise of German militarism was based not on Friedrich Nietzsche but on the following of Darwinism to its logical conclusion. Shortly thereafter, he came across *The Science of Power* (1918) by Benjamin Kidd which argued the same thesis, and he recommended it to his friends. He was also disturbed by the statistical study of Professor James H. Leuba of Bryn Mawr College which proposed to show that the faith of college students was rapidly declining. As he explained to an audience in Chicago in April, 1923:

So I began to feel a little more earnestly about the effect of Darwinism. I had found that Darwin was undermining the Christian faith, and then I found he had become the basis of the world's most brutal war, and then I found that Benjamin Kidd pointed out that he is the basis of the discord in industry.[11]

Something, obviously, had to be done about this situation.

Bryan had been out of the news since his resignation from Woodrow Wilson's Cabinet in 1915, and from that time until the Armistice, his general reputation was not high. But with his new crusade against evolution, Bryan moved once again back onto the front pages of the newspapers, where he remained until his death. In so doing he effectively captured the emerging Fundamentalist movement. Aware of the problems which the churches were having, the national press seized upon Bryan as the representative figure of Fundamentalism; this, too, helped determine the future path of the controversy. Some conservative leaders such as Episcopalian Bishop W. T. Manning and Presbyterian theologian J. G. Machen ignored the evolution issue. As soon as Bryan began his campaign against evolution, what had been primarily an interchurch controversy was suddenly brought to the attention of the entire nation. Moreover, his championing of anti-evolution so overshadowed all the other aspects of the Fundamentalist position that it alone came to be seen as the center of the movement. With this, Fundamentalism began to shift to the negative, for opposing evolution can only be viewed as the last line of defense.

Bryan first began attacking evolution seriously in the spring of 1920, but so long as he spoke chiefly to church audiences, there was little difficulty. It was not until he took this message to the college campuses that the country began to sit up and take notice. In the fall of 1920, he visited the University of Michigan and spoke there to an estimated 4,500 persons. In his address he soundly attacked Darwinism as a false and vicious mode of thinking. No sooner had he left than he began to receive letters commenting on the controversy he had stirred up there. One which particularly annoyed him came from the Reverend Arthur W. Stalker of the First Methodist Church in Ann Arbor. The Reverend Mr. Stalker criticized him for the false alternatives he had posed and claimed that until his speech, the issue of evolution had been a dead one for most of the Michigan students. He intimated that Bryan would lose his influence with the American people if he continued to speak along those lines. Hurt and annoyed, Bryan fired back an angry letter and then sat down to expand and elaborate his anti-evolution arguments. He printed up 5,000 copies of his new speech, *The Menace of Darwinism*, and sent them out for distribution. A large number were sent to Ann Arbor and he was pleased to discover when he returned that a scientist and philosopher had both taken time to denounce him.[12]

In the fall of 1921, Bryan started a similar controversy when he went to the campus of Middlebury College in Vermont. In early 1922 he caused a real uproar when he denounced evolution at the University of Wisconsin at Madison. The president of the university, E. A. Birge, was furious. After the talk the two men exchanged words and Birge denounced Bryan in the next day's papers.[13] Bryan, in turn, replied that the people of Wisconsin might like to select a new head of their university, one who would not ridicule the faith of the students' parents. He intimated Birge might be an atheist. He also claimed that the taxpayers had a right to determine what should be taught in the university and that they should not tolerate any teaching which might negate the Christian religion. Bryan's controversy with Birge dragged on for over a year and was marked by a generally low tone.[14]

The the press began calling on him. The *New York Times* asked him to present his objection to Darwinism, and he did so in the February 26, 1922, Sunday edition in the article "God and Evolution." This immediately triggered a response from the Reverend Harry Emerson Fosdick whose "Mr. Bryan and Evolution" was carried by the *Times* in March and later reprinted by the *Christian Century*. H. F. Osborn and E. G. Conklin also joined Fosdick in replying to Bryan. Osborn noted:

> Early in the year of 1922, I was suddenly aroused from my reposeful researches in paleontology by an article in the New York *Times* ... by William Jennings Bryan ... and it struck me immediately that Bryan's article was far more able and convincing than any previous utterance of his or any other fundamentalist, and that there should not be a moment's delay in replying to it.[15]

In 1923 *The Forum* asked Bryan to give them an article on "The Fundamentals" and he gladly obliged.[16] It was not long before Bryan was being credited with causing all the nation's Fundamentalist activity.[17] Moreover, with Bryan's entrance the question of the infallibility of the Bible was brought into politics. Had the issue remained in the churches, the outcry would have been considerably less. As it was, however, scientists and educators joined with the liberal Protestants to keep a group of anti-evolutionist Fundamentalists from carrying out their plans.

After 1922 Bryan increased his anti-evolution activity in a steady fashion. Except for the election campaign of 1924 (in which his brother Charles ran for vice-president on the Democratic ticket), he concentrated most of his energy on this cru-

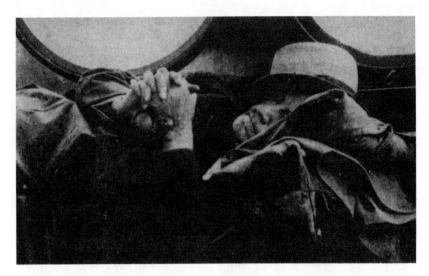

An inveterate Chautauqua speaker, William Jennings Bryan was often forced to sleep on trains with few accommodations. This picture was taken about 1920.

sade. He once wrote that he felt "called . . . to try to save young people from Darwin's false and demoralizing guesses."[18] He spoke before the state legislatures of Florida, Georgia, Kentucky, West Virginia, and Tennessee and on college campuses across the country. He even took his cause to the "enemies' country" of Dartmouth, Harvard, Yale, and Brown. But he told a University of Florida audience that "the colleges of the South [were] less affected than others," and said he expected "the South to lead in the fight against this influence."[19] His most responsive crowds were in the South, and this illustrates the increasing parochialism of his general outlook. He was successful, however, in calling the South to a renewed consciousness of its distinctive religious position.

In addition to his public speaking, Bryan kept his own Northern Presbyterian Church in turmoil over the issue of evolution from 1922 on. He almost split the denomination in two in 1923 when he was narrowly defeated for moderator by Charles D. Wishart. The prospect of the head of a major Protestant church traversing the land denouncing evolution would have caused untold controversy. The next year he helped elect the conservative Clarence E. Macartney as moderator and was, in turn, appointed vice-moderator. He agreed not to run for the head office again unless there was no chance of causing a similar disruption. He opened the 1925 general assembly with a prayer, but did little else. Although he was often seen as one of the leaders of the Presbyterian conservatives, his involvement with the inner workings of the church had never been close. By 1925 he was even moving away from the more radical wing of the Presbyterian Fundamentalists.[20]

It is fascinating to chart the road by which evolution — following in Bryan's wake — came to be seen as the ultimate enemy of evangelical Protestantism. Higher criticism was the real danger to the conservative position, but it was an awkward field for debate. One would have to be fluent in Greek, Hebrew, Latin, and Aramaic to discuss it intelligently. Here was one area gladly given over to scholars. Rationalism, often depicted as a major evil, was an old and tired enemy, and American Christianity has lacked a strong rationalist tradition with which to spar. Each age produced its own defender, but he usually stood conspicuously alone. Clarence Darrow, Joseph Lewis, and Charles Smith were prominent representatives of aggressive rationalism for the 1920's, but even the most generous estimate would label their impact as very mild. Moreover, rationalism was an old foe and could hardly be blamed for the war. Liberalism and modernism were freely bandied about as evils but everyone in the 1920's was in some sense modern, and Bryan, if anything, was a liberal. Evolution served perfectly as an explanation for what was wrong with the nation. It was new, it was different, and it had sinister connotations. Moreover, it was so loosely used that it could include complaints which all held in common and yet allow everyone to keep his own emphasis. As such, it explained Protestantism's failure to capture the 1920's. "God won the war," A. J. Brown reported to the Presbyterian General Assembly in 1924, "but thus far the devil is winning the peace. Few, if any, of the objects for which we declared we were waging war have been achieved."[21] It was through the issue of evolution that Bryan could identify with the Fundamentalist movement, and it was through this issue that they could capture him as their own.

Bryan's national prominence and his general disposition were such that many people and organizations tried to use him for their own benefit. In a sense, the main interdenominational Fundamentalist organizations were especially guilty of this. Their success might well be credited to Bryan's lack of realization as to how much their respective programs and methods differed.

As soon as they discovered their ally, organized Fundamentalism increasingly demanded his aid. T. C. Horton of the Bible Institute of Los Angeles and the Reverend J. W. Porter of Louisville bombarded him with suggestions that he lead a national organization devoted to their cause.[22] He toyed with the idea and even went so far as to design an emblem. His son, William Jennings Bryan, Jr., also talked to Horton about the possibility, but nothing was done. William B. Riley of Minneapolis, director of the World's Christian Fundamentals Association, was successful in getting Bryan to speak twice and urged him, unsuccessfully, to be allowed to arrange a series of additional addresses. Riley was eager to push Bryan to the forefront of his WCFA. In 1922 the WCFA voted to have him head a committee to organize the laymen of the country, but because of his other work and the illness of his wife, he refused.[23] The next year without his knowledge they voted him president of the entire organization. Again he turned them down.[24] Many of the Fundamentalist magazines repeatedly asked him for articles, and when he did give them something, they wanted to use his name on the masthead as an associate editor.[25]

An obvious attempt to use Bryan came in 1924 from Baptist minister J. Frank Norris of Fort Worth, Texas. The governor of Texas was thinking of introducing an anti-evolution bill into the legislature and Norris predicted that two Bryan addresses would change the minds of ten million people. Without Bryan's consent he made arrangements for special trains to bring in the faithful from the surrounding areas to hear him. Bryan toyed with the idea of journeying to Texas, since he also had an invitation from the Texas legislature, but he felt that it would not be worth his effort unless the governor decided to have his forces submit the bill on the house floor. This was not done, so Bryan did not make the trip. He was mortified when Norris offered him $1,000 if he would change his mind. "I am doing the best I can," he replied, "and those who are not satisfied with the amount of work I am doing are, I hope, in a position to do more."[26]

Perhaps the most blatant attempt to use his efforts came from the National Federated Evangelistic Committee. When Bryan expressed general favor with their work, General Secretary James H. Larson made him president of the organization. His friend, W. E. Biederwolf, assured Bryan that he would not have to take any active part, but Larson was a bit too eager. Bryan was surprised to discover that their new stationery prominently displayed his name on the masthead and that Larson had, with much publicity, scheduled an extensive tour for him across the continent. Finally, Bryan sent an angry letter withdrawing entirely from the organization. "While I feel interest in your work," he wrote, "it is not mine, and I will not allow you to decide for me what God wants me to do. . . . I have my work to do and I must do it my way. Your way and my way are entirely different and opposite." Further unctuous pleas from Larson went ignored.[27]

The variegated Fundamentalist organizations had their own ideas for planning Bryan's life, and considering the pressures placed on him, it is surprising that he resisted as well as he did. Bryan was in many cases willing to be used, of course, but one wonders if he realized how vastly different his program was from most of the other organized Fundamentalists. It is doubtful if many of them ever voted for him. The officials of Moody Bible Institute on his death admitted that they never had. Moreover, Bryan and other Fundamentalist figures had very different conceptions of evolution. The Reverend John R. Straton of New York, for example, opposed evolution primarily because it carried with it the idea of relativity in morals. William B. Riley of Minneapolis opposed it on theological grounds. Bryan opposed it, as the recent studies by W. H. Smith and L. W. Levine have shown, because he felt that by destroying regeneration, evolution would also destroy social reform. In his last speech against evolution, he noted that "by paralyzing the hope for reform, it discourages those who labor for the improvement of man's position."[28] The burden of Bryan's inspirational talks was that of service. It was the duty of man, he told countless thousands on the Chautauqua circuit, to overflow with righteous life and become a reformer. But he felt that evolution limited social reform to a slow, gradual process that man could not effect. The doctrines of the survival of the fittest and gradual improvement he saw as paralyzing individual Christian action

*Opposing lawyers at the Scopes trial in Dayton, Tennessee (1925),
were Clarence Darrow, Chicago lawyer (left), for the defence; and
William Jennings Bryan (right), for the prosecution.*

and the application of Christian principles to all aspects of society. This concern does not exist in the writings of the other
Fundamentalists.

There were other differences than those over evolution, too.
For one thing, Bryan had had no theological training and much
of the controversy went over his head. For example, he did not
believe in the pre-millennial return of Christ, a position which
was rapidly becoming a touchstone for much of Fundamentalism. Bryan once commented that there were too many people
who did not believe in the first coming of Christ to worry about
those who didn't believe in the second. While the conservative
movement in the Presbyterian Church was generally, but not

*Scene of the trial was the Rhea County
Courthouse in Dayton.*

entirely, post-millennial, the most active, best-organized Fundamentalists were all pre-millennial. They tried to deny this, but
it seems obvious that the majority held this belief. And, of
course, Bryan was not a dispensationalist. He did not view the
periodic cataclysmic entrance of the Lord into history as an
important item of belief. The burden of much of the northern
Fundamentalist message was the removal of the holy church
from the sinful world. Bryan did not want this at all. He wanted
to *merge* Christianity and the world. This, although he would
have denied it, was almost a liberal position. The issue over
which Presbyterian conservative J. G. Machen left Princeton
was Calvinism, but Bryan was no Calvinist. His comments dur-

When in Lincoln, William Jennings Bryan attended the Methodist Church in Normal, a suburb of the city near his farm. In the latter 1820's a Bryan Memorial Church was built in Miami, Florida.

ing the Scopes trial show that he was not really even a Biblical literalist. Thus, he did not hold the same doctrinal position as the conservatives among either Baptists or Presbyterians.

There were also decided differences between Bryan and the other Fundamentalists over the best tactics to be used in the campaign. Bryan did not approve of strong penalties for violation of an anti-evolution law. His faith in legality was such that he thought the simple passage of a law by a fairly elected legislature would be sufficient. Many of his comrades demanded harsh penalties, even prison terms for such violations. While Bryan had as much scorn heaped upon him in this venture as any of his companions, there was in him no anti-Semitism, no anti-Catholicism, and no ballyhoo. In fact, it was the breadth and depth of Bryan's personality that kept the movement together. When he died in 1925, the varied elements which it contained each went its own way. Glenn Frank's comment that Bryan was Fundamentalism could not have been further from the mark. He was unique, and the Fundamentalists were trying to utilize him for all his worth.

The differences between Bryan and the other Fundamentalists were lost, however, in the issue of evolution; and this connection was fixed forever in the public mind with the 1925 Scopes trial in Dayton, Tennessee. The Scopes story is a familiar one. On March 21, 1925, Governor Austin Peay signed the state's anti-evolution bill into law, and shortly thereafter John T. Scopes, a teacher in the Dayton school system, volunteered to test it. The American Civil Liberties Union, which had been closely watching the case, offered to supply Clarence Darrow as the legal defense. In turn, the WCFA, which had been actively lobbying for the bill, persuaded Bryan to represent the prosecution. Thus was the stage set for what was to become the most widely reported trial in the nation's history. After the entry of Darrow and Bryan, few people were able to see the trial in terms other than those of "the meeting of the great forces of skepticism and faith."[29]

Europe watched with astonishment as America reopened a controversy which everyone felt had been closed for sixty years. For a continent which saw itself disgraced by the recent war, the spectacle was somewhat funny and comforting.[30] Although there were many serious issues involved in the Scopes trial — such as the right of the people to control what was taught in the public schools — they were never faced directly. Instead, Dayton, Tennessee, soon became filled with religious fanatics of all persuasions. Egged on by the journalists, a mood of ballyhoo dominated the proceedings. As a result, the trial became a strange combination of both tragedy and farce. Both these themes were well illustrated on the final Monday of the proceedings when Darrow lured Bryan onto the platform. There he exposed the Great Commoner's lack of Biblical understanding to the world. Visibly upset, Bryan died in his sleep five days later.

The furor resulting from the trial had hardly begun to mount when news of Bryan's sudden death plunged the nation into mourning. The hooting of the skeptics stopped as abruptly as it started. A Broadway play making fun of him closed immediately. The *New Yorker* called back its current issue, and 20,000 copies of another national magazine were halted halfway across the continent so that a two-page section ridiculing him could be removed.[31] Editorials ranged from the extensive coverage of the *New York Times* to the compliments of the humblest weekly, and from the banal to the most perceptive. "Here in Great Britain we had almost no clue to him," remarked G. O.

Griffith. "He remained an enigma, for his mentality eluded us."[32] Lord Herbert Asquith conceded that only America could have produced such a man.[33] William Allen White, the Pulitzer-Prize-winning, small-town journalist from Kansas, called him the best political diagnostician and the worst political practitioner that the country had ever seen. Never had he been wrong on a single diagnosis; never had he been right on a single solution.[34]

Whispers abounded, however, as he was laid to rest in Arlington National Cemetery, that it was a good thing for his reputation that he died when he did. Few felt that the cause of anti-evolution, to which he appeared to be devoting his life, was worth the same degree of effort as has been his others – free silver, anti-imperialism, and prohibition. Dudley Field Malone, in a talk to a group of Unitarians, claimed that at the time of his death, Bryan was leading the most sinister movement of the day.[35] Strong evidence can be marshaled to suggest that in spite of his protestations that he had no plans after the Scopes trial, Bryan was indeed laying plans for what might well have been a new national crusade.[36]

After his death the most immediate question in Fundamentalist circles was who would take his place. "Everywhere I have been, I have been urged to take up Mr. Bryan's work," J. R. Straton of New York was quoted as saying. "It was unique and should be carried on. I would be willing to attempt it."[37] The able editor of the Baptist *Watchman-Examiner*, Curtis Lee Laws, however, noted that he could find no one authorized to select Straton to be the new leader of the Fundamentalists. "Mr. Bryan was never the leader of fundamentalism except that his prominence caused the papers to count him the leader," said Laws. "Fundamentalism has never had a leader. Any man can assume the leadership of a small or a large portion of the fundamentalists when they are willing to be led. It has been our experience and observation that the leadership of the fundamentalists is a pretty hard job."[38]

No one could take over the Fundamentalist movement from William Jennings Bryan because William Jennings Bryan had taken over the Fundamentalist movement. By his sudden increased interest in evolution, his lack of theological training, his concern for all aspects of Christianity, especially the social gospel, and the magic of his name, he had thrust himself into the center of the controversy. The newspapers kept him there until his death. Moreover, Bryan was an inclusive force whereas the other Fundamentalists were primarily exclusive. His tolerance, perspective, and genial warmth were to be found in none of his successors. In spite of their activities, none of his followers could approach the publicity which Bryan received just by being Bryan. After 1925 the Fundamentalist movement was largely limited to attempts at passing anti-evolution legislation and defections from the main line Protestant denominations.

"The newspapers ought to put up the money to build a memorial for Wm. Jennings," a friend wrote to Mark Sullivan, "because he was to the world of news what Babe Ruth is to baseball — the real drawing card, for anyone who is halfway fair has to admit that Bryan was new to his friends and enemies and the reading public 365 days a year."[39] His mantle, once he dropped it, could find no shoulders strong enough to carry it again. Bryan was unique in the Fundamentalist movement. He could never be replaced.

NOTES

1. *Richmond Times-Dispatch*, July 27, 1925; Glenn Frank, "William Jennings Bryan — A Mind Divided Against Itself," *The Century* (September, 1923), 794; *cf.*, the comments of the *Sioux City Sunday Journal*, July 26, 1925, *Dallas Morning News*, July 27, 1925, *Wheeling Register*, July 27, 1925. Many later writers such as Lewis S. Feuer, *The Scientific Intellectual* (New York, 1963), 10, also view Bryan this way. The best accounts of the Fundamentalist movement are Norman Furniss, *The Fundamentalist Controversy, 1918-1931* (New Haven, 1954) and Stewart G. Cole, *History of Fundamentalism* (Hamden, Connecticutt, 1963 — originally published 1931); See also, Ferenc M. Szasz, "Three Fundamentalist Leaders: The Roles of William Bell Riley, John Roach Straton and William Jennings Bryan in the Fundamentalist-Modernist Controversy," (Ph.D. dissertation, University of Rochester, 1969). Paul Carter's two studies, *The Decline and Revival of the Social Gospel* (Ithaca, 1954), and *The Twenties in America* (New York, 1968) are helpful, as is Kenneth K. Bailey, *Southern White Protestantism in the Twentieth Century* (New York, 1964).

2. A summary of the turmoil in this period can be found in Winthrop Hudson's comprehensive survey, *Religion in America* (New York, 1965). Proselytizers for many sects such as the Salvation Army, the Jehovah's Witnesses, Theosophy, Unity Church, etc., were also very active during these years. While the term "Fundamentalist" has often been used to include anyone from a 17th century Puritan to a modern day Christian Scientist, it should actually be restricted to a specific conservative movement within organized Protestantism from 1875 to about 1930. It will be so used in this essay.

3. *Christian Century*, XXI (June 2, 1904); *Ibid.*, October 6, 1904, quoted on 880.

4. Ernest R. Sandeen, "Towards a Historican Interpretation of the Origins of Fundamentalism," *Church History*, XXXVI (March, 1967), 66-83. My debt to Sandeen's work is obvious throughout this article He argues that the chief source for dispensationalism came from Plymouth Brethren missionaries from England. See also *A Bibliographic History of Dispensationalism* (Grand Rapids, 1965); Clarence B.

Bass, *Backgrounds of Dispensationalism* (Grand Rapids, 1960); and Harris F. Rall, *Modern Premillennialism and the Christian Hope* (New York, 1920).

5. This view was diametrically opposed to liberal post-millennialism, which argued that man would have to establish the thousand years of peace. Only then would Christ return and bring an end to Time.

6. Compare, for proof of this, the list of sermon topics for the Kansas City, Kansas, meetings held November 19-23, 1919, in the William B. Riley manuscript collection, Northwestern College, Roseville, Minn., and the sermons in *God Hath Spoken* (Philadelphia, 1919). The latter stemmed from the World's Conference on Christian Fundamentals, which was held in Philadelphia from May 25 to June 1, 1919. The Wollam quote is from *Christian Worker's Magazine* (April, 1919), 534. I have examined this liberal-conservative split more thoroughly in "Protestantism and the Search for Stability," in Jerry Israel, ed., *Building the Organizational Society* (New York, 1972).

7. WJB to *Christian Herald*, August 10, 1909, Bryan Manuscripts, Occidental College. Bryan has long been a subject of fascination for biographers. Paolo E. Coletta's *William Jennings Bryan: Political Puritan 1915-1925* (Lincoln, 1969) and Lawrence W. Levine's *Defender of the Faith: William Jennings Bryan: the Last Decade, 1915-1925* (New York, 1965) are both excellent for the period covered. For Bryan's earlier career, Paul Glad's *The Trumpet Soundeth: William Jennings Bryan and his Democracy, 1896-1912* (Lincoln, 1960) and *McKinley, Bryan and the People* (New York, 1964) are able treatments.

8. WJB to J. R. Best, January 17, 1918, Bryan Manuscripts, Library of Congress.

9. *The Commoner*, XIX (May, 1919).

10. Bryan, *The Prince of Peace*, 11.

11. The quote is from Bryan's little-known speech as printed in *The Moody Bible Institute Monthly* XXIII (April, 1923).

12. A. W. Stalker to WJB, December 1, 1920, A. W. Stalker to WJB, November 15, 1920; A. W. Stalker to WJB, January 31, 1921, Library of Congress. Levine, however, argues that his reputation was not seriously damaged by his resignation from Wilson's Cabinet, *Defender of the Faith*, 73-80.

13. *The Capital Times* (Madison, Wisconsin), February 7, 1922.

14. WJB to H. L. Hoard, December 13, 1921; WJB to Chester C. Platt, December 16, 1921, Bryan Manuscripts, Library of Congress; The story of this incident is well told in Irvin G. Wyllie, "Bryan, Birge, and the Wisconsin Evolution Controversy," *Wisconsin Magazine of History*, XXXV (Summer, 1952), 294-301.

15. Quoted in Levine, *Defender of the Faith*, 287.

16. W. J. Bryan, "The Fundamentals," *The Forum*, LXV, (July, 1923), 1665-1680.

17. *New York Times*, February 3, 1922; *Ibid.*, February 9, 1922.

18. WJB to A. W. Stalker, dated December 9, 1923, but probably 1922, Bryan Manuscripts, Library of Congress.

19. Jack Mills, *The Speaking of William Jennings Bryan in Florida, 1915-1925* (M.A. thesis, University of Florida, 1948), quoted 34.

20. *New York Times*, May 21, 1925; Edwin H. Rian, *The Presbyterian Conflict* (Grand Rapids, 1940) is the standard account of the Presbyterian troubles.

21. *New York Times*, May 24, 1924; The British journalist, S. K. Ratcliffe, writing in *Contemporary Review*, 1925, noted that when he published an article in 1922 on America's religious difficulties, evolution was not an issue of fundamentalism at all. "When the Fundamentalist movement was originally formed," wrote William B. Riley, one of the real founders, "it was supposed that our particular foe

was the so-called 'higher criticism,' but in the onward going affairs, we discovered that basal to the many forms of modern infidelity is the philosophy of evolution." *The Fundamentalist Magazine*, I (September, 1927), 8.

22.   A. C. Dixon to WJB, January 5, 1923; T. C. Horton to WJB, September 15, 1923; J. W. Porter to WJB, April 14, 1923, Bryan Manuscripts, Library of Congress.

23.   *Christian Fundamentals in School and Church* (April-May-June, 1923); WJB to W. B. Riley, March 27, 1925, Bryan Manuscripts, Library of Congress.

24.   WJB to W. B. Riley, May 3, 1923, *Ibid.*

25.   F. T. Boyer, editor, *Bible Champion*, to WJB, *Ibid.*

26.   WJB to J. F. Norris, May 1, 1923; W. B. Riley to WJB, February 7, 1923; W. J. Gray to WJB, March 2, 1923; J. F. Norris to WJB, May 30, 1923; J. F. Norris to WJB, April 24, 1923, *Ibid.*

27.   J. H. Larson to WJB, January 28, 1922; J. H. Larson to WJB, January 1922; J. H. Larson to WJB, January 10, 1922; W. E. Biederwolf to WJB, December 9, 1921; WJB to J. H. Larson, January 28, 1922, *Ibid.*

28.   From Bryan's speech as quoted in the *Christian Fundamentals in School and Church* (Oct.-Nov.-Dec., 1925), 26; W. H. Smith, "William Jennings Bryan and the Social Gospel," *Journal of American History*, LIII (June, 1966), 41-61; Levine, *Defender of the Faith*. 251-253, 358-365.

29.   R. D. Owen, "The Significance of the Scopes Trial," *Current History*, XXII (September, 1925, 811; J. R. Straton, "Is Our Modern Educational System Develop-ing a Race of Materialists, Sensualists, and Unbelievers?" Folder, Evangelical, of the John R. Straton manuscripts, to be deposited in the archives of the American Baptist Historical Society, Rochester, New York. On the trial itself, see J. T. Scopes, *Center of the Storm* (New York, 1967) and Ray Ginger, *Six Days or Forever* (Boston, 1958). The play by Jerome Lawrence and R. E. Lee, *Inherit the Wind*, is the version of the story which has received the most publicity, but it is not totally fair to Bryan. J. C. Edwards, "Bryan's Role in the Evolution Controversy" (M.A. Thesis, University of Georgia, 1966); Ferenc M. Szasz, "The Scopes Trial in Perspective," *Tennessee His-torical Quarterly*, XXX (Fall, 1971).

30.   Dennis Brogan, *American Aspects* (New York, 1964), 91.

31.   *Utica (New York) Press*, July 29, 1925.

32.   G. O. Griffith, "William Jennings Bryan," *Watchman-Examiner*, August 20, 1925, 1037.

33.   *Albany (New York) News*, July 27, 1925.

34.   *Memphis Commercial Appeal*, April 29, 1925.

35.   *The Fundamentalist*, December 6, 1925; *Washington Post*, July 30, 1925.

36.   See especially William J. Bryan to his son, June 17, 1925, Bryan Manuscripts, Occidental College.

37.   Chicago *Daily News*, August 18, 1925; *Cf.* Manila *Times*, September 8, 1928.

38.   *Watchman-Examiner*, September 3, 1925, 1131.

39.   Callahan to Mark Sullivan, August 8, 1925, in the Josephus Daniels papers, Library of Congress.

The Scopes Trial in Perspective

By Ferenc M. Szasz

In the years after World War I, the United States was afflicted with a series of spectacular courtroom trials which continually clamored for the public's attention. What the Leopold-Loeb case was to Illinois and the Mid-West, the Sacco-Vanzetti case to Massachusetts and the East, the Scopes trial was to Tennessee and the South. The Tennessee story is a familiar one. On March 21, 1925, Governor Austin Peay signed into a law a bill prohibiting the teaching of evolution in state supported schools.[1] Shortly thereafter, George Rappelyea and John Thomas Scopes of Dayton decided, as a lark, to test the law by having Scopes brought to trial. No sooner was Scopes arrested, however, than William Jennings Bryan, three-time candidate for President and prominent figure in the emerging Fundamentalist movement, offered to help prosecute. Then, Clarence Darrow, the most famous criminal lawyer of the day and a noted agnostic, offered to aid in the defense. With the entrance of these two, few people were able to view the trial in terms other than "the meeting of the great forces of skepticism and faith."[2]

When the two giants clashed in the stifling heat that July, Darrow lured Bryan onto the witness stand where he mercilessly exposed the Great Commoner's lack of Biblical understanding. When he finished, the crowd, which had been naturally pro-Bryan, rushed over to congratulate Darrow. The trial then closed and since Bryan died suddenly five days later, he never got his chance to reply.[3] Scopes was found guilty and fined $100 but the case was later overturned on a technicality by the Tennessee Supreme Court,

---

[1] The legislator who introduced the Bill, John W. Butler, said, "I never had any idea my bill would make a fuss. I just thought it would become a law, and that everybody would abide by it." Quoted in Leslie H. Allen (ed.), *Bryan and Darrow at Dayton* (New York, 1925), 1.

[2] R. D. Owen, "The Significance of the Scopes Trial," in *Current History*, XXII (September, 1925), 881; John R. Straton, "Is our Modern Educational System Developing a Race of Materialists, Sensualists and Unbelievers?" folder Evangelical, Straton manuscripts, First Baptist Church, Malden, Mass.

[3] Frank A. Pattie (ed.), "The Last Speech of William Jennings Bryan," in *Tennessee Historical Quarterly*, VI (June, 1947), 265-83, tells much of Bryan's final days. His final speech on evolution was printed by many newspapers after his death. The effect was nothing, however, to what might have been had he delivered it in person across the nation, as was his plan.

with the strong admonition that it not be reopened. Although the law remained on the books until September, 1967, no further cases were brought up under it.

The material on the Scopes trial is enormous. Accounts in the magazines and newspapers of the time were legion and more words were relayed from Dayton that July than from any other comparable event in the nation's history. Station WGN of Chicago broadcast the trial live and every afternoon the correspondent from the London *Daily News* faithfully cabled 500 words to his home office.[4]

Books on the case followed quickly. Before the year was out, Leslie H. Allen had compiled a pot boiler in *Bryan and Darrow at Dayton* (New York, 1925) and, three years later, Walter Lippmann wrote his caustic *American Inquisitors*. The protagonists for the defense, Clarence Darrow in *The Story of My Life* (New York, 1932) and Arthur Garfield Hays in *Let Freedom Ring* (New York, 1937) presented their sides while Mrs. Mary Bryan completed her husband's *Memoirs* (Chicago, 1925) from the other point of view. Jerome Lawrence and Robert E. Lee fictionalized the events in their play, *Inherit the Wind,* which later became a movie and a television special. But Matthew Harrison Brady and Henry Drummond are not convincing characters and the treatment of Brady (Bryan) is unfair. For many years, Ray Ginger's *Six Days or Forever* (Boston, 1958) was the standard historical account, but recently there have been several new works published. Lawrence W. Levine, *Defender of the Faith* (New York, 1965) and Paolo E. Coletta, *William Jennings Bryan: Political Puritan* (Lincoln, 1969) both contain excellent renditions of the trial. Jerry R. Thompkins has collected a series of essays, some by participants, in *D-Days at Dayton* (Baton Rouge, 1965) and in 1967 John Scopes finally published his own account, *Center of the Storm.* The historian does not lack factual information about the Scopes trial. The problem lies in placing it in proper perspective. This paper would like to set forth two such perspectives: the trial's position *vis-a-vis* America's anti-evolution crusade and its position *vis-a-vis* the western world's reaction to Darwinism.

On the national scene, the dramatic events of the Dayton trial are often viewed as the high point of the American Fundamentalist

---

[4] Bynum Shaw, "Scopes Reviews the Monkey Trial," in *Esquire,* 74 (November, 1970), 88ff.

and anti-evolution crusade. This, however, is open to serious doubt. Because of the reputation of the men involved, it received the most publicity, but instead of being the culmination of the movement, it was really the beginning. Before Dayton, only the South Carolina, Oklahoma, and Kentucky legislatures had dealt with anti-evolution laws or riders to educational appropriation bills.[5] With the publicity of the Dayton case, the rush began in earnest. After 1925 pressures on the state legislatures increased steadily until the peak year, 1927, when 13 states, both North and South, considered some form of anti-evolution law. All told, at least 41 bills, riders, or resolutions were introduced into the state legislatures, and some states faced the issue time and time again.[6]

Most of these bills failed to pass (Rhode Island relegated hers to the Committee on Fish and Game), but Mississippi and Arkansas both put anti-evolution laws on their books, California allowed the teaching of evolution only as "theory," and the Governor of Texas, Miriam "Ma" Perkins, personally saw to it that evolution was eliminated from the school textbooks. While it is true that there was no second Scopes case, it is still difficult to see the Dayton trial as any sort of liberal milestone. It was hardly the battle for liberal religion as the *Christian Register* saw it, nor was it successful in the battle for freedom in the classroom.[7] Furthermore, anti-evolution agitation is not dead yet. In 1964, a Baptist minister in Pheonix, Arizona, campaigned to get an amendment to the state constitution to forbid the teaching of evolution; and every issue of *The Plain Truth*, a popular southern California publication, contains an attack on the theory.[8] All this has occurred without the assistance of any major figure, and one wonders what would have happened had Bryan been around to throw the magic of his name into the controversy for another five years.

Instead of serving as the apex of the anti-evolution crusade, the Tennessee law and the Scopes trial actually provided a model for

---

[5] Elbert L. Watson, "Oklahoma and the Anti-Evolution Movement of the 1920's," in *Chronicles of Oklahoma*, 42 (Winter, 1964-65), 397.

[6] R. Halliburton, Jr., "Kentucky's Anti-Evolution Controversy," in *Kentucky Historical Society Register*, 66 (April 1968), 97-98; Virginia Gray, "Anti-Evolution Sentiment and Behavior: The Case of Arkansas," in *Journal of American History*, LVII (September, 1970), 353.

[7] *Christian Register*, CIV (August 6, 1925), 761; Irving Stone, *Clarence Darrow for the Defense* (New York, 1941), 464.

[8] Donald F. Brod, "The Scopes Trial: A Look at Press Coverage After Thirty Years," in *Journalism Quarterly*, XLII (Spring, 1965), 219.

it. The situation in the Volunteer state showed that a law could be passed and upheld. It also revealed the ingredients needed to do so. These were simply a discontented populace and an organized, aggressive group of anti-evolution speakers with a definite legislative program. That discontent existed in most parts of the country during the 1920's is obvious. Its causes lie beyond the scope of this paper, but surely two of the most important reasons were the feeling by the average man that he had lost control of the elements which were shaping his life and the feeling that World War I had failed to live up to its promises. Anti-evolution seemed to be a meaningful response to this situation. It celebrated the people by attacking an un-democratic elite of professors (Bryan called them a "scientific soviet") who were seen as replacing the Word of God, which all could understand, with a set of hypotheses open only to the educated few.[9] And by seeing evolution as the basis of German militarism, it also offered an explanation of the First World War.

But a discontented populace alone was not enough. There also had to be an organized channeling of this discontent into the area of anti-evolution sentiment. In general, one may say that the anti-evolution feeling did not arise spontaneously from the people. Evolution was not then and is not now a major part of the average man's daily life. In fact, when the Fundamentalist movement began in earnest in 1919, evolution was not an item which he was much concerned with at all.[10] It did not become a major issue until late 1921 and early 1922 when William Jennings Bryan began bringing it to the attention of the nation through the publicity which his speeches always received. The anti-evolution fights in the various states were directed agitation. Without the tireless efforts of a dedicated band of sincere, but misguided, men, the discontent would have taken another form.

Of the organized groups which led this agitation, the most im-

---

[9] William J. Bryan to Clarke, Dec. 6, 1923; to J. H. Hayne, March 15, 1924; to Mayor Hylan, June 12, 1923, in Bryan manuscripts, Library of Congress, Washington, D.C. Boxes 60, 39.

[10] S. K. Ratcliffe, "America and Fundamentalism," in *Contemporary Review*, CXXIII (September, 1925), 288; S. K. Ratcliffe, "The Intellectual Reaction in America," in *Contemporary Review*, CXXII (July, 1922). When Ratcliffe, a British journalist, wrote the first article on Fundamentalism, evolution was not yet a major issue; Ferenc M. Szasz, "Three Fundamentalist Leaders: The Roles of William Bell Riley, John Roach Straton, and William Jennings Bryan in the Fundamentalist Modernist Controversy," unpublished Ph.D. dissertation for the University of Rochester, 1969, especially Chapter 5.

portant by far was the World Christian Fundamentals Association (WCFA), organized in 1918 by William B. Riley, pastor of the First Baptist Church in Minneapolis and directed by him until 1930. The WCFA was the nation's most powerful interdenominational Fundamentalist force. Through its contacts with local Protestant ministers in every city and small town, it played a major role whenever there was discussion over evolution. It was the WCFA and Riley who got Bryan to visit Tennessee and speak in favor of the anti-evolution law. They also were the ones who secured his help in the prosecution of Scopes.[11]

Although Bryan's sudden death removed the most popular anti-evolution speaker from the field, agents of the WCFA such as T. T. Martin, J. Frank Norris, and Riley, or strong sympathizers such as John R. Straton and Mordicai F. Ham, were more than willing to take up the slack. All these men combined to give innumerable speeches against evolution, blaming it for everything imaginable. In states which did pass anti-evolution legislation, such as Arkansas and Mississippi, the agents of the WCFA were much in evidence. Riley was a prominent speaker during the Arkansas campaign and Martin, ignoring the Mississippi state lobbying regulations, harangued the legislators mercilessly. He accused them of bartering the faith of their children for a pot of gold.[12] The WCFA was also deeply involved in the bitter anti-evolution fights of Oklahoma and Minnesota, but here they were not able to overcome growing scientific and ministerial opposition. A surprisingly large number of ministers were not pleased with either their tactics or their goals.[13] Although they were not always able to get the legislation they desired, in many cases the WCFA did achieve what they set out to. Much more subtle and effective than any law

---

[11] William B. Riley traces this in "The World Christian Fundamentals Association and the Scopes Trial," article in the Riley sermon collection, Box 7, Northwestern Schools, Minneapolis, Minnesota; The files for the WCFA, unfortunately, seem to have been lost. Only a few letters remain in the Riley collection.

[12] Martin is quoted in Ed Williams, "Monkey Business," in *The Petal Paper*, 17 (May, 1970), 1.

[13] The various state studies which have been done tend to support the thesis of directed agitation. *Cf.* Halliburton, "Kentucky's Anti-Evolution Controversy," Gray, "Anti-Evolution Sentiment and Behavior: The Case of Arkansas," Watson, "Oklahoma and the Anti-Evolution Movement of the 1920's," Willard B. Gatewood, Jr., "Politics and Piety in North Carolina: The Fundamentalist Crusade at High Tide, 1925-1927," in *The North Carolina Historical Review*, XLII (July, 1965), Ferenc M. Szasz, "William B. Riley and the Fight Against Teaching of Evolution in Minnesota," in *Minnesota History*, 41 (Spring, 1969).

were the actions of the local school boards. They could choose one textbook over another on any basis they desired. Organized anti-evolution agitation was never really stopped in the 1920's. By 1928, however, the discontent which produced it was siphoned off in the successful fight against the Presidential attempt of Al Smith. After that, the depression so changed things that such questions remained meaningful to only a small minority.

The Scopes trial can also be viewed from an international perspective. Recent events in the mid-twentieth century have made us painfully aware that world history cannot be comprehended solely from within the context of American history. However, American history can best be comprehended by placing it within the context of world history—particularly that of western Europe and the British Isles. The two countries which had the most violent reactions to the writings of Charles Darwin were Great Britain and the United States. Both nations were primarily Protestant and shared similar values and traditions. The Protestant connection between the two has yet to be documented, but there is no doubt that it was extensive.[14] Methodists, Baptists, Presbyterians, and Episcopalians all had significant corresponding bodies in the British Isles. Many of the evangelical churches frequently exchanged pulpits. The first dramatic public reaction to the theory of evolution in England came when Bishop Samuel Wilberforce clashed with Thomas H. Huxley at Oxford in June of 1860. Sixty-five years later, a similar dramatic confrontation would be repeated at Dayton, Tennessee. Karl Marx opens his *The Eighteenth Brumaire of Louis Bonaparte* with Hegel's comment that all great historic facts and personages recur twice. Marx suggested, however, that Hegel forgot to add: "Once as tragedy, and again as farce."[15] When Bishop Wilberforce confronted Huxley in Oxford in 1860, it was tragedy; when Bryan confronted Darrow at Dayton in 1925, it was farce.

It was natural that England should be the first country to react. In 1859, when *Origin of Species* was first published, the United States was about to enter a Civil War which was followed by a disrupting period of industrialism. England, however, was relatively calm. She had avoided the revolutions of Europe and had

---

[14] Some of the similarities can be seen in Willis B. Glover, *Evangelical Nonconformists and Higher Criticism in the Nineteenth Century* (London, 1954).

[15] Karl Marx, *The Eighteenth Brumaire of Louis Bonaparte*, Daniel DeLeon (trans.), (Chicago, 1914), 9.

long since made her peace with industrialism. This, plus the presence of a tight-knit intellectual community, enabled English thinkers to face immediately all the religious implications in the Darwinian hypotheses. Darwin himself was careful never to become involved in these disputes. He conspicuously avoided any gathering which he thought might produce acrimony.[16] But his ideas were never lacking for defenders, the chief of which was the scientist and popularizer, Thomas H. Huxley. Huxley, who had seen Darwin's writings before they were published, claimed that the theses of *Origin of Species* came "like a flash of light to a man who was lost himself on a dark road. . . ."[17] After reviewing the book for the London *Times,* Huxley devoted much of his later life as "Darwin's Bulldog," defending evolution against all comers.

Initially, the conflict over evolution was not composed of scientists on one side and theologians on the other. Many scientists, such as Richard Owen in England and Louis Agassiz in America, expressed grave doubts about Darwin's theories. Nevertheless, it was the Anglican Bishop Wilberforce who initiated the first dramatic exchange over the issue of evolution. The scene was the annual meeting of the British Association in Oxford in late June, 1860. Both Wilberforce and Huxley were seated on the platform at one of the sessions when the American scientist J. W. Draper read a rather dull paper on "The Intellectual Development of Europe Considered with Reference to the Views of Mr. Darwin." Rumor was rife that a confrontation was in the offing and a crowd of 700, including many women and students, had packed the hall. In the discussion period after Draper's paper, Wilberforce, who knew nothing of science but had been coached by Owen, stood up and with eloquent charm began a half-hour attack on Darwin's theories. The main point of dispute hinged on man's relationship with the animals.[18] Wilberforce argued that Darwinism was obviously incompatable with the Word and works of God. During his remarks, he casually turned to Huxley and asked him if he claimed descent from the monkey through his grandfather or his grandmother. After this canard, Huxley whispered to the aston-

---

[16] Gertrude Himmelfarb, *Darwin and the Darwinian Revolution,* (Gloucester, Mass., 1967), 287.

[17] Quoted in Janet E. Courtney, *Freethinkers of the Nineteenth Century* (London, 1920), 147.

[18] Wilberforce used the same arguments in his review of the *Origin* which appeared in the *Quarterly Review* shortly after his Oxford exchange. There he spoke of "our unexpected cousinship with the mushrooms." Himmelfarb, *Darwin,* 273.

ished Sir Benjamin Brodie, seated next to him: "The Lord hath delivered him into my hands."[19] Then, when the Bishop had finished, Huxley arose to present the other side. He was listened to politely as he defended Darwinism and then stated: "If then. . . . the question is put to me would I rather have a miserable ape for a grandfather or a man highly endowed by nature and possessed of great means of influence and yet who employs those faculties and that influence for the mere purpose of introducing ridicule into grave scientific discussion—I unhesitatingly affirm my preference for the ape."[20] This shocked and thrilled the audience (one lady fainted and had to be carried out), and Huxley was the most popular man in Oxford for a short period of time. The reputation of this dramatic riposte spread quickly and A. D. White claimed it "reverberated through England, and indeed through other countries."[21]

This clash between Wilberforce and Huxley was the opening act of a tragedy. It ushered into the popular realm an aggressive conflict between science and religion which has not yet died down. The pious called the secular spirit of science heresy and the proponents call it "progress." Denunciations were prevalent on both sides. Ironically, those who attacked the "Sin of Faith" were often as dogmatic and narrow as some the creeds which they were denouncing.[22] "Neither party," noted Reinhold Niebuhr, "was able to annihilate the other as simply as it had hoped."[23] Both science and religion remained much alive as the twentieth century began.

Sixty-five years later, after the rest of the world had reached a *modus vivendi* with evolution, America revived the controversy at Dayton. And the second time, the clash emerged as farce.[24]

---

[19] Quoted in Courtney, *Freethinkers*, 150.

[20] Since there were no reporters present, the actual words of the exchange vary slightly in the several accounts of the situation. This quote comes from Cyril Bibby, *T. H. Huxley* (London, 1959), 69.

[21] Andrew D. White, *A History of the Warfare of Science with Theology in Christendom* (New York, 1896) as found in Philip Appleman (ed.), *Darwin* (New York, 1970), 423.

[22] G. M. Young, *Victorian England: Portrait of an Age* (London, 1960), 109.

[23] Reinhold Niebuhr, *Pious and Secular America* (New York, 1958), 1.

[24] The perceptive British commentator, D. W. Brogan, writes of the Scopes trial as follows: "Here was the most powerful, richest, most complacent society in the world making a fool of itself over Jonah and the Whale, over the literal accuracy of Genesis. To a continent weakened and, in its heart of hearts, thinking itself disgraced by its recent civil war, the spectacle of William Jennings Bryan, former Secretary of State, thrice candidate for President, floundering over primitive cosmology was funny and comforting." *American Aspects* (New York, 1964), 91.

To argue that the trial was farcical, however, is not to claim that no serious issues were involved. Serious issues were there, but they were quickly obscured. Clarence Darrow, Arthur Hays, and Dudley Field Malone, the mainstays of the defense, took the case with deadly seriousness, as did Bryan and his followers. But the most important aspect of the trial—the right of the people to control not only their schools but all knowledge taught within them—was never faced directly. This point was the basis of Bryan's argument. "The trial will be a success in proportion as it enables the public to understand the two sides and the reasons on both sides," he wrote a friend in 1925. "Every question has to be settled at last by the public and the sooner the subject is understood, the sooner it can be settled."[25] Scopes, however, felt that no one could have held an absolutely serious attitude toward the trial without losing his sanity; perhaps Scopes was right after all.[26] The Dayton trial was very much a product of the general atmosphere of the 1920's—a decade which produced as many moral causes as it did magnificent trivialties. The Scopes trial shared equally in both.

From the beginning the case took on an atmosphere of circus and carnival. Chattanooga tried to steal the trial away and had to be quickly stopped. Over 100 reporters poured into Dayton to rub shoulders with preachers of Armageddon, militant rationalists, and others, whom the New York *Times* suggested had "obvious mental irregularities of a religious tendency."[27] J. R. Darwin, who ran the local hardware store, had a large sign—"Darwin is Right . . . inside—" in front of his building. Looking back over the years, Scopes noted that Dayton that July contained "one of the rarest collection of screwballs" he had ever seen in his life.[28]

Of the reporters who did much to keep this atmosphere alive, H. L. Mencken, editor of the *American Mercury*, ranks at the top. While at the trial, he sent out a stream of syndicated editorials and the popular outcry against them was such that some newspapers were forced to cancel his column.[29] Mencken had al-

---

[25] Bryan to Ed Howe, June 30, 1925, in Bryan Manuscripts, Library of Congress.

[26] John T. Scopes, *Center of the Storm* (New York, 1967), 102.

[27] *New York Times*, July 10, 11, 1925 as quoted in Kenneth K. Bailey, "The Anti-Evolution Crusade of the Nineteen-Twenties," (unpublished Ph.D. dissertation, Vanderbilt University, 1953, p. 156).

[28] Scopes, *Center of the Storm*, 98; Michael Williams, "At Dayton, Tennessee," in *The Commonweal*, II (July 22, 1925), 262-65.

[29] Mencken's essays have been collected in Jerry R. Tompkins, *D-Days at Dayton* (Baton Rouge, 1965); Watson, "Oklahoma and the Anti-Evolution Movement of

ways hated Bryan and what he stood for and probably was in Dayton to ridicule him as much as possible.[30] He accused Bryan of trying to become the "Pope" of the "peasants," after three times failing to become their President.[31] His essay on Bryan, "In Memoriam: WJB," in his *Prejudices* series, is laced with vitriol and he is one of the few commentators who ever accused Bryan of being insincere.[32] The rumors were, however, that Bryan saved Mencken from being lynched, for he convinced the self-appointed committee designated for this purpose simply to ask him to leave the state. This Mencken hurriedly did. He was not present for the final day of the trial, and he never wrote much on it afterwards.

The most famous scene at Dayton came on the final Monday of the proceedings when Darrow called Bryan to the witness stand to cross-examine him on his beliefs. Scopes has suggested, however, that the real turning point of the trial came the previous Thursday when Bryan clashed in the crowded court room with the other chief defense lawyer, Dudley Field Malone. Malone was his old friend, for he had been Bryan's undersecretary in the Department of State during the Woodrow Wilson Administration. Bryan spoke for an hour about the duel to the death between science and religion and had the crowd in the palm of his hand. Then Malone arose and in twenty-five minutes totally crushed Bryan's hold on the audience. He spoke intensely, beginning with an essay on Thomas Jefferson's ideas of religious toleration which Bryan himself had written twenty years earlier. Malone made the distinction between God, the church, the Bible, Christianity, and William Jennings Bryan. "There is going to be no duel," he said. "There is never a duel with the truth. The truth always wins— and we are not afraid of it. The truth does not need the law. The truth does not need Mr. Bryan. The truth is imperishable, eternal, immortal. . . ."[33] Tremendous applause greeted his oration. When the court room cleared to leave only Malone, Bryan, and Scopes, Bryan said wearily to Malone, "Dudley, that was the greatest speech I have ever heard." "Thank you, Mr. Bryan," said

the 1920's," 400-401.
[30] William R. Manchester, *Disturber of the Peace* (New York, 1950), 164.
[31] Tompkins, *D-Days at Dayton*, 46.
[32] H. L. Mencken, *Prejudices: Fifth Series* (New York, 1926) 64ff.
[33] Quoted in John Reddy, "The Most Unforgettable Character I've Met," in *Reader's Digest*, LXIX (August, 1956), 86; The *Commoner* for May, 1903, contains Bryan's address on "Religious Freedom," which he gave to the Thomas Jefferson Memorial Association.

Malone. "I am sorry that it was I who had to make it."[34] Scopes has suggested that Malone's speech was the most dramatic event in his life, and its effect on Bryan was such as to make him desire to recoup his lost prestige on the stand with Darrow the following Monday, with its inevitable results.

Although tragedy does not necessarily involve farce, farce may well involve tragedy. This was the case in Dayton that July. It occurred in part because the terms "Darwinism" and "evolution" had long since ceased to mean what they did to Bishop Wilberforce and Thomas Huxley. In 1860, Darwinism meant a particular theory of evolution which suggested that changes in species came about through the process of "natural selection." It was on this basis that the earlier argument was conducted. By 1925, however, the terms had been drained of all precise meaning. During the 1920's, evolution was commonly used to mean: a shifting of moral certainties, the doctrine that man was continually improving and thus did not need Christian regeneration, the idea that "might makes right," and the death of all spiritual values. *The Bible Champion* entitled one article "Evolution-Devilution."[35] William B. Riley later declared that evolution was anti-Biblical, anti-social, anti-moral, anti-national, and anti-global.[36] For the generation of the 1920's, evolution had come to mean "the totality of error."[37]

As such, anti-evolution agitation such as the Tennessee law and the Scopes trial proceedings no longer had any specific theological content. Instead, to oppose evolution was a way of lashing out at what was wrong with society. To pass a law prohibiting it would be to restore the nation to peace and harmony. Anti-evolution laws became what the Single Tax was to Henry George and Free Silver to the Populists—the panacea for all that was wrong. Although the eyes of the nation were riveted on Bryan and Darrow in Dayton that July, the forces of tragedy and farce held the center stage.

---

[34] Quoted in Scopes, *Center of the Storm,* 155; a slightly different wording can be found in Scopes' article "Reflections—Forty Years After," in Thompkins, *D-Days at Dayton,* 27, and in Reddy, "The Most Unforgettable Character I've Met," 86.

[35] *The Bible Champion,* XXV (March, April, 1918), 86.

[36] Riley, "The Troublesome Apostates," Sermons, Box 7, Riley Manuscripts, Minneapolis, Minn.

[37] The quote is from L. W. Levine in his review of W. B. Gatewood, *Preachers, Pedagogues, and Politicians* in the *American Historical Review,* LXXII (January, 1967), 732.

Political Fundamentalism
and Popular Democracy
in the 1920's

Robert A. Garson

Most historical studies of the 1920's have been understandably hostile to the religious and political fundamentalism that thrived in the first half of the decade. Not surprisingly, students of the period have judged the antievolution crusade and the Ku Klux Klan by their specific activities and their declared goals. Consequently, these movements have been variously typified as "ugly," "benighted," replete with "bigotry" and "intolerant."[1] And so they were. Their contempt for educational innovation, cultural pluralism, and moral individualism seemed to contravene some basic tenets of the liberal persuasion. But there was a paradox in their overall designs that has been largely overlooked by scholars of the period. This paradox lies in the frequently contradictory conjunction between liberty and popular will. Specifically, the fundamentalists rejected several widely held social and political values on the grounds that they could not command public support in the particular areas where the antievolution crusade and the Klan flourished. The fundamentalists wished to subject knowledge and personal freedom to the scrutiny of the local community. If the community could not subscribe to a prevalent wisdom, they argued, the community was justified in regulating the activities of any person or institution that sought to disseminate unpopular beliefs. In short, the anti-Darwinists and Klansmen sought to

ROBERT A. GARSON is Lecturer in American Studies at the University of Keele in Staffordshire, England. He is the author of The Democratic Party and The Politics of Sectionalism, 1941–1948.

1. See William E. Leuchtenburg, *The Perils of Prosperity, 1914–1932* (Chicago, 1958), p. 213; George E. Mowry, ed., *Fords, Flappers and Fanatics* (Englewood Cliffs, 1963), pp. 121–22; John D. Hicks, *Republican Ascendancy, 1921–33* (New York, 1960), p. 183; George B. Tindall, *The Emergence of the New South, 1913–1945* (Baton Rouge, 1967), pp. 207–8; W.J. Cash, *The Mind of the South* (New York, Vintage, n.d.), pp. 344–46.

revive a political system that was at once more responsive and sensitive to particular local concerns, however prejudiced these might have been.

The fundamentalist movements of the 1920's exhibited a second, related characteristic, namely, the desire to exercise closer vigilance over educational, political, and legal processes stemming from a pervasive sense of social alienation and disaffection. This disenchantment did not grow merely out of a flaunted fear of immigrants, intellectuals, and minority groups. The fundamentalists believed that the fashioning of society was being undertaken by a diverse but powerful economic group that was not bound by either historic tradition or an affinity for popular consent. New social patterns had emerged after the First World War, patterns that were in part shaped by institutions that were not directly answerable to any one constituency.

Particularly in the field of communications, it was possible for a small coterie of broadcasters, educators, or movie producers to expose, or even influence, very close-knit societies to different forms of social and intellectual existence. In an earlier age, tradition-bound communities were able to immunize themselves somewhat from cultural and political change.[2] In the postwar period, however, rural and rural-minded Americans, particularly but by no means exclusively in the South, felt that they were less able to maintain control and establish standards in their neighborhoods. Changing social and sexual habits and new modes of knowledge were being publicized—and implicitly blessed and legitimized—in the media. Yet these ideas and social images were not subject to restriction or effective refutation. Although the Klan and the evangelical fundamentalists did not concern themselves specifically with the potential power of the tastemakers, they were exercised by the fact that local communities no longer seemed able to establish and regulate social conduct and convention. A new emergent culture, a culture that was at once elusive and pervasive, undermined the ability of local neighborhoods to set the tone of life in their homes.

2. Several studies of the South before the First World War have examined the varying attempts of its leaders to maintain political conformity. Recent ones include: J. Morgan Kousser, *The Shaping of Southern Politics: Suffrage Restriction and the Establishment of the One-Party South, 1880–1910* (New Haven and London, 1974); Paul Gaston, *The New South Creed: A Study in Southern Mythmaking* (New York, 1970); Bruce Clayton, *The Savage Ideal: Intolerance and Intellectual Leadership in the South, 1890–1914* (Baltimore and London, 1972); Sheldon Hackney, *Populism to Progressivism in Alabama* (Princeton, 1969).

One caveat, however, should be made at this point. Participants in social movements do not necessarily comprehend or appreciate the total configuration that stimulates those movements in the first place. They often level their critical energy at easily discernible social groups and only marginally address themselves to the fundamental problems that prevail within their society. General uncertainty and bewilderment can produce movements that focus on one particular grievance, without fathoming the roots of that grievance. For example, the sometimes singleminded preoccupation with the monetary standard in the late nineteenth century served to distract critics from examining the wider issue of the distribution of wealth and power.

Unfortunately, there is little published evidence that indicates positively that the fundamentalists were concerned with broader questions. However, they did exhibit certain tendencies that invite speculation—which is perhaps not too fanciful—about their wider, albeit latent, motives. Both the Klan and the antievolutionists revealed a common desire to establish an educational and political system that was directly answerable to the local community. They extended, although in the process they distorted, the pragmatist's view that the worth of an institution or idea should be measured by its contribution to experience and social enhancement. They believed that current educational and social practices served to undermine the cohesion of particular communities. So they determined to harness the schools, the lawmakers, and the tastemakers to the locale in which they operated. Although violence, intolerance, charlatanism, and intellectual simplicity became the hallmark of several fundamentalist leaders, these characteristics should not obscure their attempts, however inquisitorial, to inject a greater element of answerability into their institutions.

The spectacular rise of the Ku Klux Klan after the war has generated considerable historical comment. While scholars have differed somewhat in their assessments of the Klan's locus of support and its scope of activity, they have generally agreed that the hooded order was stimulated by postwar disillusionment, the witch hunts of the Red Scare years, agricultural depression, the very noticeable influx of immigrants prior to 1924, and the bewildering changes in the social climate.[3] There is no need to take issue with this general view. Rapid

3. David Burner, *The Politics of Provincialism: The Democratic Party in Transition, 1918–1932* (New York, 1968), pp. 76–88; John Higham, *Strangers in the Land* (New Brunswick, 1955), pp. 268–70, 288–99; John M. Mecklin, *The Ku Klux Klan: A*

urbanization, spurred on by the prosperity of the Progressive era and
the stimulus of war industry, had produced one of the most dramatic
changes in population distribution in American history. The demo-
graphic revolution was compounded by the innovations in technology
and communications, particularly in the spread of the radio,
magazines, and the automobile. This development narrowed the gap
between city and country, immigrant and old-stock American.
Americans were exposed to an unprecedented bombardment of
stimuli. Yet this exposure was designed and shaped by the people in
the communications industry who were frequently more concerned
with sustaining popular interest than with conveying accurate social
pictures. The broadcaster, popular writer, or film producer created an
image of a society that was often tinted, even exaggerated. But as the
fashions and life-styles of the era indicate, the media seemed to have
the facility to give birth to, or reinforce, the new cultural trends.[4]

The apparent ability of the overlords of public taste to modify,
transform, or reinforce social attitudes clearly worried certain elements of the population in both city and country. They believed that
older patterns of belief were crumbling. They determined, therefore,
to try and halt the moral disintegration they thought infected their
society. It is not necessary for the purpose of the argument to identify
these exponents of social orthodoxy who eventually joined the
crusades for political and doctrinal conformity. This essay's central
concern lies in suggesting an explanation for the proliferation of
fundamentalism in the early twenties. Their anxieties, it is proposed,
stemmed directly from their inability to influence the content and
dissemination of cultural values. While, admittedly, they did not
focus their criticisms on the instruments of communications per se,
except in the case of the schools, they did believe that a confluence
of forces existed that would destroy the integrated, corporate community neighborhoods. They responded by trying to create an effec-

---

Study of the American Mind (New York, 1924), pp. 103–25; David M. Chalmers,
Hooded Americanism: The First Century of the Ku Klux Klan (New York, 1965), pp.
28–38; Frederick L. Allen, *Only Yesterday: An Informal History of the 1920's* (New
York, 1964), pp. 52–57; Paul L. Murphy, "Sources and Nature of Intolerance in the
1920's", *Journal of American History* 51 (June 1964), 60–76.
    4. See Leuchtenburg, *The Perils of Prosperity*, pp. 158–77, 196–98; Hicks, *Repub-
lican Ascendancy*, pp. 169–74; Lewis Jacobs, *The Rise of the American Film: A Critical History* (New York, 1939), pp. 233–45; David Robinson, *Hollywood in the Twenties* (London, 1968) passim.

tive, countervailing force that would provide a counterpoint to the "agencies of mass impression."[5]

There was widespread frustration at the inability of the local community to establish standards and customs. The new modes of living appeared to owe their existence to influential, elitist forces, not to the conventions of rural or rural-minded societies. Earlier methods of social control had vanished; even the national political parties did not seem concerned with establishing a political consciousness. Extant political institutions seemed unwilling or unable to restrain these forces. Most prominent political leaders seemed blind to this growing disaffection. They appeared to make no attempt to inculcate a sense of moral purpose. The catchphrases and slogans of the era—such as "normalcy" and "the business of America is business"—suggested that society needed no political direction. The orthodox feared that an indifferent leadership would foster a purposeless, political reticence, unless it undertook at once to inject traditional local values into the social culture. They proceeded, therefore, to immunize their families and communities from forces of change that they had not themselves generated. They aimed to regulate the impact of cultural change by tightening social controls and so insulating their neighborhoods from undesired intrusions from the media and the schools.

Many Klansmen argued that laws, which should be designed to promote the cohesion of society, had not generally adjusted to new conditions. Property and life were still adequately safeguarded, but legal rules did not extend to certain modes of private behavior. For example, sexual behavior had been usually considered a private affair. But the Klansman believed that sexual convention was as vital to the cohesion and even survival of society as was public safety. In days when geographical mobility was circumscribed, the social unit was able to develop and enforce group standards informally. But urbanization had destroyed social integration and had undermined political conformity. It was legitimate and desirable, therefore, in the Klan's eyes, to enforce morality by public law—as had been the case with prohibition. So the Klan began to move into politics in the hope of extending the sphere of moral legislation. Meanwhile, it justified its extralegal use of suasion, intimidation, and violence with the

5. Report of the President's Research Committee on Recent Social Trends, *Recent Social Trends in the United States, I* (New York, 1933), p. 125; Twelve Southerners, *I'll Take My Stand: The South and Agrarian Tradition* (New York, 1962), p. 35.

claim that it promoted social cohesion.[6]

Care must be taken, of course, not to gloss over the Klansman's one-dimensional moralism, his intolerance, and the violent methods used by some to achieve their aims. Recent literature on the Klan correctly balances the macabre side of the order with its more harmless rituals. However, historians have underestimated the alienation felt by rural-minded Americans in this period from the traditional instruments of social control. The tendency of so many Americans to take the law into their own hands suggests that there was a declining faith in both law and the ability of political institutions to respond rapidly to changing local concerns. For example, in the booming southwestern oilfields, bootlegging and prostitution abounded. These illegal activities could thrive only if the people continued to drink liquor or visit brothels. However, drinking and sexual indulgence per se were not illegal. The Klan believed that moral standards would continue to decline if the purveyors and consumers of liquor and illicit sex were left untouched. The local community could justifiably act where the law could not. The Klan, claimed one member, "moves out the gangster, bootlegger, chock [beer] shop, fast and loose females, and the man who abuses and neglects his wife and children. . . . The delay in our laws is a great protective to the criminal class, and the people, taxed to the limit, are taking this method to put a stop to this cost on the town and county." In El Dorado, Arkansas, the Klan formed an appropriately named "Law Enforcement League" to combat prostitution and other forms of vice.[7] Again, it felt frustration at the inability of politicians and the police to act in this sphere. One defender of the order argued that "the limitations and technicalities that hamper the enforcement of the law are the best excuses I can give for the order."[8] An Oklahoma judge reinforced the view that law could be both cumbersome and

---

6. This analysis of the Klan draws mainly upon the following studies: Mecklin, *The Ku Klux Klan*; Chalmers, *Hooded Americanism*; Arnold S. Rice, *The Ku Klux Klan in American Politics* (Washington, D.C., 1962); Charles C. Alexander, *The Ku Klux Klan in the Southwest* (Lexington, Ky., 1965); Kenneth T. Jackson, *The Ku Klux Klan in the City, 1915–1930* (New York, 1970); Robert M. Miller, "A Note on the Relationship Between the Protestant Churches and the Revival of the Ku Klux Klan," *Journal of Southern History*, 22 (Aug. 1965), 355–68; Robert Coughlan, "Konklave in Kokomo" in Isabel Leighton, ed., *The Aspirin Age* (Harmondsworth, England 1969). Citations hereafter given for direct quotations only.

7. Quoted in Alexander, *The Ku Klux Klan in the Southwest*, pp. 34, 52.

8. Jackson, *The Ku Klux Klan in the City*, p. 68.

ineffective and would consequently promote public cynicism. "From what I've seen," he said, "I should say that the night riders averaged nearer justice than the courts do."[9] In short, a major stimulus for the spread of vigilantism was the considerable popular alienation in certain locales from a political and legal system that seemed incapable of adjusting to the changed concerns of the community.

Indeed, the Klan's racism and xenophobia can also be explained in part by the organization's apparent disaffection with political leadership. The recent large-scale wave of immigrants from Europe and the population movement from country to town had concentrated problems of social readjustment. In both urban and rural areas, the poorer, established population felt that it had been ignored by large businesses and city governments, which under the guise of "business progressivism" seemed concerned only to provide services that reinforced the economic status quo.[10] In the cities support for the Klan came from both newly arrived white immigrants and from long-established residents. The person who had just arrived in the city from the small country town had probably not shed his restricting ideas about the desirability of a homogeneous and corporate social entity. He believed that if a social group subscribed to certain beliefs, then those beliefs should be sustained by the group's institutions. In his new, more complex environment, however, social consensus became blurred. Thus the Invisible Empire, with its recognizable, reassuring paraphernalia, would give him the opportunity to reassert those fraternal values he cherished.

The more settled city dweller in turn might have felt that his status and the solidarity of his neighborhood were jeopardised by the arrival of a Negro or east European. One historian has commented: "the threatened citizen welcomed the security and respectability of a large group. Seeking to stabilize his world and maintain a neighborhood status quo, he turned to the promise of the Klan."[11] Similarly, in the countryside the Klansman was determined to defend his community against possible future immigration and the infusion of unwelcome ideas. He felt he could not look to the established leadership, since it had exhibited indifference to provincial matters. Both

9. Alexander, *The Ku Klux Klan in the Southwest*, p. 56.
10. Higham, *Strangers in the Land*, pp. 289–99; Tindall, *The Emergence of the New South*, pp. 224–33; Blaine A. Brownell, "Birmingham, Alabama: New South City in the 1920's" *Journal of Southern History*, 38 (Feb. 1972), 44–45.
11. Jackson, *The Ku Klux Klan in the City*, p. 245.

political parties were slow to impose restrictions on immigration. Furthermore, the influence of the rural South in the Democratic party was being slowly eroded and dwarfed by the growing shift to the Democrats in the cities, where immigrants clustered. According to Imperial Wizard Hiram Wesley Evans, "Every kind of inhabitant except the Americans gathered in groups which operated as units in politics, under orders of corrupt self-seeking and un-American leaders, who both by purchase and threat enforced their demands on politicians. Thus it came about that the interests of Americans were always the last to be considered by either national or city governments."

The Klan's sympathizers, then, felt that they had been shunted aside by the political system and that they were rapidly losing control over their social institutions. The only course, it seemed, was to try and reassert the doctrine of popular participation in local political affairs. Thus the Klan deliberately encouraged a spirit of solidarity in its rituals. It believed that the corporate cameraderie of the local Klan could be extended to the central institutions of the community. In particular, it held that schools should continue to reflect and advocate neighborhood values. The Klansman, who was usually a Methodist or Baptist, hoped to render the schools as susceptible to popular sentiment as his church. He was afraid that if the schools remained subject to few popular controls there would be a further decline in traditional values.

Thus in several states the Klan concentrated its political efforts on trying to secure election to local school boards. Once in control, it could scrutinize both the religious and political credentials of local teachers and the curriculum. The schools, they argued, had become remote, largely as a result of the professionalization of education. They had ceased to provide social continuity in the local communities. Schoolteachers, who often stemmed from different areas, did not necessarily subscribe to local mores and so did not fulfill their traditional function. The school was no longer responsible or answerable to the local community. "The sacredness . . . of our right to teach our own children in our own schools were [was] torn away from us," lamented Evans. "Those who maintained the old standards did so only in the face of constant ridicule."[12] The Klan could not

12. In Mowry, ed., *Fords, Flappers and Fanatics*, p. 138. See also Burner, *The Politics of Provincialism*, pp. 104–6, 136–38.

prevent the proliferation of new ideas, but it could rebuff them if the schools were amenable. "One hundred per cent American" teachers, combined with selected textbooks, could make an effective counterpoint to the packaged modernism emanating from the mass media.

This desire to harness the local school to the will of the community was also of prime concern to the religious fundamentalists who converged in the antievolution crusade. The Klan and the anti-Darwinist movement were distinct from one another, as several scholars have pointed out. Support for the fundamentalists was confined mainly to the South, and matured largely when the Klan began to decline. Their very real organizational differences, however, should not be permitted to obscure some of their common anxieties. Both movements feared that the local community was no longer in control of its own destiny. The religious fundamentalists believed that teachers antagonistic to the biblical interpretation of the Creation were implicitly persuading their pupils to reject their parents' *Weltanschaung*. Schools and colleges no longer fulfilled their prescribed function. Formerly, it was argued, the school had equipped its students with preordained intellectual skills and moral values that would enable them to assimilate smoothly into society. The fundamentalist, like the Klansman, argued that the local school should serve the local community. If the school or college developed in such a way that students would question and challenge their parents' way of life, then, according to the fundamentalists, it had abdicated from its designated role. Substantive social change was considered disruptive. The school's task was to equip new generations for painless, gradual social transition. They were, of course, mistaken in their basic assumption that social cohesion is necessarily attainable through intellectual homogeneity. But despite the fallacy and gracelessness of their argument, the fundamentalists in many respects attempted to democratize the educational system. They did, of course, hold ironic, even perverse notions of democracy—after all, they could find no room in their system for dissenters. Nevertheless, the antievolution movement was an attempt, in Jacksonian vein, to make the servants of the state answerable to the electorate.[13]

13. The discussion of the antievolution crusade is primarily based upon the following: Willard B. Gatewood, *Preachers, Pedagogues, and Politicians: The Evolution Controversy in North Carolina, 1920–1927* (Chapel Hill, 1966); Norman F. Furniss, *The Fundamentalist Controversy, 1918–1931* (New Haven, 1954); Kenneth K. Bailey, *Southern White Protestantism in the Twentieth Century* (New York, 1964); Ray

It should be recognized at the outset that this analysis is circumscribed by at least two difficulties. First, there is still insufficient evidence about the precise extent and distribution of fundamentalist support within the South. Second, and more important for the purpose of the argument, is the fact that most leaders of the antievolution crusade were committed evangelical fundamentalists, normally preachers, who frequently publicized their cause in the sweeping, ranting fashion of the revival. Little, however, is known about their followers. But it is important to attempt to account for the movement's popularity in the rural South. As is often the case, the declared reason for affiliation to a social movement may not necessarily be the only reason for that association. An adherent might be either unable to articulate his motives in sophisticated terms or indeed, unwilling, since this could erode rather than enhance support. European anti-Semitism at the turn of the century and McCarthyism in the cold war era are but two examples of social expression whose roots clearly lie beyond the immediate avowed concerns. Thus, it seems reasonable to suggest that there might have been more deep-rooted reasons for the "authentic folk movement" that flourished so dramatically, albeit ephemerally, in the twenties.[14] After all, evangelical hostility to science was not a new phenomenon, yet it took some fifty years to develop into a popular cause.

The fundamentalists reserved most of their fire for the school and college. They reasoned that, apart from parents, the educational establishment exerted the greatest influence on children and, consequently, society. If schools nurtured values contradictory to those learned at home, severe social friction would ensue with concomitant strains on communal cohesion. There had not been a significant struggle since Reconstruction to control the school curriculum in the South. Schools had reinforced regional values, and in the university

Ginger, Six Days or Forever? Tennessee v. John Thomas Scopes (New York, 1974); Jerry R. Tompkins ed., D-Days at Dayton: Reflections on the Scopes Trial (Baton Rouge, 1965); Lawrence W. Levine, Defender of the Faith William Jennings Bryan. The Last Decade, 1915-1925 (New York, 1965); Paolo E. Coletta, William Jennings Bryan, III: Political Puritan (Lincoln, 1969); Robert M. Miller, American·Protestantism and Social Issues (Chapel Hill, 1958); Kenneth K. Bailey "The Enactment of Tennessee's Antievolution Law," Journal of Southern History, 16 (Nov. 1950), 472-90; Virginia Gray, "Anti-Evolution Sentiment and Behavior: The Case of Arkansas," Journal of American History, 57 (Sept. 1970), 352-66. Citations hereafter given for direct quotations only.
14. Cash, The Mind of the South, p. 346.

the rare rebel had been easily disciplined. Teachers broadly upheld the region's racial values; and the industrial skills taught in both black and white schools were geared to the southern economy. However, the emergence of progressive education in the early part of the century appeared to challenge the traditional social purpose of the school. While the ideas of the progressive educators may not have emanated as widely in Dixie as in northern cities, a perception of change persisted.[15]

The father of progressive education, John Dewey, had argued as early as 1899 that children's education should be geared less to preconceived notions of intellectual verity than to the child's immediate environmental pressures. While the school should attempt to mirror the larger society in its organization, it should at the same time encourage the pupil to improve that society. The school, argued Dewey, should strive towards providing an atmosphere that would enable the child to change and challenge his environment. Dewey's ideas became increasingly respectable in educational circles in the 1920's, although they were somewhat refined. One of the most influential surveys of pedagogic innovations across the country, *The Child-Centered School*, found that teachers were placing greater emphasis on individual development and self-expression than ever before. Schools were increasingly loth to establish educational goals and aims. The authors of this study compared the school of the 1920's to the historic battle of the artist against society. Children were being taught to question and, if necessary, rebel.[16]

Of course, most fundamentalists would not have been consciously aware of these developments. Indeed, educational innovation affected the South, particularly in rural areas, less than in the metropolitan regions. The southern states continued to emphasize such traditional subjects as algebra and Latin. Yet it is quite plausible that even in the backwoods parents will have sensed that their children's social perspective was on the brink or in the process of transformation. The moral climate of the era, they were being constantly reminded by the

15. E.g., Henry Y. Warnock, "Andrew Sledd, Southern Methodists and the Negro: A Case History," *Journal of Southern History*, 31 (Aug. 1965), 251–71; C. Vann Woodward, *Origins of the New South, 1877–1913* (Baton Rouge, 1951), pp. 401–6, 445.

16. John Dewey, *The School and Society* (Chicago, 1899); John Dewey, *Democracy and Education* (New York, 1916); Lawrence A. Cremin, *The Transformation of the School: Progressivism in American Education, 1876–1957* (New York, 1961), p. 183; *Recent Social Trends, I*, pp. 355–58.

pulpit, the press and advertisers, had changed. In the cities, youth had adopted strange fashions in their clothes and love lives. But the schools, according to the fundamentalists, had not succeeded in discouraging the young from this perceived prodigality. They had actually imbibed the new climate. Indeed, one of the most popular intellectual figures of the period, Sigmund Freud, had preached that social convention actually inhibited personal development and was responsible for the prevalence of neurosis. Several educators endorsed Freud's view.[17]

As might be expected, such social iconoclasm alarmed the conservative residents of the Bible Belt. Pupils and students throughout the country were being encouraged to question the foundations of their social beliefs, including religion. If this development were not controlled, then religious ties would be broken. After that, the entire social edifice would crumble. The fundamentalists, therefore, rebelled against the growing professionalization of education. The curriculum, they argued, should be geared to social needs, needs that could be defined only by the community. The teacher should be the instrument of popular aspiration, not its architect.[18]

Thus in the South, rising bewilderment at the changes, real or apparent, that seemed to envelop the educational system contributed to the spread of the antievolution movement. Such anti-intellectualism was not, of course, new or even confined to Dixie. But since the slavery controversy, southerners had sought to harness educational ideas to its regional values. Educators and intellectuals either echoed the section's prejudices or, as was sometimes the case, found themselves without jobs. But rapid educational expansion and its concomitant professionalization had broken the ties between places of learning and the community. The advocates of popular control perceived that if teachers and their curriculum were not anchored more firmly, the school could become an instrument of social disintegration. They echoed the worry of John Gould Fletcher, one of the Nashville Agrarians, that "if the present system persists, in another generation nothing will remain of the local color, the diversity, the humanity, the charm of our South, and we will become assimilated

17. Cremin, *The Transformation of the School*, p. 212; *Recent Social Trends, I*, pp. 331–33, 414–22.

18. William Jennings Bryan argued that the propensity to reject religion increased in direct proportion to the duration of a student's education. Gatewood, *Preachers, Pedagogues, and Politicians*, p. 105; Levine, *Defender of the Faith*, pp. 266–67.

outwardly and inwardly to the street gangs of New York and Chicago."[19]

In the rural South fundamentalist preachers and politicians agitated against the growing rift between community and school by raising the issue of evolution. Appropriately the leadership of the antievolution crusade fell to William Jennings Bryan, who still was America's most fervent exponent of popular democracy. Bryan shared the evangelists' belief that the scriptures should act as guidelines for man's behavior. To question only one facet was to challenge the Christian basis of civilized society. Thus to Bryan, "The objection to evolution, however . . . is not, primarily, that it is not true. . . . The principal objection to evolution is that it is highly harmful to those who accept it."[20] In the early twenties, the majority of rural southerners seemed to accept the literal biblical interpretation of the origins of man. W. J. Cash thought the fundamentalist movement was an authentic folk movement, and no contemporary observers thought it commanded the support of less than half the population.[21]

Bryan thus calculated that his personal crusade was justified on two counts. First, he believed his position was doctrinally and morally correct. Second, and more important, it was incumbent upon southerners to try and control the school curriculum and screen their teachers. Since most of them were exercised by recent developments in education, they should join together to anchor the schools more firmly. The will of the majority, according to Bryan, was "the best expression of the divine will to be found upon the earth." Majoritarianism was the only sound principle of government. It might produce injustice or unwisdom, but that was considered preferable to elitist rule. "No concession can be made to the minority in this country," he proclaimed, "without a surrender of the fundamental principle of popular government. The people have a right to have what they want."[22]

Thus, despite the intellectual simplicity of Bryan's argument, his crusade must be understood in part as an attempt to shape vital local institutions in accordance with the prevailing culture. Educational

19. Twelve Southerners, *I'll Take My Stand*, p. 121.
20. Levine, *Defender of the Faith*, p. 281.
21. Cash, *The Mind of the South*, p. 346; Edwin Mims, *The Advancing South: Stories of Progress and Reaction* (New York, 1926), p. 338.
22. Levine, *Defender of the Faith*, p. 223.

developments may have produced more sophisticated high school and college graduates, but they appeared to alienate a significant proportion of the populace. Thus the *Kulturkampf* of the twenties should be seen not only as a conflict between city and country but also as one between a growing professional elite and a tradition-oriented society whose preoccupations had been ignored by the advocates of scientific rationality, objectivity, and modernism.

The story of the struggle over the teaching of Darwinian evolution is familiar and well-trodden and needs no recounting. One aspect, however, needs special emphasis, since it has often been camouflaged by the rather absurd histrionics exhibited by the evangelical crusaders who assembled at the Scopes trial in Dayton, Tennessee, in 1925. The advocates of curricular censorship argued that the schools were supported by local taxpayers, the majority of whom were opposed to the use of textbooks that encouraged children to reject their parents' beliefs. The schools, together with the churches, comprised the mainstay of the esprit de corps found in neighborhoods. If the schools were to surrender their traditionally supportive role by challenging the basis of social cohesion, then the unique *Gemeinschaft* would be terminated. So the taxpayers who supported the schools —and the churches that financed some of the denominational colleges—should be permitted to determine the way in which their money was being spent.

The North Carolina Bible League reaffirmed in its platform Jefferson's belief, " 'To compel a man to furnish contributions of money for the propagation of opinions which he disbelieves, is sinful and tyrannical.' Our State schools shall be our servants and not our masters."[23] This argument was an attractive one, especially in areas that produced few dissenters from the current evangelical wisdom.[24] And it was an argument that could be answered only by invoking ideas and concepts essentially alien and contradictory to local folkways. The fundamentalists had raised a basic question of political theory, namely the extent to which the popular will should be subordinated to received truths.

23. Gatewood, *Preachers, Pedagogues and Politicians*, p. 239.
24. Ironically, this particular view reemerged in the 1960's, although the source of advocacy had changed. Then black dissidents in the cities of the North and the champions of "free schools" also maintained the right of the local community to shape the curriculum and screen teachers. Neither group wished to employ teachers from outside the immediate social milieu.

In short, political and religious fundamentalism in the twenties must be seen not merely as a reactionary backlash to the growing cosmopolitanism of American life. It was the political expression of a populace that was genuinely bewildered by a burgeoning culture which they could not easily control or resist. Social standards appeared to be set by a new intellectual elite that permeated educational establishments and the communications industry. The Klansman and the fundamentalist perceived that political organization was necessary if the new cultural czars were to be circumscribed. The local community, they argued, should have the political ability to resist unwanted developments. It should cooperate in immunizing itself from unwelcome encroachments. The United States was still a federal nation, and local differences and customs should be respected. However, law, education, and the popular arts seemed oblivious to local distinction. It seemed that parochial peculiarities could be maintained only if teachers, lawmakers, and community leaders were answerable to the communities in which they worked. The fundamentalists reverted to the old Jacksonian belief that a public officer must serve the perceived desires of his constituency and should not dissociate himself from the social process.

Scholars who have condemned and disparaged the Klan and the Bible crusaders in some ways have been misled by their sometimes grotesque expositions. By concentrating too much on the fundamentalists' methods of expression, the critics have tended to underestimate their essentially democratic stimuli. Perhaps the disturbing feature of small-town life in the twenties does not lie purely in the admitted intolerance of its fundamentalist inhabitants. The alienated and disaffected crusaders for moral purity were the products of a society that seemed increasingly negligent of tradition. Perhaps it is time for historians to address more critical attention to those developments in the wider society that nurtured the growth of fundamentalism in the first place.

STRIDENT VOICES IN KANSAS BETWEEN THE WARS

CLIFFORD R. HOPE, JR.

I. THE KU KLUX KLAN IN KANSAS

IN THE EARLY 1920's, Kansas (especially eastern Kansas) succumbed for a brief period to the surge of the nationally revived Ku Klux Klan.[1] The Klan is mentioned herein for the purpose of relating, for the first time in print, I believe, a footnote to history, and to show some similarity between the Klan philosophy and that of Rev. Gerald B. Winrod in the 1930's.

On January 10, 1925, the Kansas supreme court rendered its decision in effect ousting the Klan from "doing business" in Kansas on the basis that it was a foreign corporation not approved to do business in the state by the then state charter board.

This article was the author's presidential address at the annual meeting of the Kansas State Historical Society, October 17, 1978, in Topeka. He wishes to acknowledge his indebtedness to the authors of the following theses on the career of Rev. Gerald B. Winrod: Ann Mari Buitrago, "A Study of the Political Ideas and Activities of Gerald B. Winrod, 1926-38" (unpublished M.A. thesis, University of Kansas, 1955); James Schrag, "Gerald Burton Winrod: The Defender" (unpublished undergraduate paper, Bethel College, 1966); Larry B. Sullivan, "Gerald B. Winrod" (unpublished M.A. thesis, Fort Hays Kansas State College, 1967); John D. Waltner, "Gerald B. Winrod and the Washington, D. C. Mass Sedition Trial of 1944" (unpublished undergraduate paper, Bethel College, 1968).

Also, the author wishes to thank Mrs. M. K. Flowers, secretary-treasurer of the Defender organization in Wichita, for lending a copy of Winrod's official biography by G. H. Montgomery, *Gerald Burton Winrod* (Wichita, Mertmont Publishers, 1965).

1. The Klan era in Kansas has been well covered in recent issues of the *Kansas Historical Quarterly*, e.g., "Kansas Battles the Invisible Empire: The Legal Ouster of the K. K. K. From Kansas, 1922-1927," Autumn, 1974, by Charles William Sloan, Jr., and "William Allen White's 1924 Gubernatorial Campaign," Summer, 1976, by Jack W. Traylor, and in other publications, e. g., "The Ku Klux Klan in Eastern Kansas During the 1920's," by Lila Lee Jones, in *The Emporia State Research Studies*, Winter, 1975.

In the 1924 election, Klan supporters had won control of the state senate and had elected a great number of state representatives. On February 25, 1925, in a sudden action, the senate passed Senate Bill 269, which measure was designed to remove the state charter board's powers of investigation and discretion in granting corporate charters.

The speaker of the house that year was Clifford R. Hope, age 31, a strong Klan opponent. Because of the Klan issue, he had won the speakership in the Republican caucus by only two votes. While Hope was delivering a speech to the Kansas Livestock Association in Wichita, the bill was rushed over to the house, where it was delivered to the chief clerk. The Klan proponents planned to have the bill passed, under emergency procedures, during Hope's absence.

O. H. Hatfield of Copeland, a former state representative from Gray county, was chief clerk of the house, an appointee of the speaker. After delivery to Hatfield, the bill became "lost," and Hatfield contacted Hope in Wichita and urged his immediate return to Topeka, meeting him at the train with a status report.

On February 27, after Hope's return, the bill was "found" by Hatfield, and an attempt was again made to emergency it through the house. Although the vote was 77 to 40 in favor of the emergency, it was not the necessary two-thirds vote and the effort hence failed. Hope thereafter referred the bill to the judiciary committee, immediately followed by an attempt to have it withdrawn from the committee and referred back to the committee of the whole. The vote on this issue was 63 to 52 in favor, but again not the necessary two thirds.

Thereupon, after Hope's promise that the bill would be neither buried in committee nor kept down on the calendar, the bill followed the usual, nonemergency legislative procedures, with the result that the furor for it diminished, and on March 6 it was defeated on its merits, 65 to 57.

Although in 1925 support for the Klan in Kansas was probably on the wane, the action of the house in defeating Senate Bill 269 was a great blow to the Klan. It is, of course, a subjective view, but I would like to think that the Klan's demise was hastened by the wisdom of Hatfield in knowing when to lose and when to find a bill, and the courage of Hope (who planned to—and did—run for congress the next year) in vigorously opposing the Klan, when it would have been easier and safer politically to have acted pleasant and not taken a decisive stand.

Lila Lee Jones has analyzed well the reasons for the Klan's popularity in Kansas:

Fanaticism and fear contributed to the Ku Klux Klan movement in eastern Kansas during the 1920's. Many Americans were disillusioned and had a deep sense of insecurity during the period following World War I. This led many leading small town citizens to the organization which appeared to preserve the status quo. . . .
There were, however, deep rooted fears and insecurities among Kansans, as elsewhere in the United States, and the Klan gave priority to those related to Catholicism, prohibition, and immorality. There were also hints in Klan literature of fears of communism, immigrants, labor organizations, Jews, and corrupt government.
Although most Klan writers began by saying that they were not anti-Catholic, they ended by condemning the Roman Catholic Church as representing the greatest single threat to Anglo-Saxon-Protestant values. There were those Protestant ministers who for years had linked the anti-Christ and the Beast of the book of *Revelation* with the Catholic Church and the Pope. It took little persuasion on the part of Klan writers to convince Protestants of all denominations that the last days had arrived and that the Pope could overthrow the government of the United States as easily as the Communists had seized Russia.
The growth of fundamentalism in protestant America also played an important role in the Invisible Empire. Fundamentalism challenged the new developments and sought to entrench traditional doctrines and practices. It was at times a bitter and divisive movement which was often militant. Its leaders were charged with stirring up conflict as much for the love of a fight as for love of truth.

The movement tended to identify Christianity with patriotism and was influenced by the stream of superpatriotism which marked the 1920's. One fundamentalist leader preached "100 per cent Americanism" and said patriotism and Christianity are synonymous terms just as hell and traitors are synonymous. Although fundamentalism declined rapidly after 1925, its influence continued in certain congregations and small denominations. Many who agreed essentially with its doctrines turned away because they did not want to be associated with its spirit of bitterness and strife. It is not difficult to see the parallels between this movement in protestant churches and the Ku Klux Klan. Like the Ku Klux Klan, fundamentalism introduced harmful tensions into the churches which offset its worthwhile contributions.

The Ku Klux Klan, then, can be viewed as an inevitable product of the 1920's. Fear and insecurity led many Kansans into the Invisible Empire, but all of those who became members of the secret order in eastern Kansas were not revolutionaries or simply chronic malcontents. Many represented honest laborers and small town businessmen—bankers, ministers, publishers of small town weeklies, lawyers, doctors, and merchants. Insecure in the present and apprehensive about the future, they turned against those things which they saw as threatening and alien to them. When the enemies failed to materialize or were eliminated, the knights no longer had anything to fear and their crusade collapsed.[2]

As above indicated, one may note great similarity between Klan "doctrine" in the 1920's and that of Winrod in the 1930's.

II. Rev. Gerald Burton Winrod

ALTHOUGH his name is little known today, Gerald Burton Winrod could be described as one of the most intriguing, strangest, and most tragic figures of Kansas history; indeed, he was regarded by some as the most likely leader of the Fascist-anti-Semitic movement in the United States in the 1930's and prior to World War II, in the remote event that all such extremist groups had ever presented a united front.

Winrod was born in Wichita on March 7, 1900. It was reported that his father, J. W.

2. Jones, "The Ku Klux Klan   .   .   .," pp. 23, 34.

Winrod, had three favorite sports—drinking, fighting, and dancing; he was a bartender in the Old Four Ten saloon until Carry Nation wrecked it. This event, coupled with his wife's miraculous cure from cancer, led to J. W.'s conversion. Young Gerald, deeply affected by these events, followed a traveling evangelist at the age of 17 and was a full-fledged evangelist at the age of 21.

During his entire career as preacher, author, and publisher, he received no religious schooling, held no pastorate, and was not a member of any denomination. He received an honorary D.D. degree from Los Angeles Baptist Theological Seminary in 1935.

His concern over modernism in religion led him to call a meeting in Salina in 1925 for the "united defense of the Christian faith." From this meeting of approximately 100 fundamentalist leaders sprang the Defenders of the Christian Faith. Winrod was elected executive secretary, "Faith of Our Fathers" was chosen as the official hymn, and the statement of objectives was derived from the book of Jude, "Contend for the faith which was once delivered unto the saints."

In 1926 Winrod established *The Defender Magazine,* and remained its editor, as well as executive secretary of the Defender organization, until his death in 1957. The Defenders' announced purpose was to "expose modernism on every hand. We will oppose by God's grace this wicked doctrine of evolution in tax-supported schools. We will meet the rationalists at every front, whether in the pulpit or elsewhere, and seek to check their work until the faith of our fathers is restored." Winrod stated the principle of "opposition without malice." "While our position is uncompromising, yet we strive to be fair and honest with the enemies of the truth for which we stand." However, the claim of a policy of tolerance soon became unsupportable. By 1933 Winrod was

During his career, Gerald B. Winrod (1900-1957), controversial Wichita radio preacher and editor who was a candidate for the U.S. senate in 1938, moved from the mainstream of fundamentalism to an anti-Semitic conspiracy view of history. "Winrodism" in the 1930's contributed to what Henry Steele Commager called a "dirty work at the crossroads" outlook in American thinking. Photograph from Wichita *Beacon*, November 12, 1957.

vigorously attacking the Roosevelt administration, the international Jewish conspiracy, and world-wide Communism.

Winrod was an ardent Prohibitionist and

soon became involved in politics. In 1928 he supported "dry" Herbert Hoover over "wet" Al Smith, but in 1932 took the stump for William D. Upham, the Prohibitionist candidate, hoping to throw the election into the house of representatives.

During the period from 1934-1937, a monthly newspaper, *The Revealer* was published; its main purpose was attacking the Roosevelt administration. In January, 1937, Winrod established the *Capitol News & Features Service* for the purpose of supplying Washington "news" to 2,000 county newspapers without charge, but this effort collapsed in April due to a lack of funds.

During this period, Winrod gave political speeches in Kansas. Concerning the campaign year of 1936, a distinguished member of this Society, Arno Windscheffel, recalls:

"Soon after I got out of law school some of the Republican leaders in Smith Center needed to raise some money. In the mid 30's money was difficult to obtain. Someone came up with the idea of having Winrod come to Smith Center and then charge admission. This was done. The Republican big wigs did not want to be identified with Winrod, so they asked Tod Reed, a quite elderly lawyer, to introduce Winrod. Tod was given a script for the introduction. Tod was almost stone deaf. He made the introduction and Winrod started blasting. He especially blasted the Republican Administration (in Kansas) and Republicans in general. Tod was an ardent Republican, but not being able to hear he politely sat smiling through the whole verbal assault on the Republicans. After the speech Tod praised Winrod and thanked him. The auditorium was packed. The next day someone told Tod about Winrod's blast against the Republicans—it was then Tod's turn to blast.

III. The Senate Race, 1938

IN EARLY 1937 Winrod mounted a nation-wide campaign in opposition to F. D. R.'s "court-packing" plan, as a forerunner to his race for the United States senate in 1938. In January that year he announced as a Republican candidate against Sen. George McGill, who was seeking his second full term.

Winrod utilized radio as his principal campaign weapon, commencing broadcasts on WIBW in Topeka on January 27, and continuing every Tuesday thereafter. He was acknowledged by all as a gifted orator.

Early on, he set forth his seven-point platform, seeking to jump the gun on other potential candidates and to capitalize upon the strong isolationist and anti-New Deal sentiment of that time:

1. Defense of Constitutional Democracy Against the encroachment of Communist, Fascist, and Nazi propaganda.
2. Rebuilding of national character by the quickening of patriotic, moral, and religious sentiments, as an antidote to possible bloodshed.
3. Rigid observance of State's Rights to reverse present trends toward centralized bureaucracy.
4. Absolute neutrality through obedience to the nationalistic impulse, that no more American blood should flow on foreign soil.
5. Restoration of the right to create money, to the United States Congress, so that duly elected representatives of the people shall control the flow of currency.
6. Repeal of experimental legislative measures which have proved themselves to be inimical to the best interests of both labor and industry.
7. An attitude on the part of the national government that will inspire public confidence, bring private capital out of hiding and create honest jobs so that a man can do an honest day's work and receive an honest day's pay.

Winrod's platform and speeches stressed the "fundamental Americanism" approach and carefully avoided, even by indirection, his previous anti-Semitic and anti-Catholic views.

It is reported he distributed more campaign literature than any other candidate and expanded his mailing list to 150,000, making use of Dr. John R. Brinkley's lists.

Winrod, a vegetarian, seldom attended party luncheons or dinners. When he did, he sometimes left immediately, without shaking hands all around. He made extensive use of a sound truck (a widely used campaign method of that time), speaking from a platform in the rear; on such occasions, he would linger after his speech to greet his listeners.

Winrod's opponents did not commence full-time campaigning until about July 1 and Winrod was conceded to have the lead as of that date. The other candidates were: Clyde M. Reed, former governor (1929-1931) and publisher of the Parsons *Sun*; Dallas Knapp, former state senator and Coffeyville lawyer; and Jesse Clyde Fisher, a Methodist minister, then superintendent of the Liberal district and a Garden City resident. Earlier in the year both Cong. Frank Carlson and Clifford R. Hope were mentioned as candidates, but both declined to run. Winrod's official biographer, G. H. Montgomery, has stated that Fisher was a tool of Winrod's enemies in an effort to split the "church vote."

Winrod's candidacy was opposed by many clergymen. One group, for example, distributed a 15-page pamphlet entitled *Drive Fascist Ideas From Kansas,* using reproductions of pages from *The Defender* and *The Revealer.* This was followed by a letter entitled "Dear Fellow Kansans."

John D. M. Hamilton, then chairman of the Republican National Committee, and former Gov. Alf M. Landon opposed Winrod, as did most of the state's newspapers, including the Wichita *Beacon* and Topeka *State Journal.*

William Allen White, true to character, lashed out at Winrod, not only in the Emporia *Gazette* editorials, but also in paid advertisements in other newspapers, charging that he was reviving the spirit of the Ku Klux Klan.

Late in the campaign, Winrod responded, both in a pamphlet, *Unmasking a Conspiracy of Lies,* and by a radio address, "Viewing the Facts." He termed his campaign "a holy crusade." He claimed he was opposed only to apostate Jews, and was not anti-Catholic, anti-

Masonic, or a Nazi. He charged that $1,000,000 was being spent by Jews and Democrats to defeat him. He threatened to expose the source of such funds, stating the disclosure "would rock the political foundations of the nation." William Allen White dared him to do so, but no such disclosures were ever made. Winrod also charged he was being harrassed by federal agents.

Winrod did receive support from a scattering of respectable Kansas leaders and newspapers, and from 26 Wichita clergymen, and undoubtedly inherited many of the old "hard-core" Brinkley supporters.

The effectiveness of the anti-Winrod campaign was evident on August 2, primary election day, especially considering that Reed and Knapp were essentially competing for the same votes. The results were: Reed 102,691, Knapp 62,418, Winrod 52,344, Fisher 25,548.

Winrod carried only Clay county (in "Brinkley country"), but ran second in several others.

Montgomery regards the senate race as Winrod's greatest mistake, believing that politics would have muted rather than amplified his "Christian influence."

As for Winrod himself, in April, 1950, he wrote in *The Defender Magazine*:

There can be no question but what my campaign frightened the *hidden masters*. They knew that I would make myself heard in the event of going to the Senate, and that they would never be able to control me. . . . I was defeated by the most powerful combination of Communist and New Deal agencies ever assembled. They had inexhaustible sums of Jewish money upon which to draw, and violated my civil rights at every turn.

IV. The Mass Sedition Trial of 1944

ON JULY 23, 1942, Winrod's name headed the list of 28 persons indicted by a federal grand jury in Washington for sedition. The

indictment alleged violations of the Espionage
act of 1917 and the Smith act of 1940, charging
that the defendants individually conspired to
disseminate Nazi propaganda for the purpose
of undermining the loyalty and morale of U. S.
armed forces. This indictment was subsequently dismissed, but another, charging 33
defendants, was issued on January 4, 1943.
This time Winrod's name was in the middle.
The indictment also listed alleged seditious
publications, including Winrod's *The Revealer*
and his short-lived *Capital News and Features
Service.* This indictment was also dismissed.

Finally, on January 3, 1944, a third indictment, based solely on the Smith act, was issued, charging 30 defendants, including both
Winrod and E. J. Garner, then the publisher of
Publicity in Wichita.

The indictment alleged that the defendants
had,

unlawfully, willfully, feloniously and knowingly conspired, combined, confederated and agreed together with
each other and with officials of the government of the
German Reich and leaders and members of the Nazi Party,
in Washington and other parts of the country, and in
Germany and elsewhere, to commit acts with intent to
interfere with, impair and influence the loyalty, morale and
discipline of the military and naval forces of this country.

Alleged subversive publications listed were
The Revealer and *The Defender;* the Defender
Publishers was listed as an alleged subversive
organization.

The trial commenced April 17, 1944, before
Judge Edward C. Eicher, a former three-term
congressman from Iowa, appointed to the
bench in 1942. (In 1950, Winrod charged that
Eicher had been appointed as a federal judge
for the specific purpose of presiding over the
trial and upon his promise to send Winrod and
others to prison.) The prosecutor was O. John
Rogge, whom Winrod charged to be a close
personal friend of the judge.

The trial began with 30 defendants, 23 lawyers, and 40 reporters present. It continued for some eight months in an atmosphere generally described as a circus.

On May 16 the jury was impaneled. Rogge presented a three-hour opening statement, alleging a worldwide Nazi conspiracy to destroy democratic governments and replace them with Nazi regimes. Similarity of the various defendants' anti-Semitic and anti-Communist statements (and opposition to U. S. entry into World War II) with those of the Nazis were cited. The defendants claimed, and probably rightfully so, that there was no evidence of a conspiracy to incite insubordination in the armed forces.

The trial dragged on and on; there was a two-week recess in July (no air conditioning then!) and thereafter the trial was held only in the afternoons. Some lawyers and defendants quit attending every day. By October, only four or five reporters were in attendance. At one point, Winrod returned to Kansas due to the illness and death of his father. The number of defendants was reduced to 26, due to the death of Garner (then 80 years old), and the ill health of others.

The trial ended abruptly on November 30, 1944, due to the death during the night of Judge Eicher. (In April, 1946, Winrod called the judge's death an act of God, in response to prayers across the nation.) On December 7 a mistrial was declared and finally, in June, 1947, the indictment was formally dismissed.

In the retrospect of history, it is clear there was little, if any, evidence of a mass conspiracy, and certainly no credible evidence that Winrod was a part of any conspiracy. Some have claimed the trial was brought about by "war hysteria"; the *Saturday Evening Post* compared the trial to Nazi and Communist propaganda trials. Alf M. Landon later told

John D. Waltner, "Guilt by association is hardly a reason to charge sedition. The issue was not calmly and cooly studied as sedition cases should be." [3]

Winrod, of course, believed he was tried because the Jews wanted him out of the way and Montgomery, his official biographer, claimed the trial was a part of F. D. R.'s plan for absolute power.

In retrospect, it is perhaps one of the ironies of history that the defendants, most of whom were so eager to charge others with various conspiracies, themselves became the victims of similar accusations.

V. THE WINROD-BRINKLEY-GARNER CONNECTION

TIME available for the completion of this paper has not permitted detailed research into the Dr. John R. Brinkley-Winrod relationship, but the evidence which has been studied indicates at least a mutual admiration society between the two.

Francis Schruben, in *Kansas in Turmoil-1930-1936* states: "Brinkley's friends ranged from the ultra-nativistic Reverend Dr. Gerald Winrod to the Jewish Levand brothers, publishers of the Wichita *Beacon*." Regarding the 1932 gubernatorial campaign (when Brinkley was running the second time, with his name on the ballot), Schruben has written:

In addition, Brinkley may have profited in September by the printed facsimile of a letter of endorsement by The Reverend Gerald B. Winrod, the controversial Wichita clergyman with nativistic tendencies. Winrod said he and his followers had their "fingers on the Kansas pulse," and personally sent Brinkley "every spiritual blessing in your stand for the highest religious and moral interests for which you are well known."

3. Waltner, "Gerald B. Winrod. . . ."

It is clear that Brinkley either owned or controlled *Publicity* in Wichita, of which E. J. Garner was the editor. While Brinkley was engaged in Kansas politics, *Publicity* was the official organ of the Brinkley-for-Governor clubs, and when Brinkley left Kansas for Texas, his "goodbye letter" was written to Garner and published in *Publicity*.[4] After Brinkley's move to Texas, *Publicity* continued to promote Brinkley's activities.

After World War II, Winrod continued his anti-Semitic campaign, opposing international Communism, the United Nations, the old Federal Council of Churches, the National Conference of Christians and Jews, and Dwight D. Eisenhower, whom he called a tool of the Jews.

In his last years he promoted repudiated forms of treatment for cancer and lambasted the medical profession. Having suffered from multiple sclerosis for years, he contracted Asian flu in the fall of 1957, refused to see a medical doctor, and succumbed on November 11 of that year.

VI. WINROD'S RELIGIOUS IDEAS AND POLITICAL BELIEFS

STUDIES of Dr. Winrod agree that he was first, last, and always a preacher. As an orator, he was compared to William Jennings Bryan.

As a fundamentalist in the late 1920's he strongly opposed modernism (defined as an attempt to restate Christian faith in light of modern scientific discoveries), the ecumenical movement, and the Catholic Church.

He possessed an almost incredible belief in biblical prophecies (as he interpreted them);

4. Topeka *Daily Capital*, December 7, 1933.

Dr. John R. Brinkley (1885-1942), well-known "goat-gland doctor," owned Kansas' first radio station and came close with his write-in campaign to becoming governor in 1930. Running again in 1932 he may have profited from the endorsement of Gerald Winrod, who sent him "every spiritual blessing in your stand for the highest religious and moral interests for which you are well known."

one author has stated his belief in prophecy was not confined to the Bible, but included the Great Pyramid of Egypt prophecies and myths. He often took passages of scripture out of context; for example, the verse: "The chariots shall race in the streets, they shall jostle one against another in the broadways; they shall seem like torches; they shall run like the lightning" was interpreted as a prophecy of the automobile with headlights. Passages in Job and Psalms referred to the radio; the lion mentioned in Ezekiel 38:13 was Great Britain.

He apparently believed the Battle of Armageddon and the end of the world would occur in his lifetime. The president of the pre-World War II World Bank he saw as the "anti-Christ" and the bank's worldwide control as "the mark of the beast." In 1933, however, he observed that the "mark of the beast" might be the NRA Blue Eagle. He traced Communism back to Cain (whose name meant "red"). Atomic power destroyed Sodom and Gomorrah. The dust storms of the 1930's were sent by God to punish the United States for the "ungodly features of the New Deal." It would appear to follow naturally that he sometimes attributed his own actions to divine inspiration.

In 1936 Winrod referred to Catholicism as "baptized paganism" and called the pope "Mr. Pius of Vatican City." By 1949, however, Winrod was urging a united Protestant-Catholic front against Communism. And the previous year he charged the World Council of Churches (which he called "the bride of the Anti-Christ") with being anti-Catholic; the reason: because the Vatican had always been anti-Communist.

Now I wish to comment in some greater detail upon Winrod's political views in the 1930's and prior to World War II, primarily as he expressed them in *The Defender* magazine. Until January, 1933, his views in general had been in the mainstream of one of the dominant and respectable traditions of our state: Religious conservatism, literal interpretation of scripture, and Prohibition.

It was a tragic day when Winrod first read (probably in 1932) *The Protocols of the Learned Elders of Zion,* a book first published in Tsarist Russia in 1905 purporting to prove a continuing Jewish plot to seize control and rule the world. One scholar has well described Winrod's reaction:

This forgery turned his old concept of Satan's unseen

hands into his new concept of THE HIDDEN HAND of Jewish Conspiracy. Winrod wrote of his new discovery in the feature article of the January, 1933, *Defender* magazine:

"There has been uncovered before my eyes, the inner workings of one of the most gigantic plots ever perpetuated in any period of world history. For months I have been sifting evidence, assembling facts, accumulating material; and step by step, fact upon fact, I have traced these destructive forces back, back, back, to their hidden sources and I am now prepared to say that I firmly believe all these horrible outbursts which we are today witnessing are simply results of intelligent causes. Behind the scenes there is what I choose to call, A HIDDEN HAND.

"I later made the amazing discovery that there is a plot to overthrow the religious, moral, and governmental systems of the earth and that it is a world conspiracy, evidently planned several hundred years ago by a self-perpetuating group of men who today control the wealth of the world. . . . They arranged to pull wires behind the scenes to precipitate political and economic upheavals."

With the aid of this "document" Winrod was able to combine his notion of demonized unseen hands with his expectation of Jewish apostasy:

"I knew that the hellish agencies which are today going out to the ends of the earth, had to have their secret octopus-roots fastened in demon-possessed brains somewhere; so I started reasoning from effects back to causes. I soon came upon the amazing discovery of what purports to be a conspiracy, which is centuries old, to overthrow the religious and governmental systems of the Gentile world."

Winrod felt it his duty to devote himself and his magazine to "unmasking" the origins and processes of this conspiracy.

"Informed persons," Winrod wrote in February, "know that the world is breaking up. We may as well face the facts. . . . A revolution is not coming, it is already here." He believed that 300 men, "conspirators against civilization," had planned and caused the social chaos of the thirties. "This group," he continued, "claiming to hold the financial destiny of the world in their hands, are wealthy Jews." Winrod addressed himself briefly, once and for all, to the question of motive: ". . . it is a religious conviction with them to create Gentile chaos. . . . Out of the chaos they believe themselves called of God to create a dictatorship built upon gold with a Superman at the head, who will be their Messiah with headquarters in Palestine." Their whole purpose, he concluded, is to set the stage for the Superman. It is not surprising that Winrod showed little interest in examining the possible motive of the alleged Elders, or in considering the authenticity of the document. The whole scheme described in the *Protocols* fit in so exactly with his prophetic views that it must have appeared to Winrod as incredible confirmation of his independent deductions. The Elders were effecting the conditions described in the Old Testament prophecies. As Winrod was fond of saying, prophecy was simply history written in advance.

The *Protocols,* Winrod explained, revealed that in carrying out this international plot, the Elders recognized no limitations. They used means legal and extra-legal; they worked on all fronts simultaneously, tearing down all institutions and rebuilding them so that all power resided in them. The outline of the activities of these Jewish Elders given in the *Protocols* became the outline of Winrod's major subject matter for the next five years. Briefly, these alleged activities were as follows: The Elders promoted immorality among Gentiles through control of the press and publishing business; attacked the Church through the device of modernism; corrupted school children by teaching jungle evolution, and immoral, godless, communistic ideas; and poisoned the general public through alcohol, cigarettes, movies, and filthy literature. One of their most powerful tools in creating national unrest was the virus of Liberalism, wʰich led to destructive experimentation with social relationships. Since this small group of Jews controlled most of the world's wealth, and all the gold, they could manipulate forces to create any sort of economic condition desired. They had, in fact, purposely created the 1929 depression. Communism was created by the Elders to foment anarchy and revolution.

Through all these avenues the Jewish conspiracy would work until such chaos would result that a world dictator would seem necessary, and in fact, be demanded as the only solution. At such a juncture the anti-Christ would be unveiled.[5]

During the period 1933-1937, Winrod, in pursuit of his conspiracy theory of history, publicized some amazing stories. An example is related by the same scholar just quoted.

[In 1776] Adam Weishaupt, a leading Jesuit teacher at Ingoldstaat University in southern Germany, created the "Illuminati," an organization purporting to embody ancient and advanced degrees of Freemasonry. The "Illuminati" patterned their secret government after the Jesuit order, by pulling wires, and by setting up spy systems to control the lodges all over Europe. Many Jesuits followed

5. Buitrago, "A Study of the Political Ideas . . .," pp. 55-58.

The Defender

An Official Statement By The Executive Committee, Of "The Defenders of the ... a Faith"

Vol. 1   Published in Wichita Kansas, June 1926   Number 2

Jesus and Modernism

By Rev. A. L. Carlton, President of "The Defenders"

The "Flying Defenders"

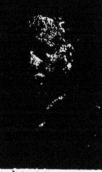

Gerald B. Winrod

Damned Souls

By Gerald B. Winrod

Gerald Winrod's concern over modernism in religion led him in 1925 to call a meeting of fundamentalist leaders for the "united defense of the Christian faith." In 1926 he established *The Defender Magazine* and he remained its editor, as well as executive secretary of the Defenders of the Faith organization, until his death in 1957. The Defenders opposed the teaching of evolution in the schools and sought to "contend for the faith" with all other enemies of truth as they saw it.

Weishaupt into the organization; many of the leaders were Jews. Weishaupt, although possibly not a Jew himself, was in the center of a Jewish circle, analogous to Stalin's position in the Soviet government.

Winrod, in his book, *Adam Weishaupt, A Human Devil,* showed how he (Weishaupt) gained control of all masonic lodge rooms in Europe and changed them into secret societies of revolutionary planning. A study of the "Illuminati," Winrod emphasized, will show where all the modern ideas of Bolshevism originated. The French and Russian revolutions were brought on by this agency. Yes, he concluded, Modern Communism and old Jewish Illuminism are one and the same thing.

This fantastic tale was based upon a four volume work written in 1798 by a Professor John Robinson entitled *Proofs of a Conspiracy Against All the Religions and Governments of Europe Carried on in the Secret Meetings of Free-Masons, Illuminati, and Reading Societies.* Here, in one neat conspiratorial package, Winrod had his three arch enemies. And some lesser ones, too, as he was swift to point out, who were responsible for the present lamentable decay in the Protestant Church. The decay was due largely, he believed, to the fact that Protestantism was being judaized. The group who held views most similar to Judaism were the modernists who were behind such nefarious organizations as the National Conference of Christians and Jews.[6]

During this period, at one time or another, Winrod attacked Jews, Communists, Catholics, international bankers, Socialists, New Dealers, Masons, Italian Fascists, and chain stores. Strange bedfellows indeed, but in Winrod's eyes they were "interconnected working parts of the Jewish plot to rule the world."

During this period also, Winrod commenced advertising and recommending the books and tracts of well-known quasi-Fascists of that time: Elizabeth Dilling, James True, Robert Edmondson, Father Coughlin, Arno Gaebelein, and Joseph Kamp, and using their writings as source material.

Another scholar has succinctly summarized:

In a little over a decade Winrod had moved from the mainstream of fundamentalism with his crusade against evolution and modernism to an isolated irrational tangent.

He had become a bigot, anti-Semitic, and hopelessly blinded by a conspiracy view of history.

One scholar has asked the question: Was Winrod a Nazi? He concluded that he was not. This conclusion would appear to be correct in the sense that there is no evidence that Winrod was a member of any Nazi organization or that his activities were directed from Berlin.

However, in response to the broader question: Was Winrod a Nazi sympathizer in the period prior to the United States' entry into World War II? The answer must be a qualified "yes." (Qualified because, as with other subjects concerning which he wrote and spoke, it is difficult to sort out the thoughts constantly churning around in his mind.) For example, in January, 1933, he listed Fascism as one of the 10 deadly enemies for that year, and in December noted that freedom of the press was gone in Germany. In December, 1936, a *Defender* article by Oswald J. Smith states, after praising Naziism and Fascism as compared to Communism, "Fascism, Naziism, and Communism are alike our foes."

At the same time, from 1933 on, Winrod expressed admiration for Hitler's "breaking the back of a Jewish plot to destroy Germany by controlling the country's finances and corrupting its morals," for purging the German motion picture industry of immorality, outlawing nudist cults, Jehovah's Witnesses, and spiritualism. Jewish persecution he dismissed as an exaggeration. He described Hitler as "poor, a vegetarian and a teetotaler, who lives very frugally."

In December, 1934, and January, 1935, Winrod traveled to Europe, spending five days in Germany. Although he always claimed his visit was for religious purposes and that he had no contact with Nazi officials, others have stated that Winrod was invited to Germany by Dr. Otto H. Vollbehr of the Propaganda Ministry.

Regardless of the facts concerning this visit, there is no question that Hitler secured even more favorable treatment by Winrod thereafter. The *Defender* published pro-Nazi articles by Oswald J. Smith (previously mentioned), Rev. M. Ballister, and Arno Gaebelein, a fundamentalist leader.

VII. CONCLUSION

It seems appropriate to conclude this paper with an assessment of Winrod, primarily as he has been seen by others, both his contemporaries and by scholars—and then attempt to state why I have chosen this subject matter.

First, his contemporaries:

His father, J. W. Winrod, wrote: "In the providence of God our son Gerald has been annointed [*sic*] and set apart by the Holy Spirit to be a battering ram for righteousness to his day and generation. . . ." There is no doubt that Gerald himself agreed with this assessment.

It would be interesting to know how William Allen White, who fought Winrod tooth and toenail during the 1938 campaign, would have assessed his life. We do know his thoughts after that primary election:

The Reverend Gerald B. Winrod is an honest man. He also displayed a certain amount of courage going against the overwhelming tide of public sentiment in Kansas. Dr. Winrod certainly exhibited a rather interesting phase of cunning intelligence, knowing the psychology of his little crowd.

The thing that was the matter with the Rev. Gerald Winrod was his gross ignorance. He did not know that all the balderdash which a tent evangelist palavers to catch emotional people would not stand him in stead when it came to appealing to a state like Kansas.

After Winrod's death in 1957, Rolla Clymer wrote:

Winrod was a strange man, with a brilliant intellect and possessed of many admirable qualities. A kindly judgment seems to indicate that he lacked the proper elements of control over his other attributes—and thus defeated the essential mission he might have carried out during his life. If it takes all sorts of people to make up the world, Winrod was in a class by himself. Despite the swirl of antagonism that he always managed to stir about him, the record indicates that he was a righteous man at heart.

Ralph Lord Roy, in his book, *Apostles of Discord,* devotes a chapter and makes other references to Winrod. He gives three alternate explanations of Winrod's career: (1) he was a sincere preacher of Christianity as he interpreted it; (2) he was a man with a perverted sense of mission which led him to seek power; (3) he was motivated by greed. Roy seems to favor the third one, stating that Winrod's former allies and employees claimed he had a miser complex and "a money-worshipping brain." He also says that in 1950 the Defenders organization was worth $276,272, and that Winrod's Wichita residence was valued at $30,000.

One Kansas graduate student (who quoted the Rolla Clymer editorial) leans to the first view, stating his belief that: "Winrod was a sincere preacher and practitioner of what he believed to be right. That he was wrong on many counts or that some of his opinions were asinine and abhorrent does not change the fact of his basic sincerity. Being sincere does not excuse him for any harm he might have done by his activities; it does, however, explain why he did them." [7]

Another Kansas graduate student presents this summary:

> In some ways Winrod embodies the American ideal of good citizenship, if being well informed, holding firm convictions, and participating in political activities are taken to make up that ideal. He was unquestionably a well-read and well-informed citizen. He was a man of convictions boldly declared; as one writer has expressed it, Winrod often called a spade a tractor, so eager was he to make his beliefs plain. Not content with agitating for his

[7]. Sullivan, "Gerald B. Winrod," pp. 125-126.

beliefs through his several propaganda organs, Winrod fulfilled the third commandment of good citizenship by organizing political action. He took to heart the admonition of his society to do something to correct the ills he saw in that society. His form seems perfect; the content badly warped. When the ideas to which he gave expression through his activities, and the resulting ills he set about to correct are regarded with an eye to their implications for a democratic society, the fallacy of constructing a concept of good citizenship without including a system of values becomes apparent.[8]

Using the Ralph Lord Roy "three alternate explanations" as a reference point, it is my personal belief that Winrod was both "a sincere preacher of Christianity, as he interpreted it," and "a man with a perverted sense of mission which led him to seek power."

A Kansas undergraduate scholar, at the conclusion of an excellent research study of Winrod and the mass sedition trial of 1944, has observed:

The threat to democracy comes not primarily from the possibility of wholesale acceptance of Winrod's views and methods. The danger lies in "Winrodism," that positive assuredness of national mission which clouds and distorts an ability to see actual causes in events that confront us. Winrod was a product of his time and currents in American life. He also helped shape the America which came after him. The effects of his fantastic efforts to see conspiracy in all that did not conform to his pattern were only amplified by the prosecution in the Sedition Trial which likewise saw a sinister plot lurking behind each bush along the rocky road of democracy. Having been bombarded from all angles in the last few decades by such viewpoints, the American people have developed a nearly paranoid interpretation of world events. Great events are not seen as the result of logical causes; if something goes wrong, be it the loss of China, the loss of eastern Europe, or the assassination of a President, the American people are receptive to conspiracy theories of explanation.

"Winrodism" in the 1930's, the Sedition Trial and Cold War of the 1940's, followed by the mad era of McCarthyism have molded what Henry Steele Commager chooses to call a

8. Buitrago, "A Study of the Political Ideas . . .," p. 147.

"dirty work at the crossroads" outlook that has
become firmly entrenched in American life and
thought.[9]

Now, why have I chosen this subject matter
for discussion? It is because the dangers of
"Winrodism," or whatever other name we may
wish to call it, continue to this day, and because of my belief that we as historians, both
professional and amateur, have a particular
duty to expose the false and to seek and tell the
truth forthrightly, incessantly, and sometimes
unpleasantly.

Robert Cecil, in his book, *The Myth of the
Master Race: Alfred Rosenberg and Nazi Ideology,* in speaking of Naziism has succinctly
and simply stated the why and the wherefor of
these dangers:

> Underlying Naziism, however, were two ideas which
> have universal application and are very much alive today.
> The first corresponds to man's basic need to think better of
> himself by thinking worse of some group of people dif-
> ferent from himself. There will always be fertile ground for
> neo-Fascist and neo-Nazi activity so long as large numbers
> of men try to compensate at the expense of others for their
> feelings of insecurity and inferiority. Secondly, the belief
> seems fated to persist among large groups that their mis-
> fortunes must necessarily be due not to individual inade-
> quacy or misdirection of effort, but to the malignant con-
> spiracy of some other groups, which may at any given time
> be labelled "the bosses," "the Unions," "the Commies,"
> "the colonial oppressors," "the blacks," or, more simply,
> "them."

Although Winrod indicated he considered
himself and his followers superior to all those
peoples and groups that he lambasted (which,
as we have seen, were many and varied), his
greatest error lay in his unshakeable belief in
the conspiracy theory of history.

That there have been some conspiracies in
history by small groups with particular objectives for a limited time is true, as Julius Caesar,
for example, well knew. But to blame our every

9. Waltner, "Gerald B. Winrod . . .," pp. 130-131.

trouble and misfortune upon the continuing machinations of a handful of "conspirators" is not only false, it greatly hinders the search for remedies for the real (and usually complex) causes for our troubles.

And so, as we plan for a new museum and new programs of service to the people of Kansas, and to those beyond our borders, let us remember that the strength of our state and the nation (indeed, of any nation) lies neither in conformity nor in the superiority of any group, however virtuous and talented it may be, nor in intellectual tranquility and stagnation. Our strength lies in diversity, in tolerance, and sometimes in turmoil.

DEFENSE OF THE FAITH

J. FRANK NORRIS
AND TEXAS FUNDAMENTALISM, 1920–1929

by

PATSY LEDBETTER

The author, a doctoral student at North Texas State University in Denton, currently is completing a dissertation on the fundamentalist controversy of the 1920s.

DURING THE 1920s a wave of reaction to liberal and scientific thought swept over the United States. Essentially it was a movement of revulsion from increased industrialization, growth of cities and loss of individualism, and a backlash from the fervor of World War I. All that America had stood for — all the traditional values of an essentially rural nation — seemed to be breaking up. One of the most important facets of this reaction was the spread of religious fundamentalism among the Protestant sects. In 1925 for the first time in its history, the Episcopal Church dethroned a bishop for preaching heresy. A few years later, ultra-conservative Presbyterians forced the nationally known preacher, Harry Emerson Fosdick, from his pulpit in New York City. During this period Texas remained a rural state, but industrialization was coming, with its attendant shift of population and influence from rural areas to the cities. Necessary readjustments to these changes proved difficult, and reactionary movements developed to defend the old against the new. Religious fundamentalism played a significant role in this climate of reaction.[1]

---

[1] Comprehensive studies of the 1920s reveal these trends. See, for example, William E. Leuchtenburg, *The Perils of Prosperity, 1914–1932* (U. of Chicago Press, 1958); David A. Shannon, *Between the Wars: America, 1919–1941* (Boston, 1965); Frederick Lewis Allen, *Only Yesterday: An Informal History of the Nineteen Twenties* (New York, 1931); and John D. Hicks, *Republican Ascendancy, 1921–1933* (New York, 1960).

The fundamentalist crusade was basically a response to the liberal theology that tried to reconcile modern science with religion. Since before the turn of the century, some leading theologians had been attempting this reconciliation by de-emphasizing those aspects of Christianity which scientific discoveries contradicted, and by emphasizing the moral teachings of religion. Rejecting especially the stories of the Old Testament, these liberal thinkers had accepted the Darwinian doctrine of evolution and made it a workable part of religion. In theological terms, evolution meant that man had risen and was still rising, rather than that he had fallen and was doomed.[2]

Having produced few advocates of this modernist approach, Texas religions remained basically untouched by the new liberal theology. As a large number of Texans began to feel the need to hold on to the past, they demanded a return to the more exacting religion of their forefathers. Rejecting liberal theology, these defenders of the faith accepted as their basic tenets the infallibility of the Scriptures, the virgin birth of Christ, His substitutionary atonement for man's sins, His resurrection, and His literal second coming. Necessary to the validity of the last four doctrines, and therefore most important in the crusade, was the belief in the infallibility of the Scriptures; the fundamentalists' most significant characteristic became their literal interpretation of the Bible. To them the crucial obstacle to such an interpretation was the theory of evolution. This concept negated the Genesis account of creation, and to the fundamentalists denial of any one part of the Bible meant rejection of the whole and the ultimate destruction of Christianity.[3]

---

[2] The tenets of modernist theology varied greatly from one individual to another, making specific summations difficult. For explanations of the modernist faith, see Harry Emerson Fosdick, *Christianity and Progress* (New York: Fleming H. Revell Company, 1922); Shailer Mathews, *The Faith of Modernism* (New York, 1924), and his "Ten Years of American Protestantism," *North American Review*, CCXVII (May 1923), 577–93. For reconciliations of science and religion, see Henry Higgins Lane, *Evolution and Christian Faith* (Princeton U. Press, 1923); Arthur Thompson, "General Aspects of Recent Advances in the Study of Organic Evolution," *Methodist Review Quarterly*, LXX (April 1921), 210–11; *El Paso Times* (Texas), February 23, 1923; *Dallas Morning News* (Texas), November 9, 1925.

[3] The most comprehensive study of the fundamentalist movement is Norman Furniss, *The Fundamentalist Controversy, 1918–1931* (Yale U. Press, 1954). For contemporary statements of fundamentalism, see William Bell Riley, *Inspiration or Evolution* (Cleveland: Union Gospel Press, 1926); William Jennings Bryan, "The Fundamentals," *The Forum*, LXX (July 1923), 1665–80, and his *In His Image* (New York: Fleming H. Revell Company, 1922), 111–16; *Houston Post Dispatch* (Texas) October 6, 1924.

The arguments of the fundamentalists were characterized by ignorance and a failure to understand scientific doctrines. As the *Honey Grove Signal,* an East Texas newspaper, proudly declared, "We don't know anything about evolution and cherish no hope of ever learning anything about it." Most Texans apparently agreed with the *Signal* that studying evolution was useless and even dangerous for the soul. As a result, throughout the state the doctrine was little studied and less understood. To most fundamentalists evolution meant that man had evolved from monkeys, and they rarely attempted to understand the process by which species developed. Some crusaders, for example, explained that dinosaurs died out, not because of evolution, but because the Ark was not large enough to transport them. Fearing that the study of evolution endangered one's soul, these people made little effort to learn the facts involved.[4]

The Protestant denominations which expressed the strongest orthodoxy predominated in the religious constituency of Texas. During the 1920s approximately three-fourths of a million Texans belonged to Baptist groups, more than' to any other denomination. Of these, contemporaries claimed that at least ninety-eight percent were fundamentalists. Other large denominations in Texas included the Methodists, Presbyterians, Disciples of Christ, and Church of Christ. The one denomination that consistently opposed fundamentalism, the Unitarian Church, had fewer than three hundred Texas members during the decade. Thus the most influential religious groups in the state were strongly orthodox.[5]

Regardless of denomination, the religious beliefs of most Texans consisted of a simple, unquestioning acceptance of the Bible as literal truth. Since, for the most part, they were indifferent to or ignorant of liberal theological developments, anti-evolution sentiment was strong

---

[4] *Tyler Daily Courier Times* (Texas), July 10, 1925; *Dallas Morning News,* December 15, 1925. For further examples of the fundamentalists' failure to understand evolution, see Thomas Theodore Martin, *The Inside of the Cup Turned Out* . . . (Jackson, Tennessee: Mercer Printing Company, 1932); Riley, *Inspiration or Evolution;* and Alfred McCann, *God or Gorilla?* (New York: Devin-Adair Company, 1922).

[5] For a breakdown of Texas church membership, see *Texas Almanac and State Industrial Guide* (Dallas: Dallas News, 1929), 220–24. See also Fort Worth *Fundamentalist* (Texas), July 7, 1922; July 9, 1926.

across the state. Although fundamentalism received support from all areas of the state, its strongest proponents hailed from Northeast Texas. Typical of comment from this area was the *Honey Grove Signal's* declaration that, "We have known several monkeys in our day and not one ever gave evidence of losing its tail and joining the pants-wearing tribe known as the genus homo." Central Texas also gave strong support to the movement, while its greatest opposition came from West and South Texas. Nationwide, the fundamentalists received their strongest support from the Old South since southern religion was more conservative than that of any other section of the nation. Northeast and Central Texas were more closely aligned with the southern Protestant groups than South and West Texas. In the southern part of the state, large numbers of Catholics were unreceptive to fundamentalist agitation. In the western regions, the wide dispersion of the ranching population helped prevent fundamentalism from winning a stronghold.[6]

In Texas, as in other states, the Protestant churches became the major battlefield for the controversy. Agitation from both sides forced the various denominations to issue statements and resolutions setting forth their official positions, which generally supported the fundamentals of the faith. Much of the agitation in Texas came from evangelist, sensationalist, and controversialist J. Frank Norris. In all probability, without him the Texas controversy would have been mild, but his activities made it a major issue in secular and religious affairs.

John Franklyn Norris was born on September 18, 1877, in the slums of Birmingham, Alabama, where his father, Warner Norris, was a poorly paid steel worker. Warner drank heavily and the family was poverty-stricken. In 1881, in an effort to start anew, the Norrises moved with their three children to Hubbard City, Texas, where the father became a tenant farmer. Neither the family's financial condition nor the elder Norris' drinking problem improved. With Warner Norris spending practically every penny he made on liquor, the family frequently lacked sufficient food and the other necessities of life.

---

[6] For an interesting contemporary analysis of Texas religious groups, see Owen P. White, "Reminiscences of Texas Divines," *American Mercury*, IX (September 1926), 95–100. For examples of sentiment in various areas of the state, see *Gilmer Daily Mirror* (Texas), July 2, 1925; *Tyler Daily Courier Times*, July 4, 1925, which quotes several area newspapers. *Austin Statesman* (Texas), March 6, 1923; *El Paso Times*, March 25, July 20, 1925; *Lubbock Morning Avalanche* (Texas), July 23, 1925; *Corpus Christi Caller* (Texas), July 6, 1925; *San Antonio Express* (Texas), January 19, 1923.

The mother, Mary Norris, found escape from her many troubles in religion and instilled a strong fanaticism in her young son. When, as a child, J. Frank Norris had recovered from a serious illness, she was so grateful to God that she carried him to the banks of a nearby river where — she told her son years later, — "I said to the music of the falls God gave this babe to me' and He snatched you from the jaws of death, and I lifted you up and said 'I give him back!' . . .and she said she heard the voice of God and He said 'You have given the world a preacher.' " At least this was Norris' version of the incident as he told it in July of 1946.

Frank was a near invalid for about three years during his teens, as a result of a gunshot wound. During this time his mother spent long hours reading the Bible to him, telling him stories of great men, and convincing him of his own exalted destiny. Even when he was a child, she had assured him that he would be a prophet, and instilled in him the ambition to become a great man. Thus, from his home life Norris acquired a hatred for liquor, a belief in the fundamentals of Christianity, and a desire to preach to huge crowds.

In spite of extreme financial difficulties Norris earned degrees from Baylor University in Waco, Texas, and the Southern Baptist Theological Seminary in St. Louis, Missouri, with high honors, and began his long, sensational career as a Baptist minister. On May 5, 1902, he married Lillian Gaddy, whom he had met at Baylor. The daughter of a Baptist missionary, her religious training and family background made her an ideal wife for Norris. Attracted by his ambitious, enthusiastic approach to religion, she gave his ministry her full support and made few demands upon him at home. They had four children: Jim Gaddy, J. Frank, Jr., George Louis, and Lillian Gaddy — but Norris was too concerned with building his ministry to be a dedicated family man.

For several years he pastored the McKinney Baptist Church of Dallas, but in 1909, he was called to the more prestigious pulpit of the First Baptist Church of Fort Worth, a position which he held until his death of a heart attack on August 21, 1952. During this time his church grew to a membership of over 10,000; his controversial church newspaper, which was variously titled *The Fence Rail, The Searchlight*, and *The Fundamentalist*, reached a circulation of over 70,000; and he influenced thousands more through his powerful radio station KTAT, later KFSL. He also established the Bible Baptist Seminary at Fort Worth, which is still in operation, and acquired a significant following of fundamentalist churches. Norris' success can be explained largely by the tactics he used, constantly engaging in some crusade against the forces of Satan — liquor, gambling, Catholicism, Sunday movies, religious modernism, or some similar evil.[7]

Few people have been more successful in winning the support of Texans than J. Frank Norris, largely because he designed his methods to appeal to the uninformed masses. Norris' appearance was attractive and appealing. He was tall and slender, light complexioned, with blue eyes that had a dreamy quality. His unusual speaking talent gave him a strong hold over his audiences. Beginning in a soft voice, almost inaudible to those in the back of the assembly, his volume increased as he became more and more excited, until he shouted and waved his arms about violently, pacing back and forth, completely carried away by what he was saying. The dominant theme of his ministry was fear, and he never failed to warn his audiences of the eternal damnation they faced if they refused to heed his message.[8]

The "Texas Tornado," as Norris was called, frequently illustrated his emotion-packed message in graphic terms. When a district attorney, who had prosecuted the pastor for arson in the burning of his own church, was killed in an automobile accident, someone found a broken liquor bottle containing a portion of the victim's brain and carried it to Norris. Taking the exhibit into the pulpit, Norris used it to illustrate a sermon titled "The Wages of Sin is Death." Although the people were terrified and some even fainted, they loved this kind of sensationalism, and Norris gave them what they wanted.[9]

Although Norris' church was located in a city, his appeal was to those who identified with the country, and his speeches were replete with references to farm animals and situations. Calling upon the "fork of the creek boys" to destroy modernism, he referred to himself as "a

---

[7] E. Ray Tatum, *Conquest or Failure? Biography of J. Frank Norris* (Dallas: Baptist Historical Foundation, 1966), *passim*. For additional biographical information on Norris, see Homer G. Ritchie, "The Life and Career of J. Frank Norris" (M. A. thesis, Texas Christian University, 1967); and William K. Connolly, "The Preaching of J. Frank Norris" (M. A. thesis, University of Nebraska, 1961).

[8] Nels Anderson, "The Shooting Parson of Texas," *The New Republic* (September 1, 1926), 35–37; Fort Worth *Fundamentalist*, January 20, 1922.

[9] Louis Entzminger, *The J. Frank Norris I Have Known for 34 Years* (n.p., n.d.), 112.

country Baptist preacher who lives in a cow town up here and fights the
devil for a living." Since Texans generally distrusted intellectuals,
whom they did not understand, Norris knew that their suspicions could
easily be aroused against "frizzled-headed professors." He also knew
that Texans loved a good fight; the same type of people who could be
aroused to hang or flog a man for little or no reason could be incited to
fight a vigorous battle against evolution, even where it was little
believed or taught. Using fear and emotion to enhance his cause, Norris
threatened all modernists with doom, boasting: "I preach old-fashioned
hellfire and damnation, 3-linked, unadulterated repentance and mourn-
ers' bench faith gospel." Evidently this was the type of religion that
appealed to Texans. As Norris pointed out he drew large crowds all
over the state, while the modernists preached to "empty woodyards."[10]

Norris caused most disturbance within the Baptist denomination,
which his agitation ultimately split apart. Actually modernists were
practically non-existent among Baptists in the state, so that the battle
was between conservatives and ultra-conservatives, rather than between
modernists and fundamentalists. No Texas Baptist leader openly
admitted belief in extreme modernist doctrines such as evolution, but a
few were willing to give a slightly less than literal interpretation to the
Scriptures. Any deviation from complete orthodoxy brought immediate
cries from the ultra-conservatives, especially from Norris and his followers. For example, when James Dawson, a Baptist minister from
Waco, contended that Sodom and Gomorrah were destroyed by volcanic
fires or natural causes, the extreme fundamentalists answered that fire
and brimstone or supernatural causes were responsible. Some of the
less conservative Baptists admitted that God might have worked through
various inspired individuals to write the Bible, but the ultra-conservatives insisted that inspiration was direct, instantaneous, and verbal
dictation from God. The only real differences among Baptists was in
degree of orthodoxy, and it was paradoxical that fundamentalist agitation became extremely intense within a denomination that agreed
almost unanimously on the fundamentals.[11]

---

[10] Fort Worth *Fundamentalist*, October 5, 1923; October 14, 1921; September 29, 1922; May
18, 1923; May 12, 1921.

[11] *Ibid.*, November 30, 1928; June 14, 1929; Martin, *Inside of the Cup, passim.*

Throughout the 1920s Norris accused such prominent leaders as Reverend George Truett, pastor of the First Baptist Church of Dallas; Lon R. Scarborough, president of Southwestern Baptist Theological Seminary in Fort Worth; and Samuel P. Brooks, president of Baylor University, of accepting modernist doctrines. These leaders, actually as orthodox as Norris himself, retaliated by proving their own fundamentalism, not by defending modernism. One leader contended that in looking for modernists among Texas Baptists Norris was "setting up men of straw to knock down," since no such sentiments actually existed. The *Baptist Standard,* the denominational periodical published in Dallas, was almost as concerned with combating modernism as Norris himself, and practically every issue condemned the doctrine of evolution. Scarborough declared that any teacher would be dismissed from Southwestern who had "a streak of modernism or Darwinian or theistic evolution in his teachings as big as the finest feather on an angel's wings." Norris obviously had little real basis for his heresy hunts among Texas Baptists.[12]

Norris' fanaticism was most destructive to the Baptist schools, especially Baylor University, which he plagued repeatedly by exposing so-called "evolution professors." His first assault on Baylor came in 1921 with an investigation of Samuel Dow, a sociology professor. Dow had written a book describing man's development from an apparently much less civilized being, and emphasizing social and family life in its various stages. Norris insisted that Adam and Eve made up the first family and that man had never been uncivilized; thus Dow's views were heretical. Finally, after prolonged, bitter attack Dow resigned, but only after publicly defending his orthodoxy and denying belief in evolution. Throughout the decade Norris constantly launched similar attacks, forcing Texas Baptists into an even more fundamentalist stand than other Southern Baptists. All of the professors whom he accused of being modernist answered him by vowing complete acceptance of the Bible. The nearest any came to modernism was when one Baylor history

---

[12] Fort Worth *Fundamentalist,* August 17, 1923; April 30, 1926; August 3, 1923; November 11, 1921; Dallas *Baptist Standard* (Texas), February 23, 1923; January 12, 26, February 3, March 30, 1922; May 17, 1923.

J. Frank Norris dominated the fundamentalist controversy in Texas during the 1920s. — *courtesy Mrs. Jean Taggart.*

SAMUEL P. BROOKS was president of Baylor University in the 1920s. — *Texas Collection, Baylor University.*

SAMUEL DOW, a sociology professor, was driven from his position at Baylor by Norris' attacks. — *Texas Collection, Baylor University.*

LULA PACE, a professor of botany, refused to resign from the Baylor faculty under pressure from Norris. — *Texas Collection, Baylor University.*

ORA C. BRADBURY resigned as a professor of zoology from Baylor under Norris' attacks. — *Texas Collection, Baylor University.*

teacher, Charles Fotergill, admitted that he did have some doubts that the ark could have been large enough to transport two of every kind of animal.[13]

The Texas Baptist General Association, a branch of the Southern Baptist Convention, constantly refuted modernism and yet opposed the extreme fundamentalist agitation caused by Norris. The Texas Association met annually in a general convention in which such matters as finances, missions, schools, and general policies were discussed. The actions and policies of this convention were the best expressions of the Southern Baptists' opinions as a denomination in Texas, and it was this group that Norris constantly opposed. On numerous occasions the convention passed, without opposition, statements giving unqualified acceptance to the Genesis account of creation. Without exception, the policies and resolutions of this convention were orthodox and fundamentalist.[14]

Their suspicions aroused by Norris, the Baptists, through their convention, frequently investigated their institutions and instructed them not to employ anyone who believed or taught anything that contradicted the literal interpretation of the Scriptures. The most significant investigation of the Baptist schools occurred in 1921 as a direct result of Norris' attacks on Baylor. Two borderline cases of heresy were discovered in the beliefs of Lula Pace and Ora Clare Bradbury, both biology teachers who accepted Genesis but thought that the language might be allegorical. Norris immediately used this deviation from total orthodoxy as a basis for bitter attack, demanding the resignation of both professors, and of President Brooks who defended them. Bradbury

---

[13] Fort Worth *Fundamentalist*, October 21, November 11, 1921; Dallas *Baptist Standard*, November 3, 1921. For Norris' similar attacks, see Fort Worth *Fundamentalist*, April 6, 13, 1923; November 24, 1922; September 26, 1924. For the defense by various professors of their own orthodoxy, see *ibid.*, December 8, 1922; November 7, 1924; Dallas *Baptist Standard*, October 16, 1924; November 3, 1921.

[14] *Annual of the* [77th] *Baptist Convention of Texas, . . . December 2–5, 1925* (n.p., n.d.), 172; *Annual of the* [73rd] *Baptist Convention of Texas, . . . December 1, 1921* (n.p., n.d.), 172; *Annual of the* [74th] *Baptist Convention of Texas, . . . November 16–20, 1922* (n.p., n.d.), 19, 159; A. Wakefield Slaten, "Academic Freedom, Fundamentalism and the Dotted Line," *Education Review*, LXV (February 1923), 74; *Annual of the* [79th] *Baptist Convention of Texas, . . . November 16–20, 1927* (n.p., n.d.), 25.

resigned but Lula Pace remained at Baylor until her death in 1925, although Norris continued to demand her resignation.[15]

Throughout the various controversies Norris managed to connect the heresies of the Baptist colleges with the financial campaigns of the denomination. He told the people through his pulpit, his newspaper, and his radio station that donations were being misappropriated, and that a large portion of their money supported institutions which taught evolution. Apparently his attacks hurt the denomination because leaders announced that churches were not paying their quotas and that the association faced a serious debt. Ecstatically taking credit for this situation, Norris even offered monetary rewards to students who would testify against the schools and help him keep the issue alive.[16]

Understandably upset at his activities, the Texas Baptist convention refused to seat his delegates in 1922 and 1923 and ousted his church from the state general association permanently in 1924. This affront failed to lessen Norris' influence or to stop his attacks. Instead, he gained a significant following of fundamentalist churches which disassociated themselves from the state convention also. Thus Norris succeeded in breaking the denomination apart, with the more fundamentalist Baptists following his lead. His separate association became known as the World Baptist Fellowship, but this group later split, primarily because of personality conflicts, and another association known as the Bible Baptists developed from it.[17]

While Norris' center of strength lay in Northeast and Central Texas, his newspaper circulated throughout the state and over much of the United States. He spoke in numerous small Texas towns, but he also drew large crowds in Houston, San Antonio, Dallas, Austin, and other cities. Conducting revivals outside the state, he frequently appeared in New York, Boston, Chicago, and similar cities.

---

[15] *Texas Baptist Annual, 1922*, 13, 17–18, 151–61; Fort Worth *Fundamentalist*, September 29, 1922; March 16, 1923; July 17, 1925; *Texas Baptist Annual, 1925*, 49.

[16] Entzminger, *J. Frank Norris*, 179–80; Dallas *Baptist Standard*, March 15, 1923; Fort Worth *Fundamentalist*, April 20, 1923; April 25, 1924.

[17] *Texas Baptist Annual, 1922*, 15–16; *Annual of the* [75th] *Baptist Convention of Texas*, ... *November 15–17, 1923* (n.p., n.d.), 18–24; *Annual of the* [76th] *Baptist Convention of Texas*, ... *November 20–22, 1924* (n.p., n.d.), 24–25 ;Tatum, *Conquest or Failure?* 260–61.

Not being satisfied with purging the Baptists, Norris contributed to the conflicts arising in other denominations also. In 1921 fundamentalism struck Texas Methodism with an attack on John Rice, a professor at Southern Methodist University in Dallas. His book, *The Old Testament in the Life of Today,* pointed out that the Old Testament was largely Hebrew folklore, which had been retold for ages before finally being recorded. Norris, along with a large number of the professor's Methodist brethren, attacked his position as being heretical, contending that he denied the direct, instantaneous inspiration of the Bible. Rice resigned under pressure, after publishing a defense of himself in which he, like the accused Baptists, insisted upon his own orthodoxy.[18]

The most explosive controversy in the Methodist church, however, erupted in 1923, when the more fundamentalist brethren, led by Reverend W. E. Hawkins, began accusing the Methodist colleges of teaching evolution. A committee of Methodist churchmen, which investigated the schools, failed to convict individual teachers but also failed to appease the fundamentalists. The climax of the agitation came in May of 1923, when the World's Christian Fundamentalist Association held its convention in Norris' church and presented a sensational two and one-half hour trial. Six young people testified that they had learned the soul-destroying heresy of evolution from Methodist teachers in Texas Women's College in Fort Worth, Southwestern University in Georgetown, and Southern Methodist University. The schools, of course, were found guilty, and much dissension within the church resulted from the meeting.[19]

Most other Texas denominations remained fundamentalist in viewpoint with few modernists speaking out. One Episcopal minister, Reverend Lee Heaton of Fort Worth, who did express liberal beliefs was threatened with being tried for heresy and suffered humiliating attacks from Norris. Finding Texas generally unreceptive to his liberal theology, Heaton resigned from the ministry and became a salesman.[20]

---

[18] Fort Worth *Fundamentalist,* May 12, 19, 26, 1921; Dallas *Texas Christian Advocate,* April 14, July 7, August 11, September 22, November 17, 1921.

[19] Fort Worth *Fundamentalist,* January 5, 26, May 4, 11, 1923.

[20] *New York Times,* January 21, December 17, 1923; Fort Worth *Fundamentalist,* July 11, 1924; January 16, 1926.

On the national level the fundamentalist controversy disturbed the Presbyterian church, but most Texans supported that denomination's strongest actions against modernism and took steps to eliminate it from their ranks. Meeting in San Antonio in 1924, the Presbyterian Church of the United States reaffirmed its acceptance of the fundamentals and voted to withdraw financial support from missions, colleges, and other institutions where modernism was taught or believed. The Cumberland Presbyterian Church also took a strong stand against modernism. Meeting in Austin in 1924, it passed a resolution, proposed by the Presbytery of Weatherford, Texas, declaring that every member of the church from "president to janitor" opposed the "ape-man" idea. It declared itself "squarely, fixedly and unmoveably against these infidelic theories ... poisoning the minds of the rising generation ... with these dangerous and soul-destroying doctrines." Texas Presbyterians expressed little opposition to such fundamentalist stands.[21]

Controversy was not intense in most of the other Protestant denominations, largely because fundamentalism was unopposed. Various Texas groups of the Missionary Baptist Association, for example, adopted resolutions opposing evolution and modernism and stating that there was not a single modernist in the denomination. The Seventh Day Adventists, holding their annual conference in San Antonio in 1925, declared that the church had no place for evolution or other modernist theories and appealed to the people to accept the infallibility of the Bible. One speaker called the evolution controversy the Christ and Anti-Christ struggle to which the Bible referred, and said that it was a sign pointing to the end of the world. The Disciples of Christ and the Church of Christ were disturbed nationally by the controversy but apparently Texas ministers either remained fundamentalist or kept their opinions quiet.[22]

In the process of upholding their beliefs, Texas fundamentalists were not satisfied to combat modernism merely within their own religious

---

[21] George Paschal, Jr. & Judith Brenner, *One Hundred Years of Challenge and Change: A History of the Synod of Texas of the United Presbyterian Church in the U.S.A.* (Trinity U. Press, 1968), 148–49; *Dallas Morning News*, July 20, 1925; *Austin Statesman*, May 11, 16, 20, 1924.

[22] *Dallas Morning News*, July 25, 30, August 3, October 16, November 14, 1925; Furniss, *The Fundamentalist Controversy*, 170–76.

denominations, but were determined to root the evil out of every institution in which it manifested itself, including public schools, colleges, and universities. Secular phases of the movement were closely tied to the religious agitation, as church leaders, like Norris, stirred their followers to fight modernism wherever it appeared.

Across the nation religious zealots attempted to use state governments to force people back to the fundamentals of the gospel. Throughout the 1920s fundamentalists inspired a number of bills and resolutions in Texas, the major goal of which was to prohibit the teaching of evolution in the public schools and colleges. In 1923, the House passed an anti-evolution bill, but it died in the Senate. In the same year the House also passed a resolution stating that teaching evolution was unconstitutional, but it too failed in the Senate. The senators, being from larger districts, were not so close to their constituents, nor so tied to the rural areas. The Senate Committee on Education reported on the 1923 bill with the recommendation that it pass, but it was allowed to die on the calendar so that the Senate was never called upon to record a definite vote. All of the various bills and resolutions proposed during the decade originated in the House but as only the 1923 bill passed the House, the others did not reach the Senate.[23]

Leading proponents of the bills in the state legislature were J. T. Stroder of Navarro County, S. J. Howeth of Johnson County, J. A. Dodd of Texarkana, James W. Harper of Mount Pleasant, and W. R. Wigg of Paris. In the debates held on the bills and resolutions, these proponents of fundamentalism identified evolution not as scientific doctrine but as religious dogma, and argued that to teach it in the public schools was to teach a religion, which the Constitution prohibited. They argued that since Christian doctrines could not be taught, neither could anti-Christian doctrine. Their major legal argument was that the majority of the people disapproved of evolution and felt that it would destroy

---

[23] H. B. No. 97, "A Bill to Be Entitled an act prohibiting the teaching of evolution...";
H. B. No. 378, "A Bill to be Entitled an act prohibiting the teaching of evolution in any of its phases..."; H. B. No. 90, "A Bill to be Entitled an act making it unlawful for any teacher...to teach as a fact that mankind evolved from a lower order of animals...," typescript copies, Legislative Library, Austin, Texas. See also the Journals of the House and Senate of the Thirty-eighth, Thirty-ninth, and Forty-first Texas Legislatures. During the 1920s five states, Oklahoma, Florida, Tennessee, Mississippi, and Arkansas, passed laws against teaching evolution in public schools.

Christianity. Since their tax money supported the schools, they reasoned that it was illegal to teach anything that opposed their beliefs and suggested that the modernists establish their own schools.[24]

Texas legislators' arguments frequently revealed the nature of the fundamentalist mind, especially their ignorance of scientific doctrine. For example, one of the strongest supporters of anti-evolution bills in the House, J. T. Stroder, denounced evolution as that "vicious and infamous doctrine . . . that mankind sprang from pollywog, to a frog, to an ape, to a monkey, to a baboon, to a Jap, to a negro, to a Chinaman, to a man." Another representative, J. A. Dodd, proclaimed:

> The state forces me to pay taxes to support schools, then forces me to send my children to those schools and there shows my children the road to hell through teaching them the hellish infidelity of evolution. We owe it to our children and to our mothers who loved their Bible and taught us its meaning to abolish forever from our schools this iniquitous fallacy which holds that the Bible is a liar and that man is a monkey.

Speaking to the legislature in support of the bills, J. Frank Norris explained that the doctrine of evolution originated in Germany, that it was more dangerous than German militarism, and that it would ultimately destroy civilization. So soon after World War I, connecting evolution with Germany was an effective scare tactic. Generally, emotionalism, ignorance, and fear characterized the arguments of the various bills' proponents.[25]

Lloyd Price of Morris County, Eugene Miller of Parker County, J. Roy Hardin of Kaufman, and Roland Bradley of Houston were the leading opponents in the House of the anti-evolution bills. These men argued that such laws would restrict teachers unnecessarily, that they were violations of academic freedom, and that such restrictions were outside the authority of the state legislature. Some argued that restrictive laws were unnecessary since Texas teachers did not teach evolution anyway.[26]

---

[24] Fort Worth *Fundamentalist*, February 23, 1923; *Dallas Morning News,* July 9, 26, 1925; March 4, 1923; *Fort Worth Star Telegram* (Texas), February 16, 1929.

[25] *Waco News Tribune* (Texas), January 19, March 4, 1923; Fort Worth *Fundamentalist,* February 23, 1923; *Austin Statesman,* March 3, 1923.

[26] *Austin Statesman,* February 24, March 3, 1923; *Fort Worth Star Telegram,* February 16, 1929; *House Journal, Thirty-Eighth Legislature, Regular Session,* 1459.

Fundamentalists enjoyed greater success in influencing the seven-man Texas textbook committee than in swaying the state legislature. Appointed by Governor Miriam A. Ferguson, the committee in 1925 ordered all references to evolution deleted from texts used in state schools, an action which proved almost as effective as the proposed laws would have been. Because of local pressures and public opinion few teachers taught evolution anyway; thus the textbooks were the primary means by which Texas young people became acquainted with the doctrine. Censorship of these books was therefore a major victory for fundamentalism.[27]

Having decided to remove all mention of evolution from textbooks, the committee, which consisted primarily of Texas educators, proceeded to a thorough accomplishment of its task. Changing the word evolution to development, the committee removed all indications that man was related to other animals or that he had ever lived in an uncivilized state. One committee member even suggested that the word evolution be stricken from dictionaries, but the group decided that these were not actually textbooks. So extensive were the revisions that the publishers had to prepare separate school books for Texas students.[28]

Little opposition to the committee's action arose. Even educators and teachers did not voice significant protest to this violation of freedom in the classroom. For the most part, Texas politicians were either fundamentalist or noncommittal. Governor Miriam Ferguson supported the committee, declaring: "I am a Christian mother . . . and I am not going to let that kind of rot go into Texas textbooks." Equally as fundamentalist, her successor Dan Moody contended, "I believe in the Bible from cover to cover. I believe that God created man in His own image and likeness, that the whale swallowed Jonah, and that the children of Israel passed through the Red Sea on dry land." This lack of opposition

---

[27] Maynard Shipley, *The War on Modern Science: A Short History of the Fundamentalist Attacks on Evolution and Modernism* (New York, 1927), 172. The Texas textbook committee consisted of Ida Mae Murray, a University of Texas graduate and San Antonio public school teacher; F. M. Black, supervisor of Houston public schools; A. W. Bridwell, president of Nacogdoches State Teachers' College; T. J. Yoe, Brownsville school superintendent; R. L. Paschall, a Fort Worth high school principal; F. W. Chudej, who had five years teaching experience in grades below the high school level; and H. A. Wroe, a businessman.

[28] *Dallas Morning News*, October 6, 1925; "No Evolution for Texas," *Literary Digest* (August 14, 1926), 30; Shipley, *War on Modern Science*, 173.

was one of the major reasons that fundamentalists were able to expunge so thoroughly the concept of evolution from the state's books.[29]

The controversy which swirled around Norris came to a climax on July 17, 1926, when he shot and killed a Fort Worth lumberman, D. Elliot Chipps, in his church study. Infuriated by some accusations Norris had made concerning Henry Clay Meacham, the mayor of Fort Worth, Chipps had come, unarmed, to the church making threats and Norris, apparently afraid for his own life, shot him. The minister was arraigned and tried for murder, but was cleared on a plea of self defense. This incident hurt Norris' influence among many Texans who thought that he was carrying sensationalism in religion too far. This was especially true when the pastor tried to use the affair to increase his popularity, saying that the evil forces in Fort Worth had hired Chipps to assassinate him.[30]

With Norris' own following, however, this incident served to increase his popularity, making him an even more sensational character. His own church welcomed him back after the trial with much singing, praying, and crying. On the other hand, the traditional Baptists of the Southern Baptist denomination were less inclined to listen to his accusations — a trend that was spreading throughout the fundamentalist movement during the latter part of the decade. Although Norris remained a very influential personality in the state's religious as well as political life, the two separate camps were definitely established and his disturbances had less effect among traditional Baptists. Maintaining a following of fundamentalist churches, Norris continued to be extremely popular with some Texans and with the ardent fundamentalists across the nation.

Beginning in 1932, Norris pastored, simultaneously with his Fort Worth church, the Temple Baptist Church in Detroit, Michigan, where many southerners had been drawn by employment in the automobile industry. He commuted by railroad between the two cities. Norris had conducted a revival in the Detroit church in 1932 and had been invited

---

[29] Maynard Shipley, "The Forward March of the Anti-Evolutionists," *Current History*, XXIX (January 1929), 582; *Literary Digest*, August 14, 1926, 30.

[30] *New Republic*, September 1, 1926, 36.

to become its pastor. He did not wish to give up his church in Fort Worth, but found in the Temple church a large number of agrarian migrants who were seeking just the kind of religion that he taught. He decided that he could serve both churches by alternating Sundays and by printing each week's sermon. The combined congregation reached a total of about 25,000, the largest in the world. Norris' appeal seemed especially strong to those people who were moving during this decade from the rural areas to the cities and who found adjustment to urban conditions difficult. He could give them the sort of emotional religion that they had known "back home," and he was fighting the forces that they despised in city life.[31]

During the latter part of the 1920s, fundamentalism gradually ceased to be a controversial issue in Texas and in the nation. By 1929 Norris had shifted his attention from evolution to other issues. In October of that year, he refused to publish an anti-evolution article entitled "The Doctor Bell Theorem vs the Gods of Evolution," and wrote the author that his paper was not printing anything on evolution at that time. Norris did not give specific reasons for ceasing his battle against evolution. However, he was always intensely aware of public opinion, and undoubtedly realized that anti-evolution was not as popular an issue as it had been in the past. After the 1925 Scopes trial in Tennessee, newspaper editors, educators, and even church groups finally began to oppose religious restrictions on teaching.[32]

---

[31] Fort Worth *Fundamentalist*, October 12, November 16, December 7, 1923; January 13, 1922; June 13, 1924. J. Frank Norris, *The Inside History of the First Baptist Church* (Fort Worth: n.p., n.d.); Entzminger, *J. Frank Norris*; and Tatum, *Conquest or Failure?* give good accounts of Norris' career.

[32] Entzminger, *J. Frank Norris*, 107–109; Arthur C. Bell to Norris, October 6, and Norris to Bell, October 9, 1929, J. Frank Norris Papers, 1927–1952, Southwestern Baptist Theological Seminary, Fort Worth, Texas. *Austin Statesman*, February 2, 11, 1927; Marian J. Mayo, "Freedom in Education," *Texas Outlook*, XII (March 1927), 9–10; Fort Worth *Record Telegram* (Texas), January 16, 1929. The Scopes trial, the climax of the anti-evolution movement, occurred in Dayton, Tennessee, after John T. Scopes, a high school biology teacher, and a group of his friends decided to test the state's anti-evolution law. When Scopes was arrested for teaching evolution, the American Civil Liberties Union engaged Clarence Darrow and other notable attorneys to defend him, while William Jennings Bryan gave his services to the prosecution. The trial attracted national attention, and seemed to be a battle to the death between the forces of fundamentalism and modernism. On July 20, 1925, Darrow called Bryan to the witness stand as an expert on religion, and under the defense attorney's searching questions the limitations of Bryan's mind were quickly revealed. Scopes was convicted, but both Bryan and fundamentalism suffered an embarrassing exposure as a result of the trial.

During the 1920s fundamentalist sentiment had been especially intense in Texas. There had been disturbances in the Protestant denominations, agitation in the state legislature, and censorship of textbooks. Most segments of the population, urban and rural, educated and uneducated, felt the effects of the controversy. In Texas, no one group could unite the forces that might have combated the movement openly, and certainly the few modernists had no leader as effective as J. Frank Norris. Texans proved a receptive audience for Norris' fanaticism, while modernist expressions were indeed rare.

OUR HOPE: AN AMERICAN FUNDAMENTALIST JOURNAL AND THE HOLOCAUST, 1937-1945

by

David A. Rausch

To the historian, the journal *Our Hope* is unique as a Fundamentalist periodical in many ways. In the first place, the longevity of its life (1894-1957) was exceptional in light of the fact that it never accepted advertisements. Thus, it was deprived of a key source of financing, but was better able to pack its pages with Bible study and world news. Secondly, the editor for most of its lifetime was Arno C. Gaebelein, a leader in the Fundamentalist movement and avid reader, who concentrated on keeping abreast of current events. Thirdly, from its inception, *Our Hope* was interested in the Jewish people and would incorporate news items concerning the Jewish community around the world. In addition, although the journal never had a readership of over 12,000 it had a broad-based Christian readership. For example, Episcopalian clergymen subscribed as well as Plymouth Brethren. Great Bible teachers within the Fundamentalist-Evangelical movement, such as M. R. DeHaan, H. A. Ironside, Wilbur M. Smith, Lewis Sperry Chafer, Donald Grey Barnhouse, F. C. Jennings, and William L. Pettingill, were avid readers of the journal. Even an editor of a large Minneapolis newspaper subscribed and thanked Gaebelein for his analysis of the news. Such a loss was felt within Evangelicalism when the journal published its last issue, that Billy Graham seriously considered reviving the periodical.[1]

To the historian of the Holocaust, there is another unique attribute of this periodical. It chronicles with unbelievable accuracy the plight of the Jews during the Nazi regime. During the 1920's, Gaebelein had pointed to the rise of dictators in Europe, emphasizing Mussolini's growing power. At the end of 1930, he wrote in his "Current Events in the Light of the Bible" section of *Our Hope*:

> Adolf Hitler—Will He Be Germany's Dictator? . . . He seems to have a wild program of leadership. He is adored by thousands of women whom he seems to captivate. He is an outspoken enemy of the Jews . . . one of the most fanatical anti-Semites of Europe. The near future will show if he will succeed.[2]

Hitler did come to power, but did not immediately begin to exterminate Jews in 1933. Rather, repressive measures were instituted against the Jewish community in an effort to force them to leave Germany. Hitler felt that they were "vermin" and if other countries wanted them, those countries could have them. Sadly, few Jews were able to find countries that would accept them.

After the 1936 Olympic games that were held in Germany, repression progressively increased, culminating in Hitler's "Final Solution" of starvation and extermination. Gaebelein traveled to Germany in 1937 and gained a firsthand view of the Nazi regime. In his published accounts sent from Germany and his subsequent detailing of persecution of the Jews in his "Current Events in the Light of the Bible" section, readers of *Our Hope* gained a unique insight into the plight of the Jewish people. In June, 1939 E. Schuyler English became Associate Editor of *Our Hope*, relieving the 80 year old Gaebelein of writing the "Current Events" section. However, Arno C. Gaebelein still received each excerpt ahead of time and authorized it for publication. In addition, he continued to contribute material.[3]

Today, it is not uncommon to hear someone remark that they "did not know" that the Holocaust was occurring or that information was "just not available." Because of this, *Our Hope*'s accounts from 1937 to 1945 are valuable and illuminating to the modern historian.

Throughout the 1930's, *Our Hope* documented the rising anti-Semitism throughout the world. By 1937, its readers had no doubt about the gravity of the situation. For example, on one page Arab opposition and Polish anti-Semitism are mentioned in consecutive excerpts. Regarding Poland, *Our Hope* noted:

> Jews have suffered severely during 1936 in Poland. Their troubles have been economic distress and many deeds of violence. The Endek National Party seems to be the leaders in spreading antagonism against the Jews. Several months ago there was a great outbreak in a Polish village. Then came uprisings against the Jews in Minsk, and almost 5,000 fled from that city to Warsaw. Warsaw, Lodz and Wilna, with many smaller places, also record similar persecutions.[4]

Our Hope's premillennial bias toward the Jewish cause in Palestine during the Arab violence is clearly seen in its evaluation of Jewish funds and migration to Palestine in 1936 when it observes: "So, in spite of the Arab opposition, Ishmael's sons rising against Isaac's

earthly offspring, Zionism continues to flourish."[5]   Of Great
Britain's new recommendation to create independent sovereign
Arab and Jewish states, *Our Hope* declared: "The suggested division of Palestine among Jews and Arabs has been a great shock to
Zionism. It is too early to say much about it. It seems to us it is
altogether against the program of prophecy." To *Our Hope*, the
separation of Jerusalem from "Israel's" rule; the lack of Jewish
control of the lighting and irrigation networks they built; and the
lack of a sound Biblical basis, made the plan appear to be absurd.[6]

Throughout the early 1930's, Gaebelein had mentioned to the
readers of *Our Hope* that he wanted to take a trip to Germany to
visit relatives. It was not until August 16, 1937 that he and his
wife sailed for Central Europe. He sent home reports for seven
issues giving his observations of Hitler, Goebbels, Streicher, Rosenberg, Hitler Youth and others in the Nazi regime. Unlike the
American anti-Semitic clergyman, Gerald Winrod, Gaebelein came
back unimpressed with Hitler's philosophy. Rather, he held the
firm conviction that the Jewish people in Europe were in for deep
trouble. While German officers had been courteous to the American couple, Gaebelein noted the trouble they caused a Jewish
woman on their train and their confiscation of foreign newspapers.
"This is done to keep undesirable information from the German
citizens," Gaebelein explained and then continued, "Well, we
appreciated our American liberties, that no policeman goes around
to confiscate newspapers. Evidently the entire German Press is
under the supervision of the government."[7]   In the next segment he
wrote:

> Our brethren in Germany need our prayers. [Alfred] Rosenberg seems
> to forge ahead; his work of destruction is unchecked. The most de-
> plorable fact is that the so-called Hitler-Youth is led astray by this mis-
> leader, led towards a terrible abyss. How long oh Lord? How long?
> *Till He comes.*[8]

Gaebelein was not pacified by Adolf Hitler's declaration that
National Socialism stood "on the Foundation of real Christianity"
and that it held religious confessions to be "valuable helps to up-
hold our people." Rather, the periodical questioned in 1938 that
if Chancellor Hitler was sincere, why had he appointed Alfred
Rosenberg, a known anti-Semite and anti-Christian to receive the
first national prize of 100,000 marks. Gaebelein believed Hitler was
actually proposing a new religion in his "New World View." He
stated to the readers of *Our Hope*:

But what is this German brand of a New World View? It might be
expressed by two sentences, "We are the people. We are superior to
other races." *The hatred of the Jew is its backbone* . . . Out of this anti-
Semitic agitation has sprung the New World View. Its prominent leader,
as stated before, is Alfred Rosenberg. Others are Wilhelm Hauer, Ernst
Bergmann, von Reventlow, and several more.[9]

Our Hope pointed out to Christians that the denial of Jewish
chosenness was logically extended beyond anti-Semitism to a denial
of the authority of the Bible, objections to Paul (because he was a
Pharisee) and the necessary Aryanization of Jesus. That Jesus was
characterized as an anti-Semite by this "arrogant" Germanic religion
gion was particularly shocking to Fundamentalist *Our Hope*. The
periodical declared that "this antichristian clique" had and would
turn against the Christian church, and therefore "believers everywhere should pray earnestly for our tested brethren and for their
continued loyalty to the faith delivered unto the saints."[10] When
Germany announced its detailed program for the German church,
Our Hope shuddered and declared that it was "viciously anti-
Semitic and therefore anti-Christian."[11]

Alas, 1938 proved to be discouraging to Gaebelein as the persecution of the Jewish people increased and spread. *Our Hope* docu-
mented the incidents, listing the anti-Jewish measures instituted in
"Roumania," Poland and Germany, while noting that Austria,
Latvia and Lithuania "also have anti-Semitic tendencies."[12] Of
Poland, *Our Hope* commented, "there has been more Jewish blood
spilled there than possibly anywhere else, except in Russia and in
Palestine."[13] The Fundamentalist periodical sadly related that
"after Hitler and his minions took over Austria," even converted
Jews were not safe from the Nazi wrath. The case of a dispossessed
physician and his wife was related and *Our Hope* exclaimed:
"This is worse than inhuman, it is a diabolic cruelty. But there is
a day of reckoning coming. God will deal with the enemies of His
people and when judgment comes, as it surely will, this modern
anti-Semitism will find its ignominious end."[14] Unfortunately, it
was not yet the end.

When Benito Mussolini made his anti-Semitic statements in
Trieste, *Our Hope* labeled him a "modern Haman." The periodical
expressed compassion for the "homeless millions" of Jews and reminded Christian readers of Jesus' compassion for the Jewish multi-
tudes. It stated: "As Christians we are moved with the same compassion; we pray for them and know the day will come when the
Lord will deal with the nations for their cruel treatment of the

Jews."[15] In a following excerpt, the fact was mentioned that refugees from Germany and Austria were arriving in the United States at a rate of one hundred a day, but the U.S. Embassy in Berlin was not accepting any more applications from German and Austrian Jews because the quota was filled for the next two years.

The acts of the German government appalled the Fundamentalist journal. "What is being done to Jewry in Germany is satinically cruel," *Our Hope declared.*[16] It related that the latest laws were against Jewish professionals. Jewish doctors' permits were canceled; one thousand Jewish dentists were purged on October 5, 1938 when the conference of the Dental Association convened. By January 1, 1939, Jews must either have a name that shows their Jewish origins or must add "Israel" or "Sarah" to the name they possess. As *Our Hope* went to press on November 14th, news of *Kristallnacht* infuriated the Fundamentalists, who passionately reported "the outrageous happenings in Germany." "The three leaders, Hitler, Goering and Goebbels, permitted and encouraged the vicious and violent outbreaks against the Jews, destroying millions of dollars worth of property," *Our Hope* asserted, "How many were killed we do not know. It is devilish from start to finish."[17]

One gains some indication of the interest of the readers of *Our Hope* in world events by the fact that some readers requested more information on the fate of Czechoslovakia, and *Our Hope* responded by publishing the ten point compromise plan "hatched out in Munich."[18] Anti-Semitic legislation in Czechoslovakia, Hungary and Romania was detailed in the early months of 1939. *Our Hope* commented, "As the march continues toward the East, Hitler's great ambitious anti-semitism develops right along the line."[19] In response to an "official" report on the increase in church membership in Germany from 1934-1936, *Our Hope* predicted that the report for the years 1937 and 1938 would show "an astonishing decline." It lamented: "The deplorable thing is that the German Youth are turning their backs on the churches; they have become 'Hitlerized.' "[20]

Our Hope continually emphasized that those who persecuted God's "Chosen People," the Jews, would come to a disastrous end. This is fervently proclaimed, for example, when Alfred Rosenberg suggested that all the Jews be settled on "reservations" in British Guiana or in Madagascar. *Our Hope* quoted him as saying, "I stress the word reservation for there can be no talk either at present or in the future about a Jewish state." *Our Hope* responded to Rosenberg's comments:

> Poor mumbling, blind Rosenberg! He hates God's Word. He despises the Old Testament! He wants to have it thrown out! German youth is warned against reading it. Soon the day will come when not a Jewish state, a Jewish reservation or anything like it, will be established, but when the kingdom will be restored to Israel, when God's covenant promises to His earthly people will all be fulfilled, when glory will dwell in "Immanuel's Land." Poor Rosenberg, Streicher, Hitler and others, you are not fighting the Jews, you are fighting God, God's Son, our Lord, God's purposes and God's Word. What a terrible defeat and fate is awaiting you![21]

In a later issue the Fundamentalist periodical noted that the Jewish people would return to Palestine, "their own land," and concluded: "Zionism, though it is not influenced one whit by God's Word, may be His chosen instrument to effect their return to Palestine."[22]

Perhaps nothing disturbed *Our Hope* more than anti-Semitism within the United States. It reported the "rumor" that leaders of American anti-Semitic organizations were combining forces, and warned its readers to "be careful" of what they read. So beguiling was the anti-Semitic literature that *Our Hope* feared Christian acceptance of it. One of the anti-Semitic propaganda leaflets being circulated in the United States was *Why Are the Jews Persecuted for Their Religion?* *Our Hope* explained that the pamphlet purported to quote from the Talmud, but the fact of the matter was that "the 'quotation' is a misquotation." The Fundamentalist periodical explained to its readers that the Talmud does *not* state that a Gentile girl can be violated by a Jew after three years of age; nor does it conclude that all property of other nations belongs to the Jewish nation. *Our Hope* sadly concluded:

> The enemies of Israel will stop at no end to accomplish their purposes to persecute them, and to drive them out. When will anti-Semitism take hold in this land? Perhaps it may be sooner than we think.[23]

Most Jews would not be happy to learn that *Our Hope* regarded this anticipated American persecution as a reason for the Fundamentalist-Evangelical to "be instant in season and out of season, to preach the Word to Israel [Jewish people], before this persecution sets in!" Nor would Jews be overjoyed with the Fundamentalist emphasis that this persecution would eventually "be a blessing to the Jews" and would cause some to "go back to their own land

[Palestine]." Nevertheless, *Our Hope* recognized the potential for anti-Semitism in the United States, dreaded it and warned Fundamentalist readers not to become a part of it.[24]

As the Nazis formed ghettos to starve out the Jewish people, *Our Hope* reported the particulars it had at hand. The starvation conditions in Lublin, Poland; armbands in Antwerp; and the spread of anti-Semitism in Spain and France are a few of the topics discussed during 1940-1941.[25] One of the early reports about the Warsaw Ghetto read:

> All the *Jews* in *Warsaw*, about one half a million of them, are now required to live within a concrete wall constructed by *German* decree. The wall is eight feet high and surrounds more than one hundred city blocks. This is not, say the *Nazis*, an *anti-Semitic* edict but a health measure, to protect the *Poles* as well as the Jews from pestilences![26]

Our Hope could not accept the "official" Nazi version. It did, however, believe that as the anti-Semitic tentacles spread a glorious future awaited the Jewish people. It declared in an earlier account of anti-Semitism: "The barbed fist of persecution is crushing the *Jews,* or trying to do so. Where will Israel go? To THE LAND—PALESTINE."[27]

During 1942, *Our Hope* documented "the ugly shadow of Anti-Semitism" arising in the British Isles. The British were discussing Jewish domination of the Black Market. *Our Hope* complained that the news coverage always headlined the few "Cohens" while saying little about the large number of "Smiths" (*Our Hope* added that these names were fictitious to illustrate the point.) At the same time *Our Hope* pointed out the consternation that Zionist leaders, such as Chaim Weizmann, experienced from the British over a "tragedy" such as the *Struma* or British intransigence to allowing Jews in Palestine to defend themselves. The Fundamentalist periodical stressed the *Struma* incident:

> Dr. Weizmann's associates pleaded with authorities to allow the transfer of 760 certificates of entry into Palestine to be assigned by and from the Jewish Agency in Britain to these unfortunates living like rats in their 200 ton ship at Istanbul. Consent was delayed, but at length permits were written, *for sixty children*—but too late! The Turkish Government by that time had ordered the *Struma* to leave port, and she went down in heavy seas.[28]

Dr. Weizmann's declaration served as the critical comment on British actions: "One is horrified to think what sort of people are those who rule the destinies of Palestine—people without under-standing, without compassion and without pity."[29]

In October of 1942 Arno C. Gaebelein, the Editor of *Our Hope*, summarized the exploits of "Monster Hitler." Gaebelein declared, "Not gradually, but suddenly Monster Hitler developed *a devilish mania to persecute and exterminate the entire Jewish race,* an am-bition which he has pushed more and more to the front and in which he persists today as never before in his bloody and despicable career." Gaebelein insisted that Hitler had instituted "the worst program of Anti-semitism known in history" and reminded his readers that Hitler also persecuted Christian believers and that he hated the Jewish book, the Bible. He informed his readers of Nazi executions of Jewish people. Of the executions in Wilno in May, 1942 he exclaimed:

> Men, women and children were taken in trucks to a suburban village where they were stripped of their clothing and then machine-gunned. Similar reports come from Latvia, the terrorism of hell continues in Czechoslovakia, Bulgaria and elsewhere.[30]

The *Our Hope* office was near the publishing office of *Jewish Con-temporary Record,* and Gaebelein suggested to his Fundamentalist readers that they subscribe to it. Gaebelein felt that this publica-tion was "trustworthy" and quoted in this particular article a long segment from it on the Nazis methodical starvation and execution of Jewish people.[31]

Gaebelein ended this article by assuring his readers that "Hitler's master, Satan, is after all a stupid fellow" and that the Nazis would make blunders that would lead to their downfall.[32] In a later article in the series (June, 1943), Fundamentalist Gaebelein refused to underestimate the Nazi atrocities. He asserted:

> The horrors of Anti-semitism continue. During the first World War many atrocious deeds were reported only to be found later untrue. But this is far from being true in this war of barbarism. If anything the number of Jews killed is underestimated. Reliable sources mention not less than two million Jews murdered since Hitler went on his devil-controlled mission.[33]

Gaebelein then probably shocked his conservative readers by relat-ing a typical incident by Hitler's "beasts of lust." He wrote:

> Crimes against women go unabated. A pathetic incident became known through a letter successfully smuggled out of Poland. It reports the suicide of 93 young Jewish girls. They were between the ages of 14

and 22, students in a religious school. Some of Herr Hitler's beasts of lust had herded them together for transportation to houses of prostitution in Germany. These 93 young Jewish heroic girls preferred mass suicide to degradation.[34]

These accounts overwhelmed Gaebelein to such an extent that he exclaimed: "But we cannot continue with such rehearsals and pass by other reports of a similar nature. How long! How long! Oh Lord!"[35]

In July of 1943, *Our Hope* used the Warsaw Ghetto as an example that there was absolutely no improvement in the treatment of European Jews. Gaebelein reminisced in "The New Great World Crisis—XXIV":

When about 50 years ago the Editor took walks through that noble city (Warsaw) he admired its stately buildings, synagogues, churches, schools, libraries, public parks, etc. The greater part of it is now a thing of the past. In 1934 the Jewish population was 336,600. When the Nazis came 5 years later they herded all Jews into a miserable Ghetto. Then came

regular reports of their increasing reduction by cold-blooded massacres and by cruel removal. Rabbi Irving Miller of New York City, the Secretary General of the World's Jewish Congress, has given horrible information. It is most shocking. The 40,000 Jews remaining in the Warsaw Ghetto had revolted against the inhuman treatment of the Nazis and killed sixty German officers and men. As a result these Jews have been murdered and forcibly removed. Nazis rolled through streets in giant tanks, leveling stores and houses and silencing the feeble guns of the defenders, in the final stage of extermination. "Every living soul was either butchered or uprooted and moved to some other part of the country," the rabbi added.[36]

Gaebelein emotionally responded: "What judgment storm is brewing and gathering against these incarnate demons!"

A few issues later, *Our Hope* published a chart entitled "Extermination of the Jews," which related to its readers that at least two million Jews were known to be dead and five million were in danger of "extermination." "That this people Israel has been the great prey of the Axis powers, Hitler in particular, and the great sufferer of World War II, is too well known to argue," *Our Hope* maintained. "Surely American Jews and Gentiles, for humanity's sake, will want to succor them. When victory is won they will be repatriated, without a doubt."[37]

With the July issue, *Our Hope* had embarked on its Jubilee year as a Fundamentalist periodical. The Associate Editors used the

November, 1943 issue as a "Jubilee Issue" to honor Arno G. Gaebelein. An assortment of articles that the Editor had written for *Our Hope* over the years were reprinted and dated. Included was an early excerpt from 1898 that stated that there was "no doubt" that the process of "Israel's restoration" to Palestine was beginning; and an excerpt from the July, 1932 issue entitled "Trembling Judaism." This latter excerpt mentioned the rise of Adolf Hitler and the increase in anti-Semitism in universities and other institutions. It stated:

> Dark clouds are gathering over Judaism in many lands. Anti-Semitism is rising. Adolf Hitler, with his rising and powerful influence, is a rabid, fanatical Jew Hater. He evidently is heading for the same end and the same fate as Haman in the book of Esther.[38]

Also published were thirty-nine "testimonials" to the importance of Gaebelein and *Our Hope*. Many of the leading Fundamentalist-Evangelicals gave high praise in this section, underlining *Our Hope*'s widespread influence on the Fundamentalist movement.[39]

Throughout 1944 and into 1945, *Our Hope* detailed the setting of "Herr Hitler's star." Blunders made by Hitler and his generals were given significant coverage. This proved to the writers of the Fundamentalist journal that those who persecute the Jewish people would be cursed by God. *Our Hope* proclaimed: "There is no place for anti-Semitism in any land which calls itself Christian, and we who are the Lord's will feel for the Jews and do all in our power to succor them in their present distress in lands across the seas."[40]

As news of the fate of the Nazi leaders reached *Our Hope,* the horrid details of the concentration camps overwhelmed the editorial staff. Under the caption "Better and Better?" was printed:

> We wonder what the evolutionists and others think who have seen the world getting better and better all the time! And what of the theologians who maintain that man will work himself into a millennium of peace and righteousness? They must have a hard time convincing themselves, after reading about and seeing the pictures of the wholesale atrocities perpetrated in German prison camps. If nothing else could persuade them, we should think that the horror scenes and graphic accounts that have come out of Buchenwald, Dachau, Belchen, Berga, Ziegenhain, Orbke, Belsen, *et al* . . . These organized murders and sadistic maltreatments were not the work of one man, or one small group of men. They were devised and carried out by thousands of Nazi beasts, the product of modern Germany.[41]

Our Hope reminded its readers that these atrocities did not occur "in the jungles of Africa" or "the uncivilized islands of the sea," but rather in what some considered the epitomy of civilization in the world. Then the Fundamentalist periodical questioned: "Is this what civilization does to man? If man evolved from some protoplasm, and then developed through the ages to his present state, what happened in Germany between 1939 and 1945?"[42]

It must be mentioned that there was a continual conflict between liberal Protestantism and Fundamentalist *Our Hope* over views toward the Jewish people and their future. The premillennial eschatology of the Fundamentalist journal led it to accept the future restoration to Palestine of Jewish people unequivocally and to *believe* the reported extermination of Jews in Europe, when Liberal Protestantism could not. Hertzel Fishman has analyzed American Liberal Protestantism's attitudes during the Holocaust in his book, *American Protestantism and a Jewish State*. He chose the *Christian Century* magazine as a major source because it had the largest number of subscribers of any Protestant weekly and because it was the only non-denominational liberal Protestant weekly in the United States at that time. He came to the conclusion that in general American liberal Protestantism had a persistently hostile attitude toward Jewish "peoplehood," minimized the Holocaust and supported restriction of immigration of Jewish refugees. Fishman was appalled to find that the *Christian Century* would not believe the daily news reports about Hitler's brutality toward Jewish people. *Christian Century* asked that "all thoughtful persons, Jews as well as Christians, [put] . . . tighter curbs upon their emotions until the facts are beyond dispute."[43] Indeed, it held this view even at the end of 1942, when Rabbi Stephen Wise told the gruesome story to the press. Fishman details the journal's skepticism and its conclusion that the statement of Nazi atrocities was "unpleasantly reminiscent of the 'cadaver factory' lie" of World War I. Rabbi Wise was aghast at such an "anti-Jewish" attitude and frankly told *Christian Century* so. At the same time, Fishman found *Christian Century* to oppose the immigration of Jews to the United States during their time of trouble (a less than "liberal" attitude).

A recent study by Professor Robert W. Ross to be published in the Spring of 1980 by the University of Minnesota Press will be of extreme interest to the scholarly community as well. In *So It Was True: The American Protestant Religious Press and the Nazi Per-*

secution of the Jews, 1933-1945, Ross details the sad performance of American Protestantism during the Holocaust. He surveyed 52 religious periodicals ranging from conservative to liberal and found that the story of the Holocaust was indeed available to American Protestants. Most editors and writers, however, would not believe the material they had. Ross documents the fear of "cadaver stories" among Protestants and supports Fishman's thesis that the *Christian Century* had a very poor record.[44] Ross did not survey *Our Hope* for his book because it was not readily accessible to him. However, he was familiar with the journal and believes that it was an important journal because of Gaebelein's "irenic" spirit within the Fundamentalist movement.[45] Comparing the quotes in *So It Was True,* I believe *Our Hope*'s extensive coverage of the Holocaust and unswerving acceptance of the events is unique when compared to American Protestantism as a whole and should be pointed out to the scholarly community for examination as well.

Our Hope was certainly not the only Fundamentalist journal to document Nazi atrocities. The periodical, *Moody Monthly,* the voice of The Moody Bible Institute of Chicago (the West Point of Fundamentalism), for example, not only believed that the atrocities were occurring, but in 1939 published an "Appeal" for money to be "used for the purpose of relieving the sufferings of Jews and Jewish Christians in Central Europe." This Appeal was signed by some of the leading Fundamentalist-Evangelicals in the United States.[46] However, *Moody Monthly* never published such occurrences with the fervent thoroughness of *Our Hope.* Such thoroughness must be commended and reevaluated. In light of our current knowledge of the details surrounding the Holocaust, the alarming questions *Our Hope* raised about the "civilized" world can not be totally dismissed by the scholar in this most modern of societies. Some "civilized" men in our world today *still question* whether the Holocaust indeed occurred.[47]

NOTES

1. For more background information on *Our Hope* and Gaebelein note my articles "Our Hope: Proto-Fundamentalism's Attitude Toward Zionism, 1894-1897," *Jewish Social Studies* 40 (Summer-Fall 1978): 239-250; and "Arno C. Gaebelein (1861-1945): Protestant Fundamentalist Zionist," *The American Jewish Historical Quarterly,* 68 (September 1978): 43-56.

2. "Adolf Hitler—Will He Be Germany's Dictator?" *Our Hope* 37 (1930): 363-364.

3. Gaebelein's youngest son, Frank, was an Associate Editor as well. I have interviewed Frank E. Gaebelein and E. Schuyler English with respect

to this time period. Note also "Hearty Welcome," *Our Hope* 45 (1939): 810.

4. "Poland's Anti-semitism," *Our Hope* 43 (1937): 552.

5. "The Jewish Population of Palestine," *Our Hope* 43 (1937): 552. *Our Hope* had no illusions about the struggle in Palestine and reported vicious attacks on Jewish settlements in Samaria. "They burned down houses and shot at people, while others broke into the stores to steal . . . vandals . . . uprooted newly planted trees in the widely known Balfour Forest." And yet, *Our Hope* gloried in the fact that friction between the two major Arab families (Husseini and Nashashibi) might destroy the Arab unity against the Jewish people in Palestine. Note "Interesting Palestine Conditions": 764-865.

6. "The Palestine Situation," *Our Hope* 43 (1937): 190-191. *Our Hope* agreed with these facts as quoted from *The Christian,* a London periodical, and in later issues reemphasized these elements.

7. "Observations and Experiences," *Our Hope* 44 (1938): 676.

8. "Observations and Experiences," *Our Hope* 44 (1938): 750.

9. "Observations and Experiences," *Our Hope* 44 (1938): 461-462. *Our Hope* was appalled that the German's "New World View" accused the Jews of defiling and ruining the Nordic race. The periodical documented the fact that it was a crime for an "Aryan" to marry a "Jewess," and the fear that Christians had of finding a Jew in their family tree.

10. *Ibid.,* p. 465.

11. "The Announcement of the Black Squadron," *Our Hope* 44 (1938): 686. *Our Hope* published the five points as follows: 1. Germany proclaims a State religion to which all citizens are obligated and which is based on the revelation of God in nature, destiny, life and death of peoples.   2. Churches are permitted to exist as private institutions if they in every way subordinate themselves to the basic doctrines of the State religion. 3. The State refuses all cooperation with the Churches. It will neither protect nor support them. 4. The State confiscates all Church property on the grounds that this property was created through the joint cooperation of all the citizens and in a period when the State and Church were still an entity. 5. Religious instruction in the Church will be conducted in the service of the State Religion by the teachers that have left the churches. In exceptional circumstances special instruction can be furnished for the children remaining in the Christian Church by teachers who are "ecclesiastical professionals."

12. "The Increasing Anti-Semitism," *Our Hope* 44 (1938): 687. Compare "Happenings in Romania," *Ibid.,* p. 552; and "Anti-semitism Rising Higher and Higher," *Ibid.,* pp. 822-823.

13. "Anti-Semitism Rising Higher and Higher," *Our Hope* 44 (1989): 822. In this same article "Jewish liquidations in Soviet Russia are scrutinized, and the Communist program in the United States against the "owning class" Jew is mentioned.

14. "The Plight of the Austrian Jews," *Our Hope* 44 (1938): 825-826.

15. "Another Modern Haman Speaks," *Our Hope* 45 (1938): 332. The following excerpt on pages 332-333 is entitled "Refugees Arriving in America."

16. "The Anti-Semitic Situation in Germany," *Our Hope* 45 (1938): 385.

17. *Ibid.,* p. 386.

18. "The Dissection of Czechoslovakia," *Ibid.*

19. "New Troubles for Jewry in Eastern Europe," *Our Hope* 45 (1939): 604.

20. "A Church Report from Germany," *Ibid.*, 605.

21. "Alfred Rosenberg, the German Anti-Christian Leader Speaks Again," *Our Hope* 45 (1939): 689.

22. "Will Britain's White Paper Keep the Jews Out of Palestine?" *Our Hope* 46 (1939): 179. As the title indicates, the article discussed Britain's White Paper, and states: "How and when the White Paper will be repudiated we do not know, but that in the end it shall be non-effective is a certainty" (p. 179). *Our Hope* is referring to the non-religious sphere of political Zionism in this excerpt.

23. "Anti-Semitism in the U.S.A.," *Our Hope* 46 (1940): 543-544.

24. *Ibid.* It is an interesting fact that *Our Hope* regarded the re-election of Franklin D. Roosevelt as a "disappointment," although the editors stressed that as Americans, they would "accept the verdict of the polls" and go on "praying for our land and for those in authority." *Our Hope* felt the third term put too much power in the hands of a few. Note "Another Shadow," *Our Hope* 47 (1940): 414.

25. Note especially "The Poor Jews," *Our Hope* 46 (1940): 625; "Anti-Semitic Tentacles Spread," *Our Hope* 47 (1940); 263; "Christian or Gentile?" *Our Hope* 47 (1940): 413; and "Armbands in Antwerp," *Our Hope* 47 (1941): 489.

26. "Ghetto in Warsaw," *Our Hope* 47 (1941): 489.

27. "Anti-Semitic Tentacles Spread," *Our Hope* 47 (1940): 263.

28. "Jewish Problems," *Our Hope* 48 (1942): 839. Compare "Two Attitudes Toward the Jews in Britain," *Our Hope,* 49 (1942): 192-193.

29. *Ibid.*

30. Arno C. Gaebelein, "The New Great World Crisis: XV" *Our Hope* 49 (1942): 240.

31. *Ibid.* Gaebelein noted to his readers that the *Jewish Contemporary Record* was published by the "Jewish American Committee." *Our Hope* sometimes interchanges the first two words of the publication's name, at times calling it "*Contemporary Jewish Record.*"

32. *Ibid.*, 242.

33. Arno C. Gaebelein, "The New Great World Crisis" XXIII, *Our Hope* 49 (1943): 815.

34. *Ibid.*, 816.

35. *Ibid.*

36. Arno C. Gaebelein, "The New Great World Crisis: XXIV," *Our Hope* 50 (1943): 22.

37. "The Jewish Problem," *Our Hope* 50 (1943): 250.

38. "Current Events in the Light of the Bible," *Our Hope* 50 (1943): 334; 343.

39. "A Handful of Testimonials," *Our Hope* 50 (1943): 324-330.

40. E. Schuyler English, "The Judgment of the Nations," *Our Hope* 51 (1945): 565. The "Editorial Notes" of this issue run the captions *Never before!* and relate the shocking details of the methodical extermination of the Jewish people. Note especially page 514. *Our Hope* also supported the opening of Palestine to Jews. "Zionist Demand to Atlee," *Our Hope* 52 (1945) comments: "The poor Jews! 6,000,000 of them died under Nazi persecution. Had Palestine been opened many of them might have escaped" (p. 189).

41. "Better and Better?" *Our Hope* 51 (1945): 851.

42. *Ibid.*, 852.

43. Note Hertzel Fishman, *American Protestantism and a Jewish Sta'e* (Detroit: Wayne State University Press, 1973), p. 54. The quote is from *Christian Century* (April 5, 1933): 443. Fishman's Chapter 4 deals with the impact of the Holocaust on *Christian Century* and his conclusion on pages 178-183 sums up his discussion on liberal Protestantism in general.

44. Robert W. Ross, *So It Was True: The American Protestant Religious Press and the Nazi Persecution of the Jews, 1933-1945* (unpublished manuscript).

45. Professor Ross was kind in letting me read his manuscript before publication. I have asked his permission to quote some of his telephone comments concerning Arno C. Gaebelein. I deeply appreciate his help. I also found it interesting that Ross had personally heard Dr. Gaebelein lecture.

46. Joseph Taylor Britan, "An Appeal for Persecuted Israel," *Moody Monthly* 39 (February 1939): 316; 345. The quote is from page 345, and the thirty men endorsing the Appeal are also on this page.

47. Fundamentalist *Our Hope*'s thorough coverage for its Christian readers of the atrocities of the Holocaust as they occurred also leads one to question the classic excuses by those who say: "We didn't know."

Baptist Fundamentalism: A Cultural Interpretation

LEON MCBETH*

A study of the heritage of Baptist thought can scarcely overlook the role of fundamentalism in shaping Baptist theology. Most Southern Baptists are not fundamentalists. Certainly, the organized fundamentalist movement of the 1920's was not able to capture the Southern Baptist Convention. Southern Baptists are conservative, Bible-believing people. While they share many of the same doctrinal beliefs of fundamentalism, Southern Baptists generally do not share in the spirit and temperament of the movement. The Convention and its agencies have never, to this point, fallen under control of extreme doctrinaire fundamentalist forces that were active in the 1920's, and that are again active today.

Southern Baptist thought, however, has been greatly influenced by fundamentalism. As in the case of Landmarkism, an earlier extremist movement, the Southern Baptist Convention fought off the threat of fundamentalism only at the cost of absorbing some of its ideas. No doubt a large number of Southern Baptists are outright fundamentalists, and others share many of their doctrines if not their spirit. There are also some organized Southern Baptist groups within and without the Convention that are fundamentalists.

There is apparently a resurgence of fundamentalism in American religion today. Southern Baptists share, to some extent, in this resurgence. Some observers think that a new fundamentalist controversy, like that of the 1920's may be shaping up.

Few movements in American religion have been so thoroughly sifted as fundamentalism. Scholars have analyzed it, sociologists have measured it, psychologists have commented on it, and educators have deplored it. Agreement may exist on what happened, where, and to whom, and on the names of the denominations involved, the leaders, and their major positions, but there is little agreement on the basic meaning of fundamentalism.

The historical events of the fundamentalist crusade of the 1920's will not be treated. That has been done, with more or less success, by many writers, such as Stewart G. Cole, Norman F. Furniss, and Ernest R. Sandeen.[1] Instead, the underlying *meaning* of the fundamentalist phenomenon will be probed. Why did fundamentalism emerge? What does it really mean? What is behind the fundamentalist phenomenon, not only of the 1920's but at other times in Baptist history?

This is no effort to put together the entire jigsaw puzzle that is fundamentalism but is simply an attempt to set one more interlocking piece of the puzzle in place in order to help make the picture more complete. Three things will be attempted in this paper: an examination of the extent of Baptist involvement in fundamentalism; a review of several familiar interpretations of fundamentalism; and then a

---

*Dr. McBeth, professor of church history, Southwestern Baptist Theological Seminary, Fort Worth, Texas, and immediate past president of the Southern Baptist Historical Society, presented this address at the annual meeting of the Historical Commission, SBC, and Southern Baptist Historical Society in Fort Worth, Texas, in April, 1978.

[1] Stewart G. Cole, *History of Fundamentalism* (New York: R. R. Smith, 1931); Norman F. Furniss, *The Fundamentalist Controversy* (Hamden, Conn.: Archon Books, 1963); Ernest R. Sandeen, *The Roots of Fundamentalism: British and American Millenarianism, 1800-1930* (Chicago: University of Chicago Press, 1970).

suggestion of one way to interpret the movement, namely, as an example of culture-conflict.

Baptists and Fundamentalism

When Jimmy Carter first burst upon the national scene, nothing about him so puzzled the general public as his religion. Some were attracted to his faith, some were repelled, and many were merely curious. Some news writers, unacquainted with Southern Baptists, had to scramble for descriptive materials to meet their daily deadlines. Their tendency often was to lump Southern Baptists all together as latter-day fundamentalists.

Some have assumed that the rural South is the home of fundamentalism, and that Southern Baptists are its primary exponents. Charles and Mary Beard, both fine historians, may be misleading when they say that fundamentalism showed "its greatest strength in rural districts where the machine process had as yet made little impression."[2] Better evidence suggests that fundamentalism probably originated in northern cities, and had its earliest impact among Presbyterians. However, Baptists have been enough involved with fundamentalism to warrant its consideration here.

Many of the early leaders of the fundamentalist movement were Baptists. One thinks immediately of William B. Riley, John Roach Straton, T. T. Shields, J. C. Massee, and, of course, J. Frank Norris. All of these were recognized as leaders of the fundamentalist crusade, and they were known throughout the nation.

Several of the organized results of the movement were Baptist. One thinks of the Fundamental Baptist Fellowship, The Baptist Bible Union, the World Baptist Fellowship, the Premillennial Baptist Missionary Fellowship, and the Baptist Bible Fellowship, among many others.[3] Baptists also exercised leadership in the interdenominational structures of fundamentalism, such as The World's Christian Fundamentals Conference. James Tull's excellent book, *Shapers of Baptist Thought*, describes nine influential Baptist thinkers who have helped determine Baptist theology. He does not include any fundamentalists, and should not have, but still both positively and negatively fundamentalism has helped to shape Baptist views.[4]

Doubtlessly, fundamentalism made serious inroads in Northern Baptist life. This agitation and schism sapped the vitality of that denomination and left it seriously crippled. Beginning with the Fundamentalist Fellowship in 1920, the ultraconservatives carried on a continuing campaign to purge Northern Baptists of unacceptable schools, professors, and missionaries. This led to several schisms, such as the General Association of Regular Baptists in 1933 and the Conservative Baptists of America in 1947. While it had long been popular to attack professors, this attack upon Northern Baptist missionaries split the fundamentalist movement.

While acquainted with fundamentalist views, Southern Baptists were less affected by agitation and schism. Probably, Sandeen was correct in writing that the Southern Baptist Convention "was scarcely troubled by the Fundamentalist controversy of the 1920's, since Liberalism had had very little impact upon the denomination to that date."[5] Sandeen then named W. O. Carver as the nearest Southern Baptists had to a liberal, thus overlooking W. L. Poteat and others.

---

[2] Charles and Mary Beard, "An Historical Interpretation," *Controversy in the Twenties*, ed. Willard B. Gatewood (Nashville: Vanderbilt University Press, 1969), p. 431.

[3] For more information on various Baptist Fundamentalist groups, see Norman W. Cox "Baptist Bodies in the United States," *The Quarterly Review*, 19:6-10, April, 1959. There are several other excellent articles in the same issue.

[4] Valley Forge: Judson Press, 1972.

[5] Sandeen, pp. 264-65.

Southern Baptists were no strangers to extremist movements, however. The Landmark movement, with its subsequent outcroppings in Gospel Missionism and the Hayden Controversy, had brought the Southern Baptist theological pot to an occasional boil. Landmarkism had much in common with later fundamentalism. Its founder, J. R. Graves, might be called a "prefundamentalist," since he espoused the same militant conservatism, dispensational premillennialism, and rigid spirit of the later movement.

Without doubt the leading Southern Baptist exponent of organized fundamentalism was J. Frank Norris of Fort Worth, a fact Sandeen almost overlooked. As the longtime pastor of First Baptist Church of Fort Worth, Norris was a tempestuous leader of the entire southern wing of fundamentalism. Norris grew up in Hill County, Texas, the victim of all kinds of harsh deprivations—economic, emotional, and spiritual.[6] With rejection toward his father and an almost unnatural emotional dependence on his mother, Frank Norris developed the personality which some observers would later describe as pathological.

Norris was a graduate of Baylor University and Southern Seminary. As a student at Baylor he led an uprising which got the president fired, thus giving him a first taste of the thrill of controversy and the glory of the spotlight. At Southern Seminary he attracted less notice, which may account for his inordinate haste to finish his studies and leave Louisville.

Norris became, for awhile, a close friend of B. H. Carroll. Through the pages of the *Baptist Standard*, Norris had perhaps more influence than any other man in the removal of Southwestern Seminary from Waco to Fort Worth. Some even speculate that he expected to succeed Carroll as president, and attribute part of his later bitterness and disappointment to the failure of the expectation.

Norris utterly transformed the old First Baptist Church of Fort Worth. He became ever more sensational, controversial, and independent. He refused to use Southern Baptist literature, refused to support the $75 Million Campaign, and refused to support the missionary work of the Convention. During his colorful career, Norris was indicted twice for arson and once for murder, but never convicted. W. W. Barnes led the move to oust Norris from Tarrant Baptist Association. Later, Norris was also excluded from the Texas and Southern Baptist Conventions. Norris carried many individuals and churches with him. Even to this day, more than half the Baptist churches in Tarrant County are not in fellowship with Southern Baptists.

Just as Baptists were involved in the old fundamentalist controversy, Baptists are also involved in the recent resurgence of this position. Perhaps the major contemporary Southern Baptist expression of fundamentalism is to be found in the Baptist Faith and Message Fellowship. This is an organization within the Southern Baptist Convention of ultraconservatives who are seeking, like Norris of old, to capture the Convention to their viewpoint. They take their name from the Southern Baptist Convention confession of faith adopted in 1963. They publish their own paper, *The Southern Baptist Journal*, and are quite supportive of (though not officially connected with) the Mid-American Baptist Seminary, now located in Memphis, Tennessee, once a center of Landmarkism. In addition, the Baptist Faith and Message Fellowship has created its own series of Sunday School literature and holds its own rallies.

Perhaps some similarity exists between the Baptist Faith and Message

---

[6] A recent scholarly study of Norris is Royce L. Measure's "Men and Movements Influenced by J. Frank Norris" (unpublished Th.D. dissertation, Southwestern Baptist Theological Seminary, Fort Worth, Texas, 1976). A more popular study is E. Ray Tatum's *Conquest or Failure: A Biography of J. Frank Norris* (Dallas: Baptist Historical Foundation, 1966). This is a fascinating book, but unfortunately it is plagued by errors.

Fellowship and the J. Frank Norris movement of two generations ago. Duke McCall has called the Fellowship a new outcropping of Norrisism, both in spirit, method, and ultimate divisiveness.[7]

Understanding Fundamentalism

Most informed people know the basic historical facts about the fundamentalist movement of the 1920's. This was a movement of ultra-conservatives who held certain basic doctrines in religion. They tried without success to take over their denominations, and later formed interdenominational structures. The movement took a body blow in the fiasco of the Scopes Trial in 1925, and went down for the count in the aftermath of the Great Depression. Fundamentalism somehow survived the awful thirties, remained mostly dormant through the forties, and then re-emerged in the 1950's in new disciplined form, with a new name—evangelicalism. Most people know all of that, but what do these facts mean? To that question attention will now be directed.

Many people have attempted to unravel the mystery of fundamentalism. Like detectives seeking clues, scholars have analyzed and dissected fundamentalism in search of its core. Like the blind men examining the elephant, observers do not agree on what they find.

Some interpret fundamentalism merely as a set of religious beliefs. They see it primarily as a religious movement, and find its core in a list of doctrines. One writer who takes this view is Ernest R. Sandeen in his excellent book *The Roots of Fundamentalism*.

Sandeen distinguishes between the fundamentalist movement and the fundamentalist controversy of the 1920's. He says the movement existed before, during, and after the controversy. Sandeen locates the core of fundamentalism not only in theology, but a particular kind of theology—British premillennialism. He traces the rise of British millennarianism, as seen in J. N. Darby and the Plymouth Brethren, and then shows how these viewpoints seeped into the United States through D. L. Moody and the Bible Institute Movement. He sees Darby, the father of dispensational premillennialism, as the archtype fundamentalist. He says, "much of the thought and attitudes of those who are known as Fundamentalists can be seen mirrored in the teachings of this man."[8] To Sandeen fundamentalism and premillennialism are one and the same thing, and he could hardly imagine one without the other.

Winthrop S. Hudson, on the other hand, takes a more sociological approach. In his standard text, *Religion in America*, Hudson interprets fundamentalism as one aspect of the rural-urban conflict in America. While Hudson does not deny that fundamentalism was a religious movement with religious issues, he emphasizes its relation to the secular environment. Hudson describes fundamentalism as an "intramural conflict" which divided Protestantism into two hostile camps as America made the transition from country to city.[9]

Viewed from this perspective, Hudson says, "Fundamentalism can best be understood as a phase of the rural-urban conflict, drawing its strength from the tendency of many who were swept into an urban environment to cling to the securities of their childhood."[10] This interpretation makes fundamentalism at least as much a social as a religious phenomenon. It stresses the environmental and non-religious factors which contributed to its emergence.

---

[7] "Schismatics and Heretics," *The Tie*, 46:8, October, 1977.
[8] Sandeen, p. xix.
[9] Winthrop S. Hudson, *Religion in America* (New York: Charles Scribner's Sons, 1965), p. 363.
[10] *Ibid.*, p. 369.

If Sandeen stresses theological and Hudson sociological factors, others stress psychological factors in explaining the rise and persistence of extremist groups like fundamentalism. In 1950 a group of scholars led by T. A. Adorno published a fascinating book called *The Authoritarian Personality*. While not specifically a study of fundamentalism, many of his findings shed light on that movement. Taking what they describe as a "socio-psychological" approach, Adorno and his associates describe the authoritarian personality as a mind-set firmly opposed to modern complexities. They describe the authoritarian as one who

seems to combine the ideas and skills which are typical of a highly industrialized society with irrational or antirational beliefs. He is at the same time enlightened and superstitious, proud to be an individualist and in constant fear of not being like all the others, jealous of his independence and inclined to submit blindly to power and authority.[11]

According to this interpretation, fundamentalism is more than religious faith. It expresses abberant or even pathological psychology. It represents a certain rigid mind-set quite apart from the content of religious belief.

Still another interpretation of fundamentalism is advanced by Gary K. Clabaugh in his book *Thunder on the Right: The Protestant Fundamentalists*. He suggests that people tend to project their social status anxieties into the realms of religion and politics. This theory holds that people who are upwardly mobile in the small towns and cities are anxious about their status. They oppose anything which threatens or diminishes that status. Such insecure people, frightened about the complexities of modern life and their place in it, form the main recruits for the fundamentalist churches, according to this view.[12]

Richard Hofstadter interprets fundamentalism as another chapter in the persistent anti-intellectualism in American religion. In his masterful book *Anti-intellectualism in American Life*, Hofstadter says that some tension between the head and the heart is characteristic of Christianity through all its history. At times the unlettered have found refuge and perhaps escape in religion as emotion, whereas they were unable to grapple with religion as rationality and creed. Hofstadter defines anti-intellectualism as "a resentment and suspicion of the life of the mind and of those who are considered to represent it; and a disposition constantly to minimize the value of that life."[13]

Others who hold basically the same view consider fundamentalism a continuation of medieval conflict between religion as faith and religion as rationalism. No one can doubt that a strong anti-intellectual element is basic to fundamentalism. Fundamentalists' preference for Bible institutes over colleges and seminaries, their suspicion of professors, and their depreciation of liberal arts education confirm this fear of learning.

Martin Marty has his own interpretation of the fundamentalist movement. He calls it the development of a "two-party system" in American religion. Drawing upon Josiah Strong, he says there are two brands of Protestantism in this country. Their difference is one of spirit, aim, point of view, and comprehensiveness. The one is individualistic; the other is social.[14] From this Marty develops his thesis of a two-party system in American religion. One he calls Private Protestantism, which has affinities with fundamentalism. The other

---

[11] T. W. Adorno and others, *The Authoritarian Personality* (New York: John Wiley and Sons, Inc., 1964), p. ix.

[12] Gary K. Clabaugh, *Thunder on the Right* (Chicago: Nelson-Hall Co., 1974), p. 162.

[13] Richard Hofstadter, *Anti-intellectualism in American Life* (New York: Alfred A. Knopf, 1966), p. 7.

[14] Martin Marty, *Righteous Empire* (New York: Dial Press, 1970), p. 177.

party he calls Public Protestantism, which would have more in common with liberalism.

Some view religious fundamentalism as but one thread in the larger fabric of American society. The extremist mentality certainly stretches beyond religion. The men in Billy Sunday's crusades and in the Ku Klux Klan had much in common, and no doubt often were the same people. Churchly fundamentalism became secularized in the 1950's and 1960's, to be expressed in political and economic extremism. Religious and secular fundamentalism are cut from the same cloth. In his own way Senator Joseph McCarthy was just as much a spokesman for fundamentalism as J. Frank Norris. Often there were connections between church and secular fundamentalism. One recalls, for example, that the John Birch Society was formed in Fort Worth at the First Baptist Church, led by J. Frank Norris. These are just a few of the ways scholars have sought to understand and interpret fundamentalism. There are others.

Perhaps more than doctrine, or sociology, or psychology, or organization, fundamentalism is primarily a spirit and temperament. Fundamentalism is a fighting faith. It is angry, militant, and narrow. Generosity, tolerance, or simple kindness have too often been foreign to fundamentalism and its major spokesmen. The result is that church people and non-church people alike often identify fundamentalism more with meanness of spirit than purity of faith. J. D. Massee, a leader of Baptist fundamentalism, said, "I left fundamentalism to save my own spirit."[15]

Fundamentalism and Culture-Conflict

With this background, yet another thesis will be advanced to help in understanding fundamentalism. This thesis, though by no means new, has never received the attention it deserves. The fundamentalist movement comes into focus and makes sense when viewed as one phase of culture-conflict in America. In times of social stress when the fabric of society threatens to rip apart, extremist thinking comes to the fore.

Usually, the fundamentalist sees little value in the world around him. He does not value its literature or art, and its drama he considers worldly and immoral. To use H. Richard Neibuhr's terms, he sees Christ *against* culture, not transforming culture. Unlike many of his Calvinist ancestors, the fundamentalist is not seeking to build God's kingdom in any sense in this world. He is building God's eschatological kingdom by snatching souls out of this world. This world is a place to fear, deplore, and eventually to escape.

Fundamentalism represents religion under seige, religion on the defensive. Behind its dogmatic affirmations of certainty lurk uncertainty and insecurity. Its pessimistic eschatology mirrors its view of the decay and hopelessness of society. Seeing its own position in the world diminished and threatened, fundamentalism responds with an angry, fighting faith. Seen from this perspective, fundamentalism represents culture-conflict by people who cannot or will not come to terms with society around them.

Basic to this interpretation is the idea that social environment affects religion. This is not to reduce religion to sociology, but it is to recognize that people live in a complex world of two-way streets. Just as religion may affect society, society may help shape religion. One sure way to miss the meaning of fundamentalism is to regard it merely as an eruption in religion and the churches, or a set of

---

[15] C. Allyn Russell, *Voices of American Fundamentalism* (Philadelphia: Westminster Press, 1976), pp. 129, 133.

doctrinal beliefs, and ignore its interaction with American culture of the 1920's, and the 1970's.

Certainly the idea of social forces shaping religion is not new. H. Richard Neibuhr's powerful little book, *The Social Sources of Denominationalism*, was published in 1929. From the New Testament itself one can see that the influx of Gentile converts modified religious emphases, and every observer acknowledges that Hellenism helped shape the ante-Nicene church. One could hardly interpret the medieval concept of God apart from the role of the medieval emperor, and such historic events as the Protestant Reformation, the modern mission movement and the modern ecumenical movement were all influenced by social forces around them. In a similar way fundamentalism was midwifed by social tensions and strains.

No one can doubt that America faced a serious culture crisis in the early twentieth century. World War I catapulted this country suddenly and somewhat unwillingly into the role of world leader. Post-war disappointment and shattered dreams took their toll, though Thorstein Veblen is probably extreme in defining fundamentalism as merely a bad case of post-war nervous prostration.[16] In that era America made vast cultural adjustments by admitting women to suffrage, and eventually to public life. Protestantism fought, won, and lost its greatest moral crusade, prohibition, and from that debacle has never again regained any semblance of its previous moral and spiritual force in this country. Morality underwent significant changes. Women's hemlines, which had swept the ground since Charlemagne's day, suddenly rose to the knees. Women's swimsuits were redesigned to require considerably less than the previous seventeen yards of material. This was the age of the flapper, the age of the automobile, the age of jazz. Some married couples even began to experiment with birth control, despite condemnation by both Protestant and Catholic leaders. One could agree with Willa Cather that "the world broke in two in 1922 or thereabouts."[17]

The decade of the twenties was capped by the greatest depression this country has ever known, a depression which plunged the nation in despair and came close to throwing several denominations, including the Southern Baptist Convention, into bankruptcy. What the depression did not destroy, the dustbowl did. Perhaps Sidney Ahlstrom is right in claiming that the great Protestant cultural synthesis that was put together during the First Great Awakening of the 1720's came apart in the cultural crisis of the 1920's.

In the aftermath of that desperate decade, it was the "hard-times" religion that flourished. Holiness and Pentecostal sects mulitiplied, and militant groups like Jehovah's Witnesses mushroomed. Their rejection of society fit well into the despair of the times, and their militant premillennialism mirrored the spirit of hopelessness in society.

The fundamentalist movement, of course, reached its heyday long before the roaring twenties whimpered to a close. Most observers agree that it lost not only its last great struggle but also its credibility at the infamous Scopes Trial in Dayton, Tennessee, in 1925.

This culture conflict motif shows up in a number of features of fundamentalism. Its militant, fighting spirit certainly reflects conflicts with culture. Even to this day anger, alienation, hostility, and rejection mark this movement. Its adherents seem to have a fondness for military metaphors; they are prepared to "do battle royal," they look for "a fight to finish," and they are always launch-

---

[16] Cited in Carey McWilliams, "The New Fundamentalists," *The Nation*, June 5, 1976, p. 687.

[17] David F. Wells and John D. Woodbridge (eds.), *The Evangelicals* (Nashville: Abingdon Press, 1975), p. 191.

ing some sort of "crusade." In his prayer before the Congress in 1918, Billy Sunday asked God to "bare his mighty right arm" and "smite down the hellish huns." Sunday later said, "I have no interest in a God who does not smite."[18]

Some observers feel this angry militance was and is related to status-anxiety. Many Americans felt threatened. Rural people were swept into the industrial cities, off the land, and in this strange and unfamiliar environment they longed for the old realities. Most historians agree that the fundamentalist movement began in the northern cities, but among people who were basically rural in orientation. As they saw their way of life slipping away, and saw their own place and influence in America diminish, they responded with anger and resentment. Some of this anger was expressed in religious ways, and became fundamentalism. In a paper on "Social Pathology," Franklin Littell describes fundamentalism as a movement "in flight from the perils and anxieties of modern secularization."[19]

Culture-conflict is also evident in the fundamentalist stress on a mentality of one hundred percent certainty. Fundamentalists tended to think in what Clabaugh calls a "Manichaean dichotomy," dividing the world rigidly into "them" and "us." This mentality looks upon the world as an arena for conflict between absolute good and absolute evil; it sees in rigid terms of black and white, right and wrong, true and false. The fundamentalist mentality seemed unable to cope with subtle differences, and compromise was not a part of its vocabulary or practice. Fundamentalists were unable to see any truth in creeds they rejected, they could find no grounds to cooperate with groups they labeled as liberal, and they saw no redeeming features in a society they had rejected. This led to an aggressive dogmatism in doctrine and practices. The fundamentalist may not have been right, but he was never in doubt. He was one hundred percent certain. This excessive dogmatism probably was, and is, a coverup for deep-down, frightening uncertainty.

This dogmatic mentality of one hundred percent certainty may also be responsible for the unshakable reputation for anti-intellectualism which clings to fundamentalism to this day. Hofstadter cites fundamentalism as a primary example of this phenomenon. While anti-intellectualism in religion certainly ante-dates the American experience, it reached a new level of crudity in the fundamentalist crusade of the 1920's. Despite their own notable scholars, such as J. Gresham Machen and B. B. Warfield, many fundamentalists tended to mock scholarship, ridicule the schools, and hold professors in low esteem.

Though militantly anti-intellectual, fundamentalists have been distinguished for founding schools. They may be anti-intellectual and even anti-education, but they support their own training schools with zeal.

Another mark of culture-conflict among fundamentalists is their extreme individualism in ethics and evangelism. Their moral emphasis is almost entirely upon personal morality. They generally oppose social ethics as tending to liberalism. They feel that the best way to deal with culture is to get individuals saved, and saved individuals will help to remake society. Part of this grows out of their millennialism, in which they neither expect nor desire any significant improvement in the world. The fundamentalists' rejection of the social gospel mirrored both their theological and social views. These views allowed Billy Sunday, for example, to refer to social ministries as merely "another pimple on the nose of modernism." Even to this day the Christian Life Commission, the

18 Hofstadter, p. 119.
19 Cited in Clabaugh, p. 139.

social action arm of the Southern Baptist Convention, is one primary target of the extreme conservatives.

A fundamentalist might reject the idea of culture-conflict, and say he was merely trying to defend the total inerrancy and infallibility of the Bible. However, even the militant defense of an inerrant Bible has social overtones.

Still another mark of culture-conflict, and the last mentioned here, is the fundamentalist's extreme pessimism with regard to the world. One vital part of his faith is that the world is done for. It will soon end and ought to. Ours is "the terminal generation." Sandeen has shown effectively that premillennialism is of absolute essence to the fundamentalist faith.

Dispensationalism is a doctrine of defeat and despair, expressing much of the pessimism of American life in the twentieth century. The end of the world has been one of the most persistent motifs across twenty centuries of church history. We may dismiss Mark Twain's laconic response that fewer than half of the predictions of the end of the World have ever worked out. This doctrine has vast social overtones. Whether or not its adherents realize it, apocalyptic millennarianism is one way to reject the world that has already rejected them. By this doctrine they withdraw fellowship from the world and proclaim their conflict with culture around them.

This article has been an effort to understand the underlying meaning of the fundamentalist movement in twentieth-century America. In many ways fundamentalists are no different from other people. They include the good and bad, the rich and poor, rural and urban, moral and immoral. One thing, however, sets them apart from other people. They seem unable or unwilling to relate to the world around them. They are divided, frightened, and insecure. They surround themselves with mental and theological fortresses, imagine themselves to be under steady attack from the world, and gird themselves for a fight to the finish. It would be difficult to imagine fundamentalism at peace and harmony. Conflict with the social environment is fundamental to fundamentalism.

Independent Baptists: From Sectarian Minority to "Moral Majority"

BILL J. LEONARD

Raymond W. Barber, Baptist pastor and president of the Baptist World Fellowship, wrote in 1982, "Fundamentalists have moved out of the storefront buildings on back alleys into beautiful sanctuaries fronting the freeways and boulevards that dissect the nation's biggest cities. No longer do fundamentalists operate from the closet of inferiority, but from the parlor of influence, affecting the spiritual and cultural life of America. The so-called "splinter-group" of yesterday has become a special vanguard of the truth whose influence is evidenced from the courthouse to the White House."[1]

Barber's assessment seems generally correct. Fundamentalists, particularly independent Baptist Fundamentalists, have come of age in Ronald Reagan's America. Sectarian congregations and fellowships long ignored or dismissed as fringe elements by analysts of American religion are bringing their political and moral agenda to the public arena. Recent studies suggest that the "greatest support" for the so-called New Religious Political Right in America comes from "independent Baptist congregations."[2]

The Moral Majority was organized primarily around pastors of local independent Baptist churches. Its founder, Jerry Falwell, stands firmly within that tradition as a graduate of Baptist Bible College, Springfield, Missouri, and pastor of the independent Thomas Road Baptist Church in Lynchburg, Virginia. Thomas Road Church is perhaps the best known of a group of independent congregations related to the Baptist Bible Fellowship, a loose-knit network of churches located primarily in the American Bible-belt, claiming a constituency of 1.4 million people.[3] Sociologist Robert C. Liebman suggests that the Baptist Bible Fellowship and other independent Baptist bodies demonstrate a "distinctive organizational character which made possible the moral majority's mobilization."[4]

---

1. Raymond W. Barber, "Fundamentalism: A Coming Together," *Fundamentalist*, November/December 1982.
2. Samuel S. Hill and Dennis E. Owen, *The New Religious Political Right in America* (Nashville, 1982), p. 15.
3. James O. Combs, "We Will Not Join the Southern Baptist Convention," *Baptist Bible Tribune*, 24 October 1986, p. 3.
4. Robert C. Liebman, "Mobilizing the Moral Majority," in Robert Liebman and Robert Wuthnow, eds., *The New Christian Right* (New York, 1983), pp. 63–66.

Mr. Leonard is professor of church history in the Southern Baptist Theological Seminary, Louisville, Kentucky.

Yet in spite of their prominence in certain New Right endeavors, the independent Baptists generally have been overlooked by students of American religion. Brief analyses of the groups have been offered only by sociologists like Liebman, by denominational "handbooks," in occasional studies of some of their more famous leaders—J. Frank Norris, John R. Rice, or Jerry Falwell—and in histories compiled by their own scholars.

This study represents an effort to identify independent Baptists historically and theologically and to ask certain questions regarding their relationship to American culture. It suggests that independent Baptists have modified their earlier sectarian separatism to varying degrees in an effort to promote their particular moral and spiritual agendas within the broader, and in their view, increasingly humanistic culture. While continuing to use the rhetoric of separatism with their constituency, they have found it necessary to become more "civilized" (to use John Murray Cuddihy's motif) in their response to heretical and worldly outsiders.[5] Their calling to bring moral and spiritual renewal to America sometimes has required them to join forces with individual groups they earlier had declared theologically and ethically corrupt. Given their strong emphasis on separatism, such new alignments may create serious identity crises for the independent Baptist constituency.

These questions provide direction for this article: (1) Who are the independent Baptists? (2) Where do they come from? (3) What do they believe? (4) What are the implications of their transition from sectarian, separatist minority to leadership of an American "moral majority"?

The independent Baptist movement may be described as a collection of fiercely autonomous local congregations, fundamentalist in theology, Baptist in polity, and separatist in their understanding of ecclesiastical relationships. Although most independent Baptist churches are small to moderate in size (100–500 members), the movement often is identified with certain "mega-churches," characterized by huge memberships and presided over by a charismatic senior pastor (often founder) who is the primary authority figure within the community of faith. In 1982, for example, America's largest independent Baptist congregations ranged from the mammoth First Baptist Church, Hammond, Indiana (67,267 members) to the Anchorage Baptist Temple, Anchorage, Alaska (membership 4,800).[6]

Many of these churches maintain a type of denominationalism in miniature, developing local programs and services which mirror those of national denominations. Such congregations frequently establish an extensive system of parochial schools, which may provide education from kindergarten through seminary. They send out their own missionaries and "church planters" for service at home and abroad. They often maintain media

5. John Murray Cuddihy, *No Offense: Civil Religion and Protestant Taste* (New York, 1978), pp. 6–7.
6. "Ten Largest Independent Baptist Churches," *Fundamentalist Journal* 1 (1982): 58.

resources which may include a printing house, radio and television production, and a cassette-tape ministry.[7]

Some congregations originally were related to other Baptist denominations, North and South, but "came out" in response to denominational liberalism, worldliness, compromise, and bureaucratic "hierarchy."[8] Others originated as independent churches. In fact, one of the primary aims of the movement is "church-planting," the founding of new indigenous congregations throughout America in the independent Baptist tradition.[9]

Although most Baptist churches historically have reflected a high degree of congregational autonomy, even independence, the independent Baptist movement is a fairly recent phenomenon. It demonstrates a particular antidenominational, fundamentalist interpretation of Baptist ecclesiology. Eschewing denominational structures as unbiblical, corrupt, and tainted by modernism, many of these churches maintain varying degrees of "fellowship" through such loosely organized groups as the Baptist Bible Union, the World Baptist Fellowship, the Baptist Bible Fellowship, the Southwide Baptist Fellowship, and the General Association of Regular Baptist Churches.

The Baptist Bible Union was organized in 1923 by such prominent Fundamentalists as W. B. Riley, pastor of First Baptist Church, Minneapolis, T. T. Shields, Canadian Baptist pastor, and J. Frank Norris, infamous Fundamentalist pastor of First Baptist Church, Fort Worth, Texas. The Baptist Bible Union, Riley said, was "a continent wide fellowship," founded to give "open, determined opposition to menacing modernism" in the Northern and Southern Baptist Conventions.[10] The union was of "inspirational character" for those who sought to purge liberalism from Baptist denominations.

Today's World Baptist Fellowship is the direct descendant of several groups formed by the charismatic and controversial J. Frank Norris. In 1931 he led his Fort Worth congregation out of the Southern Baptist Convention to found the Premillennial, Fundamental Missionary Fellowship, and later the World Fundamental Baptist Missionary Fellowship (1948).[11] By 1939 he had established a school, the Fundamental Baptist Institute, "to fight liberalism, save souls, train young men and women and turn out pastors,

7. George Dollar, *A History of Fundamentalism in America* (Greenville, S.C., 1973), pp. 218–220.
8. Combs, "We Will Not Join the Southern Baptist Convention"; "Two Notable Conventions in Kansas City," *Searchlight*, 25 May 1923; and R. T. Ketcham, *The Answer* (Chicago, n.d.), pp. 47–49.
9. R. O. Woodworth, "The BBF and Baptist Fundamentalism," *Baptist Bible Tribune*, 3 January 1986, p. 3.
10. William B. Riley, "Why the Baptist Bible Union?" *Searchlight*, 4 May 1923.
11. James O. Combs, ed., *Roots and Origins of Baptist Fundamentalism* (Springfield, Mo., 1984), p. 62; and C. Allyn Russell, *Voices of American Fundamentalism* (Philadelphia, 1976), pp. 37–40.

evangelists and missionaries." The curriculum was based on the English Bible and on a pragmatic approach to "practical Christian work."[12] The institute was founded, Norris said, according to "the New Testament method," by and in a local church, and was funded by local churches. Thus its orthodoxy was "guaranteed."[13]

The Baptist Bible Fellowship was organized on 24 May 1950 as a result of a schism between Norris and certain of his followers over control of the Baptist Bible Seminary. G. Beauchamp Vick, the seminary's president, claimed that Norris was maintaining veto power using an unauthorized set of by-laws. Although organized in Fort Worth, the Baptist Bible Fellowship's geographical center is Springfield, Missouri, where its school, the Baptist Bible College, and its periodical, the *Baptist Bible Tribune,* are located. Fellowship leaders continue to repudiate any effort to become a denominational convention. They insist that the Baptist Bible Fellowship was organized "to preserve the sanctity and sovereignty of local churches and provide an opportunity for local churches to labor together in supporting missionaries and establishing churches."[14] Membership in the fellowship is limited to pastors of local churches.

The General Association of Regular Baptist Churches was founded in 1932, primarily in reaction to liberalism in the Northern Baptist Convention and to the "elaborate machinery" of denominational systems.[15] The association developed numerous independent missionary agencies through which funds are channeled from local churches directly to missionary volunteers. Every agency is independent based on that "thoroughly Baptistic" principle of authority "from the local churches up and not from a convention board down."[16] Its confession of faith is the standard New Hampshire Baptist Confession with a premillennial interpretation of eschatology.

These are but a few of the many groups and alliances of independent Baptists which share numerous similarities. Most were born in reaction to what one supporter called "worldliness, modernism, apostasy and compromise," often under the leadership of a charismatic pastor/authority figure. They generally maintain a loose confederation of local congregations, promoting and funding independent missionaries and "church-planters, periodicals and Bible schools."[17] Sociologist Nancy Ammerman observes that despite their independence, "there is remarkable uniformity among Funda-

---

12. R. O. Woodworth, "Baptist Bible Seminary," *Fundamentalist,* 27 May 1949.
13. *Fundamentalist,* 19 August 1949; and Dollar, *History of Fundamentalism in America,* pp. 33–34.
14. Granville LaForge, "Has the BBFI become Another Convention?" *Baptist Bible Tribune,* 5 December 1986, p. 2.
15. Dollar, *History of Fundamentalism in America,* p. 221; and Ketcham, *The Answer,* p. 47.
16. Ketcham, *The Answer,* p. 49.
17. "Regular Baptists Hold 55th Annual Conference in Grand Rapids," *Baptist Bible Tribune,* 18 July 1986.

mentalist churches" which makes cooperation, exchange of members, and "organizational networks" possible.[18]

What do independent Baptists believe? Frequently, their own self-descriptions typify their theology: "Independent, Fundamental, Premillennial and Baptistic."[19]

Independent Baptists are, above all, Fundamentalists. Raymond Barber writes: "to state the matter succinctly, we are compatriots in the cause of Fundamentalism. We are a living organism helping to shape the cause of Fundamentalism in the twentieth century."[20] Indeed, independent Baptists consistently define their Fundamentalism in terms of the classic "five points": the infallibility and inerrancy of holy scripture; the virgin birth and full deity of Jesus Christ; the substitutionary atonement of Christ; the bodily resurrection of Christ; and the literal, premillennial, second advent of Christ.[21]

By inerrancy they mean the plenary, verbal inspiration of the scripture, "chapter by chapter, verse by verse, line by line, word by word, syllable by syllable, and letter by letter." The Virgin Birth, they believe, was a "supernatural union" resulting in a "supernatural conception that produced a supernatural Savior." Without it, Raymond Barber declares, "Christianity is only a myth that will succumb to its own deception." "Blood atonement" is necessary for human salvation since "apart from the saving efficacy of His blood there is no remission of sin." Likewise, they affirm that Jesus rose from the dead "bodily, physically, really, literally."[22]

The premillennial and, in most cases, pre-tribulational return of Christ remains an important and popular doctrine among independent Baptists. It involves a "rapture of the saints," who go immediately to be with Christ, and, following a period of intense tribulation, the establishing of Christ's kingdom on earth for a thousand years.[23] These five doctrines provide a basis for fellowship and a means of identifying modernism and apostasy.

The departure of independents from Baptist conventions North and South is a direct result of a perceived liberal drift in these denominations and the inability of Fundamentalists to impose the dogma of the five points on denominational confessions of faith. Thus defending Fundamentalist orthodoxy also means attacking modernist heresy. Modernism and its contempo-

18. Nancy Ammerman, *Bible Believers: Fundamentalists in the Modern World* (forthcoming from Rutger's University Press, 1987), p. 21.
19. Ibid.
20. Raymond W. Barber, "Who Are We and Why Are We Here?" *Sword of the Lord,* 24 April 1981, p. 1.
21. Ibid., p. 15; Ed Dobson, "Fundamentalism—Its Roots," *Fundamentalist Journal* 1 (1982): 27; "A Look at Ourselves," *Fundamentalist,* November 1976, p. 2; and Jerry Faiwell, "Why I Am a Fundamentalist," *Fundamentalist Journal* 1 (1982): 6. There are more elaborate lists of doctrines; see "An Affirmation of the Biblical Faith of Baptists," *Sword of the Lord,* 10 January 1964; and *Fundamentalist,* 4 May 1956.
22. Ibid.
23. Ibid.

rary offspring, secular humanism, are continuing objects of independent Baptist denunciation. J. Frank Norris wrote that he broke with the Southern Baptist Convention because "modernism had laid hand on the great body of Southern Baptists and something had to be done. There was no alternative."[24] John R. Rice, publisher of *The Sword of the Lord,* perhaps the most influential independent Baptist periodical, warned against "yoking up with modernists, having modernists on the platform, and calling on them to lead in prayer, and sending 'inquirers' to modernist churches."[25]

Who is a modernist? Any one who denies the five points of Fundamentalism. Rice wrote: "So any time one goes on record denying the deity of Christ, His blood atonement, etc.—we have good reason to know such a person is not saved."[26] Rice's own list of "unsaved" modernists included Bishop Gerald Kennedy, Henry Pitney Van Dusen, Gerald McCracken, even Quaker Elton Trueblood.[27] Rice "renounced denominational ties" with the Southern Baptist Convention in the 1930s when he could not defeat "modernist" leadership and stop convention funds from going to "modernist" seminaries.[28]

These Fundamentalist Baptists also identify themselves as "independent." Independency involves several issues: the sovereignty of the local congregation as the basic source of ecclesiastical authority; the authority of the pastor as the "undershepherd" of the flock of God; a strong antidenominational polemic; and a doctrine of "biblical separatism," involving separation from all sin, worldliness, and compromise with modernism in both the church and the world.

In their view of local church authority and independence, these Baptists reflect the Old Landmark interpretation of Baptist history. Landmark Baptists believe that Jesus Christ established the church as a succession of local congregations which extend from the New Testament era to the present. They reject any concept of a universal or invisible church, believing that only local congregations possess the "marks" of the true church of Christ. Local churches are the source of all ecclesiastical authority and alone fund and authorize missionary activity.[29] This high view of the local congregation has influenced the strong antidenominational bias which prevails among independent Baptists.

J. Frank Norris was a notorious antidenominationalist who challenged the denominational "machine" for years before he finally became an indepen-

24. *Fundamentalist,* 22 October 1948.
25. John R. Rice, "Bible Questions Yoking With Modernists," *Sword of the Lord,* 1 November 1957.
26. Ibid.
27. Ibid.; also G. Archer Weniger, *Sword of the Lord,* 3 January 1964; and *Sword of the Lord,* 10 January 1964, p. 3.
28. Ibid.
29. Norman Wells, "The Battle of the Baptists," *Fundamentalist,* 26 July 1956, pp. 1–2.

dent. He repudiated any suggestion that he would found a new denomination, however, concluding: "That is the curse of the hour now, denominationalism. I have yet to see where there is anything else in the New Testament but churches and how these churches cooperated together, but not one single time was there any overhead, overlord centralized hierarchy."[30]

The distinction between independent and denominational Baptist churches is evident in the following description from the independent Central Baptist Church, Denton, Texas, in 1949. The church claimed to tolerate: no "holy budget," every member canvass, pledge cards, suppers, or bazaars but "scriptural," "voluntary" tithes and offerings; no drives for schools, colleges, or hospitals to educate lawyers, doctors, school teachers or football coaches, "all of which may be good," but are not part of the "business of the church"; no waste of "sacred mission funds" on bureaucratic organization; no "quarterlies" replacing the Bible in classrooms "but a study of the Bible itself," verse by verse, "not by any hop-skip-and-jump" method; no "soft-pedaling" the church's missionary calling; no misinterpretation of the Great Commission; and, finally, no "side-stepping" on premillennial prophecy.[31]

The failure of other Christian bodies to follow that New Testament program and their abiding compromise with modernism influenced one of the most prominent characteristics of independent Baptists, "biblical separatism." Like many sectarian groups, these independents demand separation from any hint of worldliness or moral compromise. John R. Rice was a powerful spokesman for this kind of separation, warning that true Christians should have no fellowship with evil. Rice was convinced that merely tolerating card tables in the home produced gamblers; that "wine at meals" or "eggnog at Christmas" "turned out the drunkard." And, he asked, "How many dances in nice homes, patronized by church members, have aroused lust, seduced innocent youth, set the fires of hell raging in the breasts of manly young men and turned innocent girls into prodigals and harlots?"[32]

Yet moral separation was not enough. Michigan pastor Tom Malone reflected the views of most independent Baptists when he insisted, "We are to be separated doctrinally as well as from the stand point of conduct." He condemned those who "very freely mix and mingle with people who do not believe in the virgin birth of Christ and the entire verbal inspiration of the original writings of the Word of God."[33] Any association with liberals, any tolerance for modernism in denominational life, was unacceptable for true believers.

This doctrine of separation provides the key for understanding indepen-

---

30. *Fundamentalist*, 8 January 1932, p. 1; and J. Frank Norris, "The New Testament Method of Financing," *Fundamentalist*, 4 November 1927.
31. *Fundamentalist*, 13 May 1949, p. 5.
32. *Sword of the Lord*, 30 March 1945; and Norris, *Fundamentalist*, 4 May 1956, p. 3.
33. Tom Malone, "Empty Churches—Why?" *Sword of the Lord*, 31 October 1958, p. 6.

dent Baptists and their relationship to American culture. It is this doctrine, not a hesitancy to speak out on questions of politics or morality, which until recently kept them out of the political mainstream. It is the theory of separation which made them more difficult to "civilize," as Cuddihy calls it. Separation kept them away from the denominationalizing pluralism of American religious life. It is also this doctrine which frequently turned independent Baptists against themselves, as one faction challenged the orthodoxy of another because of the violation of certain separatist norms.

These independents also claim the name Baptist and identify their heritage with that historical and ecclesiastical tradition. They reflect what might be called a type of Baptist scholasticism, born of certain legalistic, propositional segments of the diverse Baptist heritage. In many respects, twentieth-century independent Baptists are heirs of earlier eighteenth and nineteenth-century landmark, successionist, antimission factions within Baptist life.[34]

Most independents believe that Baptists represent true New Testament Christianity preserved since the first century. One 1955 apologetic states it succinctly: "Baptists are not Protestants. They never belonged to the Roman Catholic institution and never came out of it. Baptists existed seven hundred years before the papacy, and much more than that before Protestants."[35] Noel Smith, longtime editor of the *Baptist Bible Tribune,* insisted that independent Baptists are the true representatives of the historic Baptist heritage. He wrote, "We have a continuity of doctrines, principles, and practices that go back to the apostolic age. Our continuity is the longest of any Christian group in the world."[36]

While affirming the standard litany of "historic Baptist principles"— believers' church, baptismal immersion, missions, religious liberty, and freedom of conscience—independent Baptists interpret those doctrines in light of fundamentalist dogmatism and independent/separatist ecclesiology. Thus they represent the scholastic wing of the broader, more diverse, Baptist heritage. Theirs is a doctrine informed by Calvinist, Arminian, and Land-mark theologies but united in a common fundamentalist and separatist ideology. Thus Pastor George Norris declared in 1956 that the term Fundamentalist Baptist is synonymous with orthodox Christian. "In fact," he wrote, "one cannot truly be a Fundamentalist without being a Baptist with just as much emphasis on the 'Baptist' end of the name as on the 'Fundamentalist' end."[37]

Given these doctrinal attitudes, early independent Baptists occupied a

---

34. Robert A. Baker, *The Southern Baptist Convention and Its People, 1607–1972* (Nashville, 1974), pp. 150–159.
35. "John's Baptism," *Fundamentalist,* 10 June 1955.
36. Noel Smith, "The Dignity and Relevancy of the Baptist Bible Fellowship," *Baptist Bible Tribune,* 12 September 1986, p. 3.
37. George Norris, *Fundamentalist,* 19 July 1956, p. 3.

paradoxical position in relation to American culture. On the one hand, they claimed the dissenting tradition of the Baptists, speaking out against those issues which they believed contrary to scripture, conscience, and Christian morality. On the other hand, their concern for doctrinal order and authority made them sound like an establishment, protecting, enforcing, and maintaining the Fundamentalist status quo against various religious minorities (Catholics, Mormons) and against challenges to their concept of the American way of life. Their vision of world evangelism and Christian commonwealth created a tendency toward cultural domination, a call to subdue the land for Christ. But their separatist fear of being tainted by the world kept them from involvement in compromising political alliances. They gloried in their remnant status but longed for a day when they would evangelize the world into the image of Christ's kingdom. Premillennialism made them hesitant to change the world; evangelical conversionism compelled them to try. In this attitude they reflect George Marsden's contention that Fundamentalists show that "paradoxical tendency to identify sometimes with the 'establishment' and sometimes with the 'outsiders.' "[38]

Independent Baptist preachers have been outspoken in matters of politics and public morality since the beginning of their movement in the 1920s and 1930s. Their approach to politics and morality may be summarized as follows.

First, they insist that the primary mission of the church "is not to 'reform the world,' but to preach and teach the Gospel of salvation to each individual soul."[39] Political involvements are secondary, sometimes incidental, to evangelism.

Second, they have not hesitated to address such issues of personal and public morality as alcohol, divorce, sexual immorality, card playing, gambling, smoking, and movies.[40] In this area they freely assume the prophet's role in denouncing the sins of the age.

Third, it is in the role of jeremiad that independent Baptist preachers have addressed political issues. They view America as a chosen nation, blessed by God, but, like ancient Israel, in constant danger of divine retribution because of immorality within and the forces of anti-Christ (Romanism, Communism, Liberalism) without. John R. Rice declared that "when moral questions enter into politics, the preacher ought to express himself and help people to know what is right." Such matters, Rice believed, involved opposing the New Deal, the New Frontier, socialism, welfare, racial intermarriage, and civil rights legislation.[41]

38. George Marsden, *Fundamentalism and American Culture* (Oxford, 1980), pp. 6–7.
39. *Fundamentalist*, 4 May 1956; also John R. Rice, "Moral Principles in National Politics," *Sword of the Lord*, 24 July 1964.
40. John R. Rice, "Are Christian Films Wrong?" *Sword of the Lord*, 8 February 1952.
41. Rice, "Moral Principles in National Politics," p. 7.

As early as 1922, J. Frank Norris was uniting politics and religion in his attack on Roman Catholics, declaring that "in the name of the American Flag and of the Holy Bible I defy the Roman Catholic machine of New York."[42] In 1937 Norris charged the Federal Council of Churches with "communistic connections," insisting that its purpose "is to overthrow the present government" while denying "the fundamentals of Christian religion."[43]

Rice wrote in 1952 that "every Christian American" should work to free the country from "the wicked, corrupt Democratic administration." "A Republican President" was the only chance to make America safe again. Rice endorsed Senator Robert Taft, not Eisenhower, for the presidency.[44] In the 1964 presidential election, Rice declared that the Democratic party "is against fundamental Christianity and fundamental Americanism" due to its support for Russia, Cuba, and the United Nations, and its encouragement to lawlessness and civil disobedience by "negroes and carpetbagger whites." He noted that "the better informed fundamentalist Christian leaders generally feel as I do, that they ought to support Sen. Goldwater for President."[45] By 1968 Rice was supporting "the convictions and character of Gov. George Wallace" but did not think he could be elected. If not Wallace, then one should support Nixon, Rice declared.[46]

The civil rights movement, integration, the Supreme Court rulings on prayer and abortion, rising divorce rates, and accompanying social upheavals set the stage for the increasing involvement of independent Baptists in the public arena. While they repudiated the participation of liberal churches in civil and political matters, they expressed increased concern for the decline of American morality. During the 1960s and 1970s independent Baptist periodicals were filled with editorials, sermons, and articles which respond to civil rights issues, civil disobedience, the expanding role of the federal government, and the assassinations of the Kennedys and Martin Luther King.

While deploring the murders, many preachers suggested that God "permitted" the Kennedys' assassinations due perhaps to judgment on their "liquor-selling" father, their socialistic tendencies, the sinfulness of a "communistic" assassin, and their efforts to build a political "dynasty."[47] Rice

42. *Searchlight,* 14 April 1922; and "Six Reasons Why Al Smith Should Not Be President," *Searchlight,* 18 November 1927.
43. Norris, "Federal Council of Churches Brought Out Into the Open," *Fundamentalist,* 8 January 1937; and *Fundamentalist,* 22 January 1937.
44. *Sword of the Lord,* 18 July 1952.
45. Rice, "Why We Will Vote for Goldwater," *Sword of the Lord,* 16 October 1964.
46. *Sword of the Lord,* 26 January 1968, p. 6.
47. "Why Did God Allow Kennedy's Death?" *Sword of the Lord,* 24 January 1964; "What Was Back of Kennedy's Murder?" ibid., 31 January 1964; Tom Malone, "What Means the Death of President Kennedy?" ibid.; Rice, "Senator Robert F. Kennedy's Death," ibid., 28 June 1968.

wrote that if Martin Luther King could promote violence among negroes, why should not Arabs feel the same way? "Is it really any worse for an Arab to shoot at Senator Kennedy, than for a negro in Chicago to shoot a policeman?"[48] Most writers concluded that Martin Luther King died, as one sermon title declared, "by the lawlessness he encouraged."[49] Noel Smith, editor of the *Baptist Bible Tribune,* charged that King, though a Baptist pastor, was "not a Christian at all."[50]

Independent Baptists have not hesitated to speak out on political issues, usually reflecting sectarian, right-wing positions in American religio-political life. They willingly addressed such questions, but because of their strong separatism were reluctant to become involved in the broader political context lest they compromise their distinctive witness. In entering that arena they were forced to reassure their constituents that supreme moral principles were at stake and that the organizations to which they related were political, not ecclesiastical.

In 1980 Rice's associate editor, Curtis Hutson, defended his decision to join the Moral Majority by insisting that it was a political, not a religious, organization and therefore involved no violation of the principle of biblical separation. He insisted, "I would not knowingly yoke up with a modernist or liberal in a religious movement."[51] But involvement with liberals or unbelievers in a nonreligious movement was not a compromise of the separation doctrine. Such a position represents a major reinterpretation from earlier days.

Independent Baptists also were pushed toward such rightist political alignments because they were courted by right-wing political action leaders and because they believed that American society was near total moral collapse. Pastor Gerald Thompson is among those who catalogued the moral depravity of America in the 1980s. The sins he lists include: abortion, the public school system, criminal violence, sexual immorality, drinking, homosexuality, drugs, rock music, adultery, fornication, communism, divorce, "and on and on and on." "America," he concluded, "has lost a consciousness of sin."[52]

Innumerable studies of Fundamentalist involvement in the New Religious Political Right document the fact that during the 1980s diverse elements within the movement coalesced in ways not known before. Sources within independent Baptist life illustrate and corroborate that thesis. The right-wing Republican leaders moved to bring independent Baptists into the party as a

48. Rice, "Senator Robert F. Kennedy's Death."
49. Bob Spencer, "Dr. Martin Luther King Died by the Lawlessness He Encouraged," *Sword of the Lord,* 14 June 1968.
50. Noel Smith, *Baptist Bible Tribune,* 23 April 1965; and Archer Wenger, "The Faith of Martin Luther King," *Sword of the Lord,* 6 September 1968.
51. Curtis Hutson, *Sword of the Lord,* 17 October 1980.
52. Gerald Thompson, "The Real Child Abusers," *Sword of the Lord,* 29 August 1980.

new voting block through organizations such as Moral Majority and Religious Roundtable.[53]

Independent Baptist leaders, long outspoken on the total depravity of American cultural life, became convinced that they could effect a revival of morality and were compelled to do so in order to save the Republic. Millennial expectations, while affirmed rhetorically, were no longer reason enough to withdraw from the arena of public morality and civil order. In a sense, Jesus may have tarried, but independent Baptists did not, in their efforts to influence social and political renewal in American life. In so doing, independent Baptists have maintained the rhetoric of sectarian separatism while increasing their involvement with certain groups (Catholics, Mormons, charismatics) whom they earlier repudiated as apostate.

In this shift, Jerry Falwell and his colleagues have led the way. Articles in Falwell's *Fundamentalist Journal* consistently redefine separatism in ways which Norris, Rice, Noel Smith, and other early independent Baptists surely would have repudiated. Falwell himself insists that he is a biblical separatist. Yet he also acknowledges that in his earlier days he was "somewhat insulated and isolated by my Fundamentalist teaching." This separatism led him to oppose such things as anthems and choir robes in worship, beards and long hair on men, and make-up and slacks on women as signs of liberalism and moral compromise. Now he insists that loyalty to the fundamentals of the faith may permit moderation on such lesser issues. He concludes, "let us not be known for how many believers we have rejected or how many extraneous issues we have added to the gospel."[54]

Others of Falwell's staff defend such moderate separatism by insisting that true Fundamentalists are best identified by their loyalty to the historic "five points": infallibility of scripture; virgin birth; substitutionary atonement; bodily resurrection; and literal second advent. A "pseudo-Fundamentalist," therefore, "is one who *subtracts from or adds to* these fundamentals."[55] Those who develop seven, ten, twenty, or more fundamentals are as guilty as those who deny the five. "Anyone who demands more is denying the historic roots of the movement."[56] They also warn that separation has led independent Baptists to "isolation," negating their influence and effectiveness in the world. "On the doctrinal absolutes of the faith (the 'fundamentals') there can be no compromise." Other issues of "personal preference" need not divide the faithful.[57]

53. Jerome L. Himmelstein, "The New Right," in *The New Christian Right*, pp. 13–30; and James L. Guth, "The New Christian Right," in *The New Christian Right*, pp. 31–45.
54. Jerry Falwell, "Moderation: A Biblical Command," *Fundamentalist Journal* 2 (1983).
55. Ed Dobson and Ed Hindson, "Who Are the 'Real' Pseudo-Fundamentalists?" *Fundamentalist Journal* 2 (1983): 10–11. Emphasis mine.
56. Ibid.
57. Dobson and Hindson, "Guilt By Association or Burned by the Second Degree," *Fundamentalist Journal* 2 (1983).

But unity, even cooperation, remains elusive for independent Baptists. One pastor wrote recently: "The Jews stand together, the Catholics stand together, the Modernists stand together, but not Fundamentalist Baptists." The result, he concluded, is "fanaticism," egotism, and "ambitious, independent, self-centering" movements.[58]

Independent Baptists are facing a dilemma: how to work together to secure their social and political agenda while remaining true to their most distinctive traditions, fundamentalist doctrine and ecclesiastical separatism. At this time, conclusions regarding the outcome involve the following questions and observations.

First, might we describe early independent Baptists as the "theological Amish" of the Baptist tradition? They, like other Fundamentalists, in Martin Marty's words, "encountered modernity, did not like what they saw and regrouped or refashioned their faith."[59] The early independents believed that fundamentalist faith could be preserved best by means of the ecclesiological and ethical separatism characteristic of nineteenth and early twentieth-century Baptist sectarians. Much of the Baptist identity they seek to preserve is of relatively recent vintage. At the same time, they increasingly have reflected a technological modernism in their willingness to utilize numerous modern methods, particularly the media, for propagating their faith and organizing their churches. Has that tendency contributed to a kind of anti-worldly worldliness within their fellowships?

Second, independent Baptists are experiencing a significant transition in their churches and fellowships as well as in their response to "the world." Jerry Falwell recently acknowledged this transition when he wrote that the "Patriarchal Period" of the modern Fundamentalist (Baptist) movement is over. The controversial and authoritarian founder-leaders (Norris, Rice, and others) are "gone-forever" and cannot be replaced. The movement is now in its "Pastoral Period," Falwell observed, which means that "no single individual will ever again speak for all of us. . . . Every pastor is 'a voice for Fundamentalism.' "[60] Clearly, many of these pastors are seeking ways to work together as a powerful political, even ecclesial, coalition while avoiding the appearance of denominational affiliation and compromise with the "civilizing" tendencies of American religious pluralism. In light of this transition, are the independent Baptists already a fundamentalist nondenominational denomination? Is their entrance into the public square a brief incursion or a sustained assault? Many of the politicians who once encouraged them are now somewhat less than supportive.

58. R. O. Woodworth, "The BBF and Baptist Fundamentalism," *Baptist Bible Tribune,* 3 January 1986, p. 2.
59. Martin E. Marty, "Modern Fundamentalism," *America,* 27 September 1986, p. 134.
60. Jerry Falwell, "Who is the Voice of Fundamentalism Today?" *Fundamentalist Journal* 3 (1984): 8-9.

Finally, while they may have modified in their separatism, independent Baptists are by no means moderate in their theological, moral, or political agendas. They continue to affirm fundamentalist ideology, rejecting apostasy and modernism wherever it may appear. Thus their dilemma. A people schooled in separatism may find it difficult to follow leaders who push them into too many uneasy alliances with apostasy. Compromise, even for political and moral ends, is difficult for those who have been ta ight that all compromise is equivocation from the truth. At present, "moderates" like Falwell must maintain a fragile contract with their own constituency. They must remain sectarian at home while eating and drinking with assorted Republicans and sinners in order to accomplish certain political and social goals. Falwell's recent involvement with the so-called PTL scandal is a case in point. Will his leadership of a charismatic media movement enhance or undermine his influence among independent Baptists? Can he continue to call himself a "biblical separatist" while closely associating with well-known charismatic Christians?

Independent Baptists remain an important contemporary phenomenon in American life. Their continuing struggle with modernity makes them an important case study in American civil and "civilizing" religion.

FUNDAMENTALISM AT HARVARD:
THE CASE OF
EDWARD JOHN CARNELL

RUDOLPH L. NELSON

Why would a promising Fundamentalist student choose to study at that citadel of Liberalism, Harvard Divinity School? The answer to this question suggests that, in theological education, differences in religious ideology do not necessarily pose insurmountable barriers to learning.

In March, 1940, Willard L. Sperry, dean of Harvard Divinity School since 1922, referred in a "Dean's Letter" in the *Divinity School Bulletin* to the recent appearance at Harvard of a type of student quite new to that institution—"a man who has already had one theological course in a conservative-to-fundamentalist seminary, and who is now anxious to begin all over again another three years of theological re-education."[1] It is not clear exactly how many such students there were. Dean Sperry does number "a half-dozen in the three regular classes" but is vague about those on upper levels ("to say nothing of the Graduate group"). More important, he seems to have some anxiety as to how the divinity school constituency is going to react to this new development. Acknowledging that these men present an academic problem "in that they have seldom had the four years of regular college work, which we technically require for admission here," he points out that some of them have come with an A.B. in theology. And whatever the value of that degree, they do

Rudolph L. Nelson is assistant professor of English at the State University of New York at Albany.

have certain qualities which highly commend them: "They already know their way around the major biblical and historical fields and their store of relevant facts-in-advance, awaiting reinterpretation, is much in excess of that of the ordinary college graduate." The experience with these men, he says, has on the whole been encouraging.

It was only the beginning. For whatever reasons, Fundamentalists kept on applying to Harvard Divinity School, and through the 1940s and 1950s the school kept accepting them, giving them what Dean Sperry called a "theological re-education" and sending them out with baccalaureate and graduate degrees. While classification by theological label is always suspect and especially hard to decipher in a rear-view mirror, the number of self-acknowledged Fundamentalists who were granted Th.D.'s or Ph.D.'s from Harvard in the remaining thirteen years of Sperry's deanship was at least twelve. Two others who received their degrees in the 1960s took most of their classwork within that earlier time. In addition, perhaps another ten or twelve men of similar background and conviction were matriculating at the divinity school during this time, working at the master's or bachelor's level or subsequently transferring to other institutions. And although these figures may not at first seem significant, it should be noted that the average number of annual graduates from all programs at the divinity school during the 1940s was less than twenty.

These were not barbarian hordes overrunning the citadel of learning. The roster of Harvard Fundamentalists now reads like an honor roll of mid- to late-twentieth-century American evangelicalism,[2] including names like Edward John Carnell, philosophical apologist and late president of Fuller Theological Seminary (Th.D., 1948); Kenneth Kantzer, theology professor at several evangelical colleges and seminaries before becoming editor of the movement's unofficial house organ *Christianity Today* (Ph.D., 1950); Merrill Tenney, long-time dean of Wheaton College Graduate School (Ph.D., 1944); John Gerstner, church historian at Pittsburgh Theological Seminary (Ph.D., 1945); Harold Kuhn, professor of philosophy of religion at Asbury Theological Seminary since 1944 (Ph.D., 1944). No fewer than three current Fuller Seminary faculty appear on the list: theologian Paul King Jewett, interpreter of the works of Emil Brunner and author of the controversial *Man as Male and Female* (Ph.D., 1951); George Eldon Ladd, author of several books on the kingdom of God and moderate evangelical voice in the sensitive area of New Testament criticism (Ph.D., 1949); and Provost Glenn Barker, a New Testament scholar who was granted his Ph.D. in 1962 but did most of his class work within the period on which we are focusing. The remainder of the list: Burton Goddard, professor and dean emeritus, Gordon-Conwell Theological Seminary (Th.D., 1943); Roger Nicole, professor of theology at Gordon-Conwell (Ph.D., 1967); Samuel Schultz, professor of Bible at Wheaton College (Th.D., 1949); George Turner, professor of biblical literature, Asbury Theological Seminary (Ph.D., 1946); J. Harold Greenlee, formerly professor of Greek at Asbury, now with OMS International (Ph.D., 1947); Jack P. Lewis, professor of Bible at Harding Graduate School of Religion (Ph.D., 1953); Lemoine Lewis, professor of Bible and church history at Abilene Christian University (Ph.D. granted in 1959 but all class work done in 1940s).

Almost certainly, the name on this list best known within the framework of American Christianity is Edward John Carnell. Author of eight books, several of which together constitute a major effort toward a conservative Christian apologetic for our time, and president of Fuller Theological Seminary in the crucial five-year period when it gained accreditation, Carnell gained further credibility as a spokesman when in the late 1950s Westminster Press, having conceived the idea of a trilogy of theological books from three different perspectives, selected Carnell to write the book representing conservative theology.[3] And although Carnell was far too much of an individual to be designated as typical, his experience at Harvard does shed light on the experiences of others.

What brought a young man like Carnell to Harvard in the first place? He seemed an unlikely Harvard applicant, as the son of a midwestern Baptist minister whose only post-high school training was two years at Moody Bible Institute, and as a graduate himself of Wheaton College in Illinois, which, despite its reputation for academic respectability, was a fortress of Fundamentalist theology and moralistic pietism. So too with Kantzer, Tenney, Ladd—all of them unlikely Harvard men. What lay behind this curious symbiotic relationship between students and an institution theologically so far apart?

Let's look at the institution first.[4] Whereas in the minds of most American church people Harvard Divinity School in the 1940s was generally thought to be affiliated with the Unitarians, its denominational status was by that time a thing of the past. In 1879 Dean Francis Greenwood Peabody had argued before the board of overseers for an unsectarian school whose system would be determined by belief in "sound methods, broad knowledge, and quickened interest." Noting that this kind of school or faculty had long been known in Germany, he felt that Harvard Divinity School should be such a school in the United States. In its undenominational classification, Harvard was to be virtually alone in this country for some time; as Levering Reynolds points out, the *New Schaff-Herzog Encyclopedia of Religious Knowledge* in 1911 lists only Harvard as undenominational, "although Union Theological Seminary in New York could have fairly laid claim to the distinction." More significant, however, is the educational philosophy and methodology that underlay Dean Peabody's pioposal to the board. His stress on "scientific" study essentially meant the historical method, an approach which remained in effect at the divinity school through the era which we are discussing. Of six new faculty appointed between 1880 and 1883, five had studied in Germany. Their influence was decisive in making Harvard Divinity School "the most advanced expression of the liberal religio-historical Protestant scholarship of which Germany was the homeland, but in which most of the leading scholars in all major American Protestant seminaries participated."[5]

Although for a time the institution forged ahead, the

second quarter of the twentieth century proved to be a time of trouble. For one thing, the president of the university from 1932 to 1953, Dr. James B. Conant, a scientist whose broader interests were in general education both within the university and in the society at large, had little interest in and gave little support to the divinity school, since he believed theology was a divisive force in education. Another problem was the depressed state of American religion in general. Historian Robert Handy has referred to the years 1926 to 1935 as "the American religious depression," a period when, in all but the rural areas of the nation, the Protestant churches had lost most of their influence.[6] Under the Conant administration, needing students to justify its very existence, Harvard Divinity School showed no substantial increase in student enrollments. Like so many of America's factories operating at less than full capacity through the years of economic depression, the divinity school, with its impressive array of learned scholar-teachers, was falling far short of its potential contribution to the university, the church, and society. When interest in theology did begin to awaken, it was a Barthian or Niebuhrian Neo-orthodoxy which carried the day, at a considerable remove from Harvard historicism, and the seminaries which were more favorably oriented ideologically to this position tended to attract the growing numbers of theology students. In short, Harvard Divinity School needed students. In the lean years of the late thirties and early forties, applications from a new potential reservoir of students, even from within the ranks of Fundamentalism, were not to be dismissed out of hand.

What about those Fundamentalist students? What were *their* motives?[7] In one sense we hardly need ask. Young men who had progressed far enough on the academic ladder to contemplate seriously graduate education could hardly have avoided at least thinking of Harvard. Why not the best? Or at least what they perceived to be the best. Some, like Merrill Tenney, had additional personal reasons, since Harvard was conveniently near Gordon College and divinity school, where Tenney taught. Several found Harvard's offer of scholarship money decisive. At a deeper level there had to be mythic overtones—encountering the Beast of Scholarly Unbelief in its own labyrinth and emerging with new confidence and new powers. As Roger Shinn remarked, reviewing a subsequent book by Carnell, "In some circles he is described, somewhat glibly, as one of the new generation of brainy fundamentalists who have studied at Harvard in order to learn the arguments they will spend the rest of their lives attacking."[8] Shinn is correct in his insistence that the description is less than fair, for the mythic truth which underlies the statement has been too easily parodied. But it was a factor. John Gerstner of Pittsburgh Theological Seminary says: "I went there because it was ultraliberal and academically competent, desiring my conservatism to be put to its test." And Samuel Schultz of Wheaton: "Although I was quite fully informed about the perspective in the Harvard Divinity School prior to enrollment, I valued this opportunity of first-hand exposure to a naturalistic viewpoint on the Bible which confronted me with a sincere examination of my approach to biblical scholarship."

There is another, more significant, perfectly sensible reason why young Fundamentalist scholars chose Harvard in the 1940s. The divinity school's thorough dedication to historicism as its educational philosophy and methodology is the decisive fact. The institution's position at the far left of the theological continuum—perhaps better described as a refusal to assume a position—not only was not a deterrent to the prospective student coming all the way from the right end of the continuum but was its most attractive characteristic. The young Fundamentalists were convinced that if they went to the nation's prestigious Liberal and Neo-orthodox schools of religion they would be pressured to conform to a certain theological mold. At Harvard they knew that they could continue to keep the orthodox faith at the same time they devoted themselves to a historically oriented, theologically neutral program of study. All that Harvard would ask of them was academic excellence. It was a fair bargain all around.

In order to bring into sharper focus Edward Carnell's personal decision to attend Harvard and to illuminate what he did when he got there, it is important to examine a chain of decisions extending back to the summer of 1937, when he sent in his application to Wheaton College in Illinois. Many of Carnell's classmates at Wheaton chose the college because, within Fundamentalist groups across the entire country, it had the reputation (somewhat overinflated) of providing the very best Christian college education available anywhere. As a member of a Fundamentalist Baptist church in Albion, Michigan, geographically well within Wheaton's primary range of influence, Carnell knew about that reputation. But as a high school student with a record that was mediocre at best, he simply didn't care. It was not until he discovered that as the son of a Baptist minister he would be eligible at Wheaton for a partial tuition scholarship that he allowed himself to be persuaded to apply at the last minute.

Getting in proved easier than staying in. Even in those days a hundred-dollar scholarship did not go far, and he found it necessary to work an average of fifty hours a week in the college dining halls. But Carnell found himself intellectually at Wheaton, discovering a love and capacity for learning that set the direction for the rest of his life. The evidence from his Wheaton transcript clearly indicates that it was philosophy which captured his mind. While he was a good enough student to be graduated with honors in 1941, he ranked no higher than forty-eighth in a class of 205. But his grades in philosophy courses were outstanding. The faculty member most responsible for igniting Carnell's intellectual fires was Gordon Haddon Clark, who had come from the University of Pennsylvania in 1936 as visiting professor of philosophy and joined the permanent faculty the next year. Clark was a powerful force on campus. In a school that often seemed to place more importance on chapel sessions and semi-annual evangelistic campaigns than on academic concerns, Clark held out for the primacy of the classroom. Many Wheaton alumni, even some who in subsequent years have strayed far from Clark's philosophical rationalism and dogmatic theology, still think of him as the one person most responsible for rousing them from intellectual slumber. In Carnell's case,

Clark was to remain, in varying ways, a determining influence in his life. Unquestionably, the decision to apply to Harvard—to settle, that is, for nothing less than what he perceived to be the best—was attributable in large part to standards inculcated by Gordon Clark.

But there was also something in Carnell's own nature which made him peculiarly receptive to Clark's rigorous academic ideals—a cluster of traits which add up to a tendency towards perfectionism. Partly this was a reaction against the emotional stigmata of his Fundamentalist background. Having moved during the twenties and thirties from place to place across the Midwest, knowing always that wherever they went his father, with his two years of education at Moody, was at the bottom rung of the local ladder of clerical prestige, Carnell entered college with a strongly negative image of himself as a member of a more or less despised minority. When, under Clark's tutelage, he began to see Christianity as intellectually respectable, he was appalled at the gulf between his new understanding of the historic Christian faith and the severely limited version of it represented by his father. One can easily imagine him convincing himself that he would never get a fair hearing for the Christian faith unless he could display academic credentials second to none. Only some such idealistic vision as that can explain the commitment necessary for Carnell to have earned in succeeding years a doctor of theology degree from Harvard and a doctor of philosophy degree from Boston University.

However, on this academic pilgrimage, before Cambridge and Boston there would be Philadelphia. Before Harvard, there would be Westminster Theological Seminary. During the spring semester vacation of his senior year at Wheaton, Carnell went on a 1700-mile hitchhiking tour which took him to four seminaries on the eastern seaboard: Eastern Baptist, Princeton, Westminster, and Faith. The first two, of course, were affiliated with major denominations; Westminster (in suburban Philadelphia) had split off from Princeton in the late 1920s as a result of internal Fundamentalist-modernist dissension among the Presbyterians; and Faith Seminary (in Wilmington, Delaware) came into being as a result of still further dissension in that extended separation process.

Carnell's letter, in late April, 1941, accompanying his application for admission to Westminster, clearly defines goals much more likely to be met at Westminster than at Harvard:

I feel, after sitting in on the classes of the various schools, that you offer the most scholarly defence of the Gospel of Jesus Christ, and so I am prepared, if I am accepted, to join with the student body and faculty at Westminster to fight against all forms of anti-Christian systems, and to preach to the world as consistent as possible, the Whole Council (*sic*) of God.

But Carnell did not feel completely at home in the Westminster atmosphere. It seems evident that he went there hoping to get a thorough grounding in traditional Calvinistic theology, and there is no reason to believe that in that respect he was disappointed. Academically he did well, and was graduated after three years with bachelor's and master's degrees in theology and the William Brenton Greene, Jr., prize in apologetics. But along with these came a burden of dissatisfaction with what seemed to him a lifeless orthodoxy. A friend from Wheaton, Jim Tompkins, also a Westminster student, recalls that they both visited and considered transferring to Eastern Baptist.

The warmth and enthusiasm of Wheaton was so noticeably absent at Westminster that we really felt isolated. What we seemed to be searching for was the theological rigor of Calvinism joined to the spiritual exaltation of Fundamentalism. Eastern was theologically superficial; Westminster was spiritually dead. With such a choice, we settled for the corpse.

Even in ideological matters, especially apologetics, Carnell was increasingly disenchanted with Westminster and spoke in later years of having rejected Cornelius Van Til, the seminary's leading force in theology and apologetics.

In the spring of 1944 he sent a letter to Harvard's Dean

Sperry outlining an entire Ph.D. program in history and philosophy of religion. It reads in part:

I wish to relate very specifically the question of epistemology to each of the six fields in which I shall engage myself in study. The first three fields, which are required, will provide me with an opportunity to apply methods of epistemology to Christianity and to some other chosen religion through a study of the contents and literary history of each. Also, I shall be able to do research work in the abstract question of the methods of knowledge and truth when related to religion and religious convictions. This probably will constitute my critical study of a phase of Philosophic Thought. The fourth field will be in the field of mysticism, conversion, and religious experience, a branch of epistemology viewed subjectively, a field which has always intrigued me because of the discrepancies in convictions, yet all appealing to objective reality. The fifth field will be in the study of ethics, its demands and relation to religion; yet always I shall tease out the theory of knowledge which each theory appeals to for its basis and verity. The last field will cover the problem of metaphysics involved in a religious view of the world. This, finally, will bring the theory of knowledge into relation with science and other phenomena of reality, and religious convictions to the structure of the objective universe. Thus, through this tentative schedule, I expect to apply my time and talents to this one critical problem as it manifests itself in these six fields of study.

One cannot help but be impressed by the contrast between this letter and the one three years earlier to Westminster. The difference in tone and substance cannot be fully explained by the normal intellectual growth of a graduate student or by an appropriate switch in rhetorical strategy. The writer is like an aircraft about to take off from a carrier deck, its engine at high pitch, waiting for the restraining wires to be released. The admissions committee, however, was not overwhelmed. A letter in Carnell's permanent Harvard file states the committee's agreement that he should be admitted but eased out at the end of the first semester if he did not measure up. There is no evidence to suggest the question was even raised again.

Whereas Harvard was as rigorous academically as the Fundamentalists expected, its atmosphere was more benign than they could have hoped. ''I think the conservative went to

Harvard," says Jack P. Lewis, "expecting to meet the Devil
and instead encountered gentlemen of the highest character
who were far kinder to *us* than *we* would have been to them
had the case been reversed." More than any other faculty
member, Henry J. Cadbury, New Testament scholar and
Hollis Professor of Divinity, stands out in their minds as the
exemplar of rigor and fairness. Glenn Barker determined
early in his own teaching career that he would always want to
treat those who differed from his own theological position
with the same complete respect showed toward him by
Professor Cadbury. Referring to Cadbury as "a man of complete intellectual honesty," Carnell recalled how he would
chide those nonfundamentalists in his class who were handling
the Greek text irresponsibly. "We may not agree with Paul,"
Cadbury would insist, "but let us at least be honest with what
he says." Cadbury was memorable also for his Socratic teaching
method, for which the Fundamentalists were a perfect
pedagogical foil. He seemed better able to provoke confrontation with the issues he wanted to pursue if some of them were
present in class, and if they were slow to join the interchange he
was not the least bit reluctant to ask one of them for the
Fundamentalist view on a given question. And he was not
above a playful jab now and then. "If you have praying
mothers," he would say at the beginning of a class in which he
would be taking a radical position on a sensitive issue, "they'd
better be on their knees this morning."

But this study is concerned with more than nostalgic
memories of revered teachers and the usual triumphs and
tragedies of graduate school days—sacred detritus lodged
permanently in the lore and legends of all alumni of all
graduate schools. Something more important was happening
to the Harvard Fundamentalists. It must not be overlooked
that this group of students inherited all the emotional wounds
of the Fundamentalist-modernist wars of the 1920s, one result
of which was a tendency in Fundamentalist ranks to discredit
the value of the free pursuit of truth in the educational
process. These young men were all aware that they were the
advance guard of a new interest in education within a

tradition which for a generation or more had been distrustful of it. This realization did not mean, however, that all of them responded identically to their graduate study at Harvard. In the eyes of certain of their nonfundamentalist fellow students, some of them appeared to be interested less in learning than in gaining the prestige that went along with a Harvard degree. It is also true that some of the Fundamentalists began their Harvard programs relatively late, having already settled into an understanding of the Christian faith which satisfied their intellectual demands. And whereas, as previously mentioned, some came to Harvard precisely because they eagerly anticipated the prospect of putting their faith to the test, others majored in areas highly technical rather than philosophical (such as ancient languages) and thus, whatever their personal interests, had less time for an inclination toward the struggle with conceptual problems. Consequently, while the fires of the academic crucible burned hot for some of them, others survived like Shadrach, Meshach, and Abednego, with not a hair of their heads singed. Except for Carnell, the *angst* level seems to have been surprisingly low. D. Elton Trueblood, who was teaching at the divinity school in the fall of 1944, noted that Carnell had written some poor papers, concluded that he was "greatly inhibited by narrow dogmatism," and sensed "some emotional disturbance" which he attributed to the theological problems he was facing. There is reason to think Trueblood analyzed the problem correctly, for Carnell told a fellow Fundamentalist student that he had come to Harvard with a sack full of arguments in defense of the faith, only to find to his dismay when he reached into it one day that the sack was empty.

We must balance that remark, however, with other evidence. In 1958, during his tenure as president of Fuller Seminary, speaking at a fund-raising dinner in Chicago, Carnell reflected on some of his reactions at Harvard: "The more I exposed myself to the competing ideologies, the more convinced I became that the Christian world view can be accepted with the consent of all our faculties. A conviction grew in my mind as I cast myself on the perils of graduate study that any fair-minded individual who is open before the facts, if he pursued a course carefully and with patience, would arrive at the biblical, theistic position."[9]

If we allow for some fraternal exaggeration in the empty sack remark and some presidential rhetoric in the Chicago speech, these two statements are not so far apart as initially they seem, certainly not irreconcilable. What was missing in Carnell's sack in graduate school was not a confident Christian faith but an apologetic stance. He had been thoroughly immersed for three years in the consistent Calvinism of Van Til at Westminster Seminary but grew to believe that Van Til, in his unwillingness to acknowledge that the unbeliever is capable of arriving at *any* valid truth, was eliminating every potential point of contact between the believer and the unbeliever, thus undercutting the task of apologetics and leaving the faith without defense. For his own part, Van Til charged that by insisting that faith must have a rational foundation, Carnell was making man autonomous, the creature in effect setting up a standard that the Creator was being forced to meet. The rift between the men was a permanent one. In 1971, four years after Carnell's death, Van Til said: "Everything he wrote in his first book of apologetics, and in all those to follow, he wrote with full consciousness of the differences which arose between us during his days at Boston."[10]

But an even more powerful influence in Carnell's intellectual development was his college philosophy professor, Gordon Haddon Clark. Was there nothing left in the sack of arguments from the formidable Clark? There was indeed. In a 1953 letter to James Tompkins, an old Wheaton and Westminster classmate who had drifted from orthodoxy, Carnell explicitly declared his continuing indebtedness to Clark: "Like yourself, I rejected Van Til; but unlike yourself, graduate studies presented no option superior to that which Clark taught me."

Then why the remark to his seminary friend that the

apologetic sack was empty? Because although Carnell never turned his back on the law of contradiction, which was the formal basis of Clark's system of deductive rationalism, he became convinced that Clark, like Van Til, was severely limited in his apologetic usefulness. The law of contradiction was the means by which one could ferret out the inconsistencies and illogicalities of opposing systems—and that, according to Clark, is the main task of apologetics—but Carnell was looking for something more positive. In Clark's view there is no evidence which can certify the God of Calvinistic orthodoxy, for if such evidence existed it would be more foundational than God himself and therefore undermine His status as a first principle. However logical, this point of view struck Carnell (as he took philosophical inventory at Harvard) as unnecessarily constricted, doing justice to neither the full dimensions of human life nor the breadth of Holy Scripture.

So he wrote a book. To be sure, Carnell is not the first graduate student to have published a scholarly volume while still a degree candidate, but not many such books are greeted with quite the acclaim received by *Introduction to Christian Apologetics* in the spring of 1948.[11] One of fifty manuscripts submitted to the William B. Eerdmans Publishing Company in an Evangelical Book Award contest, it won the first prize of $5,000 and is generally cited as one of the signs of an intellectual shift in Fundamentalism at that time.

Appearing in the spring just prior to his reception of Harvard's Th.D., *Apologetics* gave a big boost to Carnell's career. Having established himself as a teacher of some brilliance at Gordon College in Boston for the preceding three years, he was tapped by the new but already prestigious Fuller Seminary as professor of apologetics. But before departing Harvard he ran into some unexpected trouble. In December he had submitted copies of his doctoral thesis to both his adviser, Professor Auer, and the second reader, Prof. Nels F. S. Ferré of Andover-Newton. Johannes Abraham Christoffel Fagginger Auer had come to the divinity school in 1929 and been designated Parkman Professor in 1930. In 1942 he had taken over the responsibility of directing the programs of graduate students particularly interested in theology. In fact, during his last dozen years on the faculty, he was the *only* professor of theology, a fact that takes on added significance when one knows that his own theological position was humanism. But he was a warm, genial, fair-minded man, and Levering Reynolds's claim that "he never failed in his respect for other men's opinions, however much he might disagree with them"[12] is substantiated by the group of Fundamentalists.

Carnell's thesis was on Niebuhr—"The Concept of Dialectic in the Theology of Reinhold Niebuhr."[13] Auer approved it, which is not surprising since as adviser he would hardly have allowed Carnell to submit it to the formal academic approval process if he had not. But Ferré did not approve it. On December 30 Ferré wrote Prof. Arthur Darby Nock to the effect that he found the thesis inadequate in historical background, deficient in relating Niebuhr's dialectic to contemporary thought, and therefore unacceptable. He further accused the writer of superficial generalities and of unfairly caricaturing Niebuhr's ideas in order to get the better of an argument. We can get just an inkling of how much of a blow this rejection was to Carnell in a letter he wrote to Dean Sperry on January 10. Emphasizing that he had already talked with Ferré by telephone, he said, in part: "I want you to know that I am so anxious to do things right at Harvard that I will rewrite the entire dissertation with the criticisms of my readers in mind." He asked Sperry, if possible, to keep the matter confidential, "since there is a certain academic disgrace connected with the experience that is difficult to bear."

Carnell was overreacting a bit perhaps. It is not a disgrace to have to revise a doctoral thesis. As a matter of fact, the revisions required by Ferré cannot have been very extensive, for he signed the completed dissertation later that same semester. But Carnell's reaction in January has to be seen in the context of his emerging career as a Fundamentalist scholar. The agony caused by Ferré's negative response to his thesis was balanced by the ecstasy of hearing that he had won the Eerdmans prize for *Introduction to Christian Apologetics.*

After the initial high feeling had worn off, Carnell must have come down hard when he realized what a cruel irony it would be if word got around that the author of this acclaimed book could not even get his doctoral thesis approved. Such a conjecture may have motivated in part his appeal for confidentality.

A brief critical review of the content of his prize-winning book will bring into better focus the theology and the experience of the Harvard Fundamentalists and provide some support for concluding generalizations. In this earnest inquiry into metaphysical realms, the basic issues are epistemological. How can we *know* what the truth is? When Carnell says, "Metaphysics and epistemology go together like Scarlatti and the harpsichord," we recall his letter to Dean Sperry accompanying his Harvard application, a letter which tentatively organized an entire graduate program around the issue of epistemology. I think it is accurate to see *Apologetics* as a partial fulfillment of that agenda.

What is immediately evident is the grand scope of Carnell's purpose—nothing less than a demonstration of how conservative Christianity (which he explicitly equates with Fundamentalism) "is able to answer the fundamental questions of life as adequately as, if not more adequately than, any other world-view." He does not minimize the seriousness of the task, pointing out that "the core of man's dilemma is the problem of relating his insatiable desire for self-preservation to the realities of a death-doomed body and an impersonal universe." The stakes are high: five times in the first ten pages it is stated or implied that if the Christian philosophy of life is rejected the only alternatives are either suicide or else an indifference or despair which offers a person no convincing reason for *not* committing suicide.

For all of its author's willingness to tackle the tough questions, however, what lingers in the mind of today's reader is the book's curious ineffectiveness as apologetics. It may in its day have buttressed the crumbling intellectual defenses of Fundamentalism (the fact that there were four editions by 1952 suggests something of how it was perceived by its readers), but to the unconvinced it carries little weight of conviction. One reason for this is Carnell's presuppositional methodology. He goes about the task of showing how, if one presupposes the existence of the God who has revealed himself in the Bible, the Christian religion best fulfills the requirements of "systematic consistency" (i.e., negatively, it does not violate the law of contradiction; positively, it adheres to the facts of experience). There is nothing inherently wrong with presuppositionalism, but Carnell's use of the method is troublesome. He begins by convincingly portraying the ultimate dilemma of humanity, as if he is going to proceed step by step to show how Christian theism is the answer to this dilemma. But we soon realize that the argument is not developing in this manner; rather we are riding a set of rails which were laid down long before the first words were put on paper. As early as page 47 we are confidently assured that "there is no reality apart from the eternal nature of God Himself and the universe which He has created to display His glory." Moreover, his assumptions are the ones really worth being troubled about—especially the question of the existence of God—and without confronting them straightforwardly the book comes close to being 359 pages of begging the question. Carnell is adept and sometimes merciless in attacking the assumptions of others but behaves much too charitably toward his own.

A second serious flaw, one with far-reaching implications, is the book's much too easy bifurcation of thought systems into Christian and non-Christian, a trait which shows up most vividly in Carnell's discussion of "the problem of common ground."

The very nature of Christianity demands that there be no common ground between the system of the godly and the system of the ungodly, for a man's attitude toward what he considers to be the highest logical ultimate in reality determines the validity of his synoptic starting point, his method, and his conclusion.

And because "the reach of metaphysics is absolute," starting with the wrong assumptions can lead one astray at every point along the way. This kind of either/or thinking permits Carnell to claim, for example, that only a Christian has the right to make statements and expect to be understood (because only a God whose very nature is the guarantee of meaning can provide the basis for intelligible communication).

With no sovereign God to set the course of reality and to give promises of hope to man, there is a 50/50 chance of anything happening. In five minutes, not only may elephants fly and roots grow up, but doors may have only one side, spinach may grow in patches of square circles, the sun may turn to silk, the moon to mink, up may be down, right may be left, and good may be bad.

The choice is clear: either the Fundamentalist interpretation of historic orthodox Christianity or absurdity and nihilism. We can see the same bifurcation in Carnell's doctoral thesis on Reinhold Niebuhr, the concluding lines of which read as follows:

Niebuhr must either turn to revelation in Scripture seriously, in which case he must go all the way with the problems which attend special revelation; or he must break with the appeal to special revelation and take up empiricism seriously, in which case he must go all the way with the scientific method. There does not seem to the author any stopping point between these two termini.

Finally, it is clear that Carnell perceived these same two options as the only live competing ones available to him in the educational milieu at Harvard: either the Fundamentalism with which he came or the scientific method whose end was nihilism. And whereas the divinity school cannot be blamed for the tendency of Carnell's logic to rule out any middle ground between alternatives, it is a fact that theology was not Harvard's strong suit in these years. The 1947 "Report of the Commission to Study and Make Recommendations with Respect to the Harvard Divinity School" expressed grave concern that Harvard offered a total of only three courses in theology and philosophy of religion and none in Christian ethics, adding that "the tendency to stress the historical rather than the constructive aspects of theology is in itself a symptom of theological decline."[14]

Carnell left Harvard still a Fundamentalist, albeit a more intellectually sophisticated one. And although he later explicitly disowned Fundamentalism and delivered a harsh critique of the movement in *The Case for Orthodox Theology*, his published books and articles reveal that as a thinker he never completely freed himself of much of Fundamentalism's ideological baggage. But there was another side to Carnell. In his inaugural address as president of Fuller Seminary, he developed three characteristics which constitute "the glory of a theological seminary." After resolutely reaffirming the mandate to preserve and propagate faithfully the theological distinctives to which the institution was committed, he went on to expound two further points not historically common to Fundamentalism: (1) that in fulfilling the first objective, the seminary "make a conscientious effort to acquaint its students with all the relevant evidences—damaging as well as supporting—in order that students may be given a reasonable opportunity to exercise their God-given right freely to decide for or against claims to truth"; and (2) "that the seminary inculcate in its students an attitude of tolerance and forgiveness toward individuals whose doctrinal convictions are at variance with those that inhere in the institution itself."

It may be a measure of some kind of progress that even though the inaugural address evoked a serious protest from a sizable segment of the seminary's old guard faculty, and though Charles E. Fuller, the seminary's founder, ordered the impounding of all copies, Fuller Seminary today distributes an elegantly printed twenty-page brochure including the entire text of the address.[15] In any case, in fostering an academic atmosphere in which students are encouraged to venture into that ambiguous territory between a simplistic unquestioning acceptance of the faith once delivered and an equally simplistic rejection of that faith, whether in the name of humanism or nihilism, Edward Carnell not only transcended the Fundamentalism of his heritage but left his most enduring legacy.

NOTES

1. "Dean's Letter," *Harvard Divinity School Bulletin*, March 15, 1940, pp. 42, 43.

2. It is not my purpose here to explore the subtle distinctions between the terms "Fundamentalist" and "evangelical," but rather to accept the labels by which these men have identified themselves, then and more recently. For an illuminating recent discussion of the subject, see Joel A. Carpenter, "Fundamentalist Institutions and the Rise of Evangelical Protestantism, 1929-1942," *Church History* 49 (March 1980): 62-75.

3. Edward J. Carnell, *The Case for Orthodox Theology* (Philadelphia: Westminster Press, 1959).

4. I am indebted for most of my general information about Harvard Divinity School to George Huntston Williams, ed., *The Harvard Divinity School: Its Place in Harvard University and in American Culture* (Boston: Beacon Press, 1954), especially the chapter "The Later Years (1880-1953)" by Levering Reynolds, Jr.

5. Williams, pp. 168, 186, 10.

6. Robert T. Handy, "The American Religious Depression, 1926-1935," *Church History* 29 (March 1960): 3-16.

7. Material quoted or cited concerning the Fundamentalist students at Harvard is based on either personal correspondence or interviews, in some cases both.

8. Roger Shinn, *Theology Today* 15 (July 1958): 278.

9. Unpublished notes for address, Fuller Theological Seminary files.

10. E. R. Geehan, ed., *Jerusalem and Athens: Critical Discussions on the Theology and Apologetics of Cornelius Van Til* (Grand Rapids, Mich.: Baker Book House, 1971), p. 368.

11. Edward J. Carnell, *An Introduction to Christian Apologetics: A Philosophic Defense of the Trinitarian-Theistic Faith* (Grand Rapids, Mich.: Eerdmanns, 1948).

12. Williams, ed., p. 225.

13. Edward Carnell, "The Concept of Dialectic in the Theology of Reinhold Niebuhr," unpublished Th.D. thesis, Harvard Divinity School, 1948. The quotes that follow are found in the thesis on pp. 7, 23, 212, 54, and 417.

14. Report of The Commission to Study and Make Recommendations with Respect to the Harvard Divinity School, July 1947, pp. 15, 16. Used by permission of Andover-Harvard Theological Library.

15. Edward J. Carnell, "The Glory of a Theological Seminary," Fuller Theological Seminary Alumni Association, 135 North Oakland Avenue, Pasadena, California 91101.

The Creationists

Ronald L. Numbers

Scarcely twenty years after the publication of Charles Darwin's *Origin of Species* in 1859 special creationists could name only two working naturalists in North America, John William Dawson (1820–1899) of Montreal and Arnold Guyot (1806–1884) of Princeton, who had not succumbed to some theory of organic evolution. The situation in Great Britain looked equally bleak for creationists, and on both sides of the Atlantic liberal churchmen were beginning to follow their scientific colleagues into the evolutionist camp. By the closing years of the nineteenth century evolution was infiltrating even the ranks of the evangelicals, and, in the opinion of many observers, belief in special creation seemed destined to go the way of the dinosaur. But contrary to the hopes of liberals and the fears of conservatives, creationism did not become extinct. The majority of late-nineteenth-century Americans remained true to a traditional reading of Genesis, and as late as 1982 a public-opinion poll revealed that 44 percent of Americans, nearly a fourth of whom were college graduates, continued to believe that "God created man pretty much in his present form at one time within the last 10,000 years."[1]

Such surveys failed, however, to disclose the great diversity of opinion among those professing to be creationists. Risking oversimplification, we can divide creationists into two main camps: "strict creationists," who interpret the days of Genesis literally, and "progressive creationists," who construe the Mosaic days to be immense periods of time. But even within these camps substantial differences exist. Among strict creationists, for example, some believe that God created all terrestrial life—past and present—less than ten thousand years ago, while others postulate one or more creations prior to the seven days of Genesis. Similarly, some progressive creationists believe in numerous creative acts, while others limit God's intervention to the creation of life and perhaps the human soul. Since this last species of creationism is practically indistinguishable from theistic evolutionism, this essay focuses on the strict creationists and the more conservative of the progressive creationists, particularly on the small number who claimed scientific expertise. Drawing on their writings, it traces the ideological development of creationism from the crusade to outlaw the teaching of evolution in the 1920s to the current battle for equal time. During this period the leading apologists for special creation shifted from an openly biblical defense of their views to one based largely on science. At the same time they grew less tolerant of notions of an old earth and symbolic days of creation, common among creationists early in the century, and more doctrinaire in their insistence on a recent creation in six literal days and on a universal flood.

THE LOYAL MAJORITY

The general acceptance of organic evolution by the intellectual elite of the late Victorian era has often obscured the fact that the majority of Americans remained loyal to the doctrine of special creation. In addition to the masses who said nothing, there were many people who vocally rejected kinship with the apes and other, more reflective, persons who concurred with the Princeton theologian Charles Hodge (1797–1878) that Darwinism was atheism. Among the most intransigent foes of organic evolution were the premillennialists, whose predictions of Christ's imminent return depended on a literal reading of the Scriptures. Because of their conviction that one error in the Bible invalidated the entire book, they had little patience with scientists who, as described by the evangelist Dwight L. Moody (1837–1899), "dug up old carcasses . . . to make them testify against God."[2]

Such an attitude did not, however, prevent many biblical literalists from agreeing with geologists that the earth was far older than six thousand years. They did so by identifying two separate creations in the first chapter of Genesis: the first, "in the beginning," perhaps millions of years ago, and the second, in six actual days, approximately four thousand years before the birth of Christ. According to this so-called gap theory, most fossils were relics of the first creation, destroyed by God prior to the Adamic restoration. In 1909 the *Scofield Reference Bible*, the most authoritative biblical guide in fundamentalist circles, sanctioned this view.[3]

Scientists like Guyot and Dawson, the last of the reputable nineteenth-century creationists, went still further to accommodate science by interpreting the days of Genesis as ages and by correlating them with successive epochs in the natural history of the world. Although they believed in special creative acts, especially of the first humans, they tended to minimize the number of supernatural interventions and to maximize the operation of natural law. During the late nineteenth century their theory of progressive creation circulated widely in the colleges and seminaries of America.[4]

The early Darwinian debate focused largely on the implications of evolution for natural theology; and so long as these discussions remained confined to scholarly circles, those who objected to evolution on biblical grounds saw little reason to participate. But when the debate spilled over into the public arena during the 1880s and 1890s, creationists grew alarmed. "When these vague speculations, scattered to the four winds by the million-tongued press, are caught up by ignorant and untrained men," declared one premillennialist in 1889, "it is time for earnest Christian men to call a halt."[5]

The questionable scientific status of Darwinism undoubtedly encouraged such critics to speak up. Although the overwhelming majority of scientists after 1880 accepted a long earth history and some form of organic evolution, many in the late nineteenth century were expressing serious reservations about the ability of Darwin's particular theory of natural selection to account for the origin of species. Their published criticisms of Darwinism led creationists mistakenly to conclude that scientists were in the midst of discarding evolution. The appearance of books with such titles as *The Collapse of Evolution* and *At the Death Bed of Darwinism* bolstered this belief and convinced antievolutionists that liberal Christians had capitulated to evolution too quickly. In view of this turn of events it seemed likely that those who had "abandoned the stronghold of faith out of sheer fright will soon be found scurrying back to the old and impregnable citadel, when they learn that 'the enemy is in full retreat.'"[6]

For the time being, however, those conservative Christians who would soon call themselves fundamentalists perceived a greater threat to orthodox faith than evolution—higher criticism, which treated the Bible more as a historical document than as God's inspired word. Their relative apathy toward evolution is evident in *The Fundamentals*, a mass-produced series of twelve booklets published between 1910 and 1915 to revitalize and reform Christianity around the world. Although one contributor identified evolution as the principal cause of disbelief in the Scriptures and another traced the roots of higher criticism to Darwin, the collection as a whole lacked the strident antievolutionism that would characterize the fundamentalist movement of the 1920s.[7]

This is particularly true of the writings of George Frederick Wright (1838–1921), a Congregational minister and amateur geologist of international repute. At first glance his selection to represent the fundamentalist point of view seems anomalous. As a prominent Christian Darwinist in the 1870s he had argued that the intended purpose of Genesis was to protest polytheism, not teach science. By the 1890s, however, he had come to espouse the progressive creationism of Guyot and Dawson, partly, it seems, in reaction to the claims of higher critics regarding the accuracy of the Pentateuch. Because of his standing as a scientific authority and his conservative view of the Scriptures, the editors of *The Fundamentals* selected him to address the question of the relationship between evolution and the Christian faith.[8]

In an essay misleadingly titled "The Passing of Evolution" Wright attempted to steer a middle course between the theistic evolution of his early days and the traditional views of some special creationists. On the one hand, he argued that the Bible itself taught evolution, "an orderly progress from lower to higher forms of matter and life." On the other hand, he limited evolution to the origin of species, pointing out that even Darwin had postulated the supernatural creation of several forms of plants and animals, endowed by the Creator with a "marvelous capacity for variation." Furthermore, he argued that, despite the physical similarity between human beings and the higher animals, the former "came into existence as the Bible represents, by the special creation of a single pair, from whom all the varieties of the race have sprung."[9]

Although Wright represented the left wing of fundamentalism, his moderate views on evolution contributed to the conciliatory tone that prevailed during the years leading up to World War I. Fundamentalists may not have liked evolution, but few, if any, at this time saw the necessity or desirability of launching a crusade to eradicate it from the schools and churches in America.

THE ANTIEVOLUTION CRUSADE

Early in 1922 William Jennings Bryan (1860–1925), Presbyterian layman and thrice-defeated Democratic candidate for the presidency of the United States, heard of an effort in Kentucky to ban the teaching of evolution in public schools. "The movement will sweep the country," he predicted hopefully, "and we will drive Darwinism from our schools."[10] His prophecy proved overly optimistic, but before the end of the decade more than twenty state legislatures did debate antievolution laws, and four—Oklahoma, Tennessee, Mississippi, and

Arkansas—banned the teaching of evolution in public schools. At times the controversy became so tumultuous that it looked to some as though "America might go mad." Many persons shared responsibility for these events, but none more than Bryan. His entry into the fray had a catalytic effect and gave antievolutionists what they needed most: "a spokesman with a national reputation, immense prestige, and a loyal following."[11]

The development of Bryan's own attitude toward evolution closely paralleled that of the fundamentalist movement. Since early in the century he had occasionally alluded to the silliness of believing in monkey ancestors and to the ethical dangers of thinking that might makes right, but until the outbreak of World War I he saw little reason to quarrel with those who disagreed. The war, however, exposed the darkest side of human nature and shattered his illusions about the future of Christian society. Obviously something had gone awry, and Bryan soon traced the source of the trouble to the paralyzing influence of Darwinism on the human conscience. By substituting the law of the jungle for the teaching of Christ, it threatened the principles he valued most: democracy and Christianity. Two books in particular confirmed his suspicion. The first, Vernon Kellogg's *Headquarters Nights* (1917), recounted firsthand conversations with German officers that revealed the role Darwin's biology had played in persuading the Germans to declare war. The second, Benjamin Kidd's *Science of Power* (1918), purported to demonstrate the historical and philosophical links between Darwinism and German militarism.[12]

About the time that Bryan discovered the Darwinian origins of the war, he also became aware, to his great distress, of unsettling effects the theory of evolution was having on America's own young people. From frequent visits to college campuses and from talks with parents, pastors, and Sunday-school teachers, he heard about an epidemic of unbelief that was sweeping the country. Upon investigating the cause, his wife reported, "he became convinced that the teaching of Evolution as a fact instead of a theory caused the students to lose faith in the Bible, first, in the story of creation, and later in other doctrines, which underlie the Christian religion." Again Bryan found confirming evidence in a recently published book, *Belief in God and Immortality* (1916), by the Bryn Mawr psychologist James H. Leuba, who demonstrated statistically that college attendance endangered traditional religious beliefs.[13]

Armed with this information about the cause of the world's and the nation's moral decay, Bryan launched a nationwide crusade against the offending doctrine. In one of his most popular and influential lectures, "The Menace of Darwinism," he summed up his case against evolution, arguing that it was both un-Christian and unscientific. Darwinism, he declared, was nothing but "guesses strung together," and poor guesses at that. Borrowing from a turn-of-the-century tract, he illustrated how the evolutionist explained the origin of the eye:

> The evolutionist guesses that there was a time when eyes were unknown—that is a necessary part of the hypothesis. . . . a piece of pigment, or, as some say, a freckle appeared upon the skin of an animal that had no eyes. This piece of pigment or freckle converged the rays of the sun upon that spot and when the little animal felt the heat on that spot it turned the spot to the sun to get more heat. The increased heat irritated the skin—so the evolutionists guess, and a nerve came there and out of the nerve came the eye!

"Can you beat it?" he asked incredulously—and that it happened not once but twice? As for himself, he would take one verse in Genesis over all that Darwin wrote.[14]

Throughout his political career Bryan had placed his faith in the common people, and he resented the attempt of a few thousand scientists "to establish an oligarchy over the forty million American Christians," to dictate what should be taught in the schools.[15] To a democrat like Bryan it seemed preposterous that this "scientific soviet" would not only demand to teach its insidious philosophy but impudently insist that society pay its salaries. Confident that nine-tenths of the Christian citizens agreed with him, he decided to appeal directly to them, as he had done so successfully in fighting the liquor interests. "Commit your case to the people," he advised creationists. "Forget, if need be, the high-brows both in the political and college world, and carry this cause to the people. They are the final and efficiently corrective power."[16]

And who were the people who joined Bryan's crusade? As recent studies have shown, they came from all walks of life and from every region of the country. They lived in New York, Chicago, and Los Angeles as well as in small towns and in the country. Few possessed advanced degrees, but many were not without education. Nevertheless, Bryan undeniably found his staunchest supporters and won his greatest victories in the conservative and still largely rural South, described hyperbolically by one fundamentalist journal as "the last stronghold of orthodoxy on the North American continent," a region where the "masses of the people in all denominations 'believe the Bible from lid to lid.'"[17]

The strength of Bryan's following within the churches is perhaps more difficult to determine, because not all fundamentalists were creationists and many creationists refused to participate in the crusade against evolution. However, a 1929 survey of the theological beliefs of seven hundred Protestant ministers provides some valuable clues. The question "Do you believe that the creation of the world occurred in the manner and time recorded in Genesis?" elicited the following positive responses:

Lutheran	89%
Baptist	63%
Evangelical	62%
Presbyterian	35%
Methodist	24%
Congregational	12%
Episcopalian	11%
Other	60%

Unfortunately, these statistics tell us nothing about the various ways respondents may have interpreted the phrase "in the manner and time recorded in Genesis," nor do they reveal anything about the level of political involvement in the campaign against evolution. Lutherans, for example, despite their overwhelming rejection of evolution, generally preferred education to legislation and tended to view legal action against evolution as "a dangerous mingling of church and state." Similarly, premillennialists, who saw the spread of evolution as one more sign of the world's impending end, sometimes lacked incentive to correct the evils around them.[18]

Baptists and Presbyterians, who dominated the fundamentalist movement, participated actively in the campaign against evolution. The Southern Baptist Convention, spiritual home of some of the most outspoken foes of evolution, lent encouragement to the creationist crusaders by voting unanimously in 1926 that "this Convention accepts Genesis as teaching that man was the special creation of God, and rejects every theory, evolution or other, which teaches that man originated in, or came by way of, a lower animal ancestry." The Presbyterian Church contributed Bryan and other leaders to the creationist cause but, as the above survey indicates, also harbored many evolutionists. In 1923 the General Assembly turned back an attempt by Bryan and his fundamentalist cohorts to cut off funds to any church school found teaching human evolution, approving instead a compromise measure that condemned only materialistic evolution. The other major Protestant bodies paid relatively little attention to the debate over evolution; and Catholics, though divided on the question of evolution, seldom favored restrictive legislation.[19]

Leadership of the antievolution movement came not from the organized churches of America but from individuals like Bryan and interdenominational organizations such as the World's Christian Fundamentals Association, a predominantly premillennialist body founded in 1919 by William Bell Riley (1861–1947), pastor of the First Baptist Church in Minneapolis. Riley became active as an antievolutionist after discovering, to his apparent surprise, that evolutionists were teaching their views at the University of Minnesota. The early twentieth century witnessed an unprecedented expansion of public education—enrollment in public high schools nearly doubled between 1920 and 1930—and fundamentalists like Riley and Bryan wanted to make sure that students attending these institutions would not lose their faith. Thus they resolved to drive every evolutionist from the public-school payroll. Those who lost their jobs as a result deserved little sympathy, for, as one rabble-rousing creationist put it, the German soldiers who killed Belgian and French children with poisoned candy were angels compared with the teachers and textbook writers who corrupted the souls of children and thereby sentenced them to eternal death.[20]

The creationists, we should remember, did not always act without provocation. In many instances their opponents displayed equal intolerance and insensitivity. In fact, one contemporary observer blamed the creation-evolution controversy in part on the "intellectual flapperism" of irresponsible and poorly informed teachers who delighted in shocking naive students with unsupportable statements about evolution. It was understandable, wrote an Englishman, that American parents would resent sending their sons and daughters to public institutions that exposed them to "a multiple assault upon traditional faiths."[21]

CREATIONIST SCIENCE AND SCIENTISTS

In 1922 William Bell Riley outlined the reasons why fundamentalists opposed the teaching of evolution. "The first and most important reason for its elimination," he explained, "is the unquestioned fact that evolution is not a science; it is a hypothesis only, a speculation." Bryan often made the same point, defining true science as "classified knowledge . . . the explanation of facts."[22] Although creationists had far more compelling reasons for rejecting evolution than its alleged unscientific status, their insistence on this point was not merely an obscurantist ploy. Rather it stemmed from their commitment to a once-respected tradition, associated with the English philosopher Sir Francis Bacon (1561–1626), that emphasized the factual, nontheoretical nature of science. By identifying with the Baconian tradition, creationists could label evolution as false science, could claim equality with scientific authorities in comprehending facts, and could deny the charge of being antiscience. "It is not 'science' that orthodox Christians oppose," a fundamentalist editor insisted defensively. "No! no! a thousand times, No! They are opposed only to the theory of evolution, which has not yet been proved, and therefore is not to be called by the sacred name of *science*."[23]

Because of their conviction that evolution was unscientific, creationists assured themselves that the world's best scientists agreed with them. They received an important boost at the beginning of their campaign from an address by the distinguished British biologist William Bateson (1861–1926) in 1921, in which he declared that scientists had *not* discovered "the actual mode and process of evolution." Although he warned creationists against misinterpreting his statement as a rejection of evolution, they paid no more attention to that caveat than they did to the numerous proevolution resolutions passed by scientific societies.[24]

Unfortunately for the creationists, they could claim few legitimate scientists of their own: a couple of self-made men of science, one or two physicians, and a handful of teachers who, as one evolutionist described them, were "trying to hold down, not a chair, but a whole settee, of 'Natural Science' in some little institution."[25] Of this group the most influential were Harry Rimmer (1890–1952) and George McCready Price (1870–1963).

Rimmer, Presbyterian minister and self-styled "research scientist," obtained his limited exposure to science during a term or two at San Francisco's Hahnemann Medical College, a small homeopathic institution that required no more than a high-school diploma for admission. As a medical student he picked up a vocabulary of "double-jointed, twelve cylinder, knee-action words" that later served to impress the uninitiated. After his brief stint in medical school he attended Whittier College and the Bible Institute of Los Angeles for a year each before entering full-time evangelistic work. About 1919 he settled in Los Angeles, where he set up a small laboratory at the rear of his house to conduct experiments in embryology and related sciences. Within a year or two he established the Research Science Bureau "to prove through findings in biology, paleontology, and anthropology that science and the literal Bible were not contradictory." The bureau staff—that is, Rimmer—apparently used income from the sale of memberships to finance anthropological field trips in the western United States, but Rimmer's dream of visiting Africa to prove the dissimilarity of gorillas and humans failed to materialize. By the late 1920s the bureau lay dormant, and Rimmer signed on with Riley's World's Christian Fundamentals Association as a field secretary.[26]

Besides engaging in research, Rimmer delivered thousands of lectures, primarily to student groups, on the scientific accuracy of the Bible. Posing as a scientist, he attacked Darwinism and poked fun at the credulity of evolutionists. To attract attention, he repeatedly offered one hundred dollars to anyone who could discover a scientific error in the Scriptures; not surprisingly, the offer never cost him a dollar. He also, by his own reckoning, never lost a public debate. Following one encounter with an evolutionist in Philadelphia, he wrote home gleefully that "the debate was a simple walkover, a massacre—murder pure and simple. The eminent professor was simply scared stiff to advance any of the common arguments of the evolutionists, and he fizzled like a wet fire-cracker."[27]

George McCready Price, a Seventh-day Adventist geologist, was less skilled at debating than Rimmer but more influential scientifically. As a young man Price attended an Adventist college in Michigan for two years and later completed a teacher-training course at the provincial normal school in his native New Brunswick. The turn of the century found him serving as principal of a small high school in an isolated part of eastern Canada, where one of his few companions was a local physician. During their many conversations, the doctor almost converted his fundamentalist friend to evolution, but each time Price wavered, he was saved by prayer and by reading the works of the Seventh-day Adventist prophetess Ellen G. White (1827–1915), who claimed divine inspiration for her view that Noah's flood accounted for the fossil record on which evolutionists based their theory. As a result of these experiences, Price vowed to devote his life to promoting creationism of the strictest kind.[28]

By 1906 he was working as a handyman at an Adventist sanitarium in southern California. That year he published a slim volume entitled *Illogical Geology: The Weakest Point in the Evolution Theory*, in which he brashly offered one thousand dollars "to any one who will, in the face of the facts here presented, show me how to prove that one kind of fossil is older than another." (Like Rimmer, he never had to pay.) According to Price's argument, Darwinism rested "logically and historically on the succession of life idea as taught by geology" and "if this succession of life is not an actual scientific fact, then Darwinism . . . is a most gigantic hoax."[29]

Although a few fundamentalists praised Price's polemic, David Starr Jordan (1851–1931), president of Stanford University and an authority on fossil fishes, warned him that he should not expect "any geologist to take [his work] seriously." Jordan conceded that the unknown author had written "a very clever book" but described it as

> a sort of lawyer's plea, based on scattering mistakes, omissions and exceptions against general truths that anybody familiar with the facts in a general way cannot possibly dispute. It would be just as easy and just as plausible and just as convincing if one should take the facts of European history and attempt to show that all the various events were simultaneous.[30]

As Jordan recognized, Price lacked any formal training or field experience in geology. He was, however, a voracious reader of geological literature, an armchair scientist who self-consciously minimized the importance of field experience.

During the next fifteen years Price occupied scientific settees in several Seventh-day Adventist schools and authored six more books attacking evolution, particularly its geological foundation. Although not unknown outside his own church before the early 1920s, he did not attract national attention until then. Shortly after Bryan declared war on evolution, Price published *The New Geology* (1923), the most systematic and comprehensive of his many books. Uninhibited by false modesty, he presented his "great *law of conformable stratigraphic sequences* . . . by all odds the most important law ever formulated with reference to the order in which the strata occur." This law stated that "*any kind of fossiliferous beds whatever, 'young' or 'old,' may be found occurring conformably on any other fossiliferous beds, 'older' or 'younger.'*"[31] To Price, so-called deceptive conformities (where strata seem to be missing) and thrust faults (where the strata are apparently in the wrong order) proved that there was no natural order to the fossil-bearing rocks, all of which he attributed to the Genesis flood.

A Yale geologist reviewing the book for *Science* accused Price of "harboring a geological nightmare." But despite such criticism from the scientific establishment—and the fact that his theory contradicted both the day-age and gap interpretations of Genesis—Price's reputation among fundamentalists rose dramatically. Rimmer, for example, hailed *The New Geology* as "a masterpiece of REAL science [that] explodes in a convincing manner some of the ancient fallacies of science 'falsely so called.'"[32] By the mid-1920s Price's byline was appearing with increasing frequency in a broad spectrum of conservative religious periodicals, and the editor of *Science* could accurately describe him as "the principal scientific authority of the Fundamentalists."[33]

In the spring of 1925 John Thomas Scopes, a high-school teacher in the small town of Dayton, Tennessee, confessed to having violated the state's recently passed law banning the teaching of human evolution in public schools. His subsequent trial focused international attention on the antievolution crusade and brought William Jennings Bryan to Dayton to assist the prosecution. In anticipation of arguing the scientific merits of evolution, Bryan sought out the best scientific minds in the creationist camp to serve as expert witnesses. The response to his inquiries could only have disappointed the aging crusader. Price, then teaching in England, sent his regrets—along with advice for Bryan to stay away from scientific topics. Howard A. Kelly, a prominent Johns Hopkins physician who had contributed to *The Fundamentals*, confessed that, except for Adam and Eve, he believed in evolution. Louis T. More, a physicist who had just written a book on *The Dogma of Evolution* (1925), replied that he accepted evolution as a working hypothesis. Alfred W. McCann, author of *God—or Gorilla* (1922), took the opportunity to chide Bryan for supporting prohibition in the past and for now trying "to bottle-up the tendencies of men to think for themselves."[34]

At the trial itself things scarcely went better. When Bryan could name only Price and the deceased George Frederick Wright as scientists for whom he had respect, the caustic Clarence Darrow (1857–1938), attorney for the defense, scoffed: "You mentioned Price because he is the only human being in the world so far as you know that signs his name as a geologist that believes like you do. . . . every scientist in this country knows [he] is a mountebank and a pretender and not a geologist at all." Eventually Bryan conceded that the world was indeed far more than six thousand years old and that the six days of creation had probably been longer than twenty-four hours each—concessions that may have harmonized with the progressive creationism of Wright but hardly with the strict creationism of Price.[35]

Though one could scarcely have guessed it from some of his public pronouncements, Bryan had long been a progressive creationist. In fact, his beliefs regarding evolution diverged considerably from those of his more conservative supporters. Shortly before his trial he had confided to Dr. Kelly that he, too, had no objection to "evolution before man but for the fact that a concession as to the truth of evolution up to man furnishes our opponents with an argument which they are quick to use, namely, if evolution accounts for all the species up to man, does it not raise a presumption in behalf of evolution to

include man?" Until biologists could actually demonstrate the evolution of one species into another, he thought it best to keep them
on the defensive.[36]

Bryan's admission at Dayton spotlighted a serious and long-
standing problem among antievolutionists: their failure to agree on a
theory of creation. Even the most visible leaders could not reach a
consensus. Riley, for example, followed Guyot and Dawson (and
Bryan) in viewing the days of Genesis as ages, believing that the
testimony of geology necessitated this interpretation. Rimmer favored
the gap theory, which involved two separate creations, in part because
his scientific mind could not fathom how, given Riley's scheme, plants
created on the third day could have survived thousands of years
without sunshine, until the sun appeared on the fourth. According
to the testimony of acquaintances, he also believed that the Bible
taught a local rather than a universal flood. Price, who cared not a
whit about the opinion of geologists, insisted on nothing less than a
recent creation in six literal days and a worldwide deluge. He regarded
the day-age theory as "the devil's counterfeit" and the gap theory as
only slightly more acceptable. Rimmer and Riley, who preferred to
minimize the differences among creationists, attempted the logically
impossible, if ecumenically desirable, task of incorporating Price's
"new geology" into their own schemes.[37]

Although the court in Dayton found Scopes guilty as charged,
creationists had little cause for rejoicing. The press had not treated
them kindly, and the taxing ordeal no doubt contributed to Bryan's
death a few days after the end of the trial. Nevertheless, the antievolutionists continued their crusade, winning victories in Mississippi
in 1926 and in Arkansas two years later. By the end of the decade,
however, their legislative campaign had lost its steam. The presidential election of 1928, pitting a Protestant against a Catholic, offered
fundamentalists a new cause, and the onset of the depression in 1929
further diverted their attention.[38]

Contrary to appearances, the creationists were simply changing
tactics, not giving up. Instead of lobbying state legislatures, they
shifted their attack to local communities, where they engaged in what
one critic described as "the emasculation of textbooks, the 'purging'
of libraries, and above all the continued hounding of teachers." Their
new approach attracted less attention but paid off handsomely, as
school boards, textbook publishers, and teachers in both urban and
rural areas, North and South, bowed to their pressure. Darwinism
virtually disappeared from high-school texts, and for years many
American teachers feared being identified as evolutionists.[39]

During the heady days of the 1920s, when their activities made front-page headlines, creationists dreamed of converting the world; a decade later, forgotten and rejected by the establishment, they turned their energies inward and began creating an institutional base of their own. Deprived of the popular press and frustrated by their inability to publish their views in organs controlled by orthodox scientists, they determined to organize their own societies and edit their own journals.[40] Their early efforts, however, encountered two problems: the absence of a critical mass of scientifically trained creationists and lack of internal agreement.

In 1935 Price, along with Dudley Joseph Whitney, a farm journalist, and L. Allen Higley, a Wheaton College science professor, formed a Religion and Science Association to create "a united front against the theory of evolution." Among those invited to participate in the association's first—and only—convention were representatives of the three major creationist parties, including Price himself, Rimmer, and one of Dawson's sons, who, like his father, advocated the day-age theory. But as soon as the Price faction discovered that its associates had no intention of agreeing on a short earth history, it bolted the organization, leaving it a shambles.[41]

Shortly thereafter, in 1938, Price and some Seventh-day Adventist friends in the Los Angeles area, several of them physicians associated with the College of Medical Evangelists (now part of Loma Linda University), organized their own Deluge Geology Society and, between 1941 and 1945, published a *Bulletin of Deluge Geology and Related Science*. As described by Price, the group consisted of "a very eminent set of men. . . . In no other part of this round globe could anything like the number of scientifically educated believers in Creation and opponents of evolution be assembled, as here in Southern California."[42] Perhaps the society's most notable achievement was its sponsorship in the early 1940s of a hush-hush project to study giant fossil footprints, believed to be human, discovered in rocks far older than the theory of evolution would allow. This find, the society announced excitedly, thus demolished that theory "at a single stroke" and promised to *"astound the scientific world!"* But despite such activity and the group's religious homogeneity, it, too, soon foundered—on "the same rock," complained a disappointed member, that wrecked the Religion and Science Association, that is *"pre-Genesis time for the earth."*[43]

By this time creationists were also beginning to face a new problem: the presence within their own ranks of young university-trained scientists who wanted to bring evangelical Christianity more into line with mainstream science. The encounter between the two generations often proved traumatic, as is illustrated by the case of Harold W. Clark (b. 1891). A former student of Price's, he had gone on to earn a master's degree in biology from the University of California and taken a position at a small Adventist college in northern California. By 1940 his training and field experience had convinced him that Price's *New Geology* was "entirely out of date and inadequate" as a text, especially in its rejection of the geological column. When Price learned of this, he angrily accused his former disciple of suffering from "the modern mental disease of university-itis" and of currying the favor of "tobacco-smoking, Sabbath-breaking, God-defying" evolutionists. Despite Clark's protests that he still believed in a literal six-day creation and universal flood, Price kept up his attack for the better part of a decade, at one point addressing a vitriolic pamphlet, *Theories of Satanic Origin*, to his erstwhile friend and fellow creationist.[44]

The inroads of secular scientific training also became apparent in the American Scientific Affiliation (ASA), created by evangelical scientists in 1941.[45] Although the society took no official stand on creation, strict creationists found the atmosphere congenial during the early years of the society. In the late 1940s, however, some of the more progressive members, led by J. Laurence Kulp, a young geochemist on the faculty of Columbia University, began criticizing Price and his followers for their allegedly unscientific effort to squeeze earth history into less than ten thousand years. Kulp, a Wheaton alumnus and member of the Plymouth Brethren, had acquired a doctorate in physical chemistry from Princeton University and gone on to complete all the requirements, except a dissertation, for a Ph.D. in geology. Although initially suspicious of the conclusions of geology regarding the history and antiquity of the earth, he had come to accept them. As one of the first evangelicals professionally trained in geology, he felt a responsibility to warn his colleagues in the ASA about Price's work, which, he believed, had "infiltrated the greater portion of fundamental Christianity in America primarily due to the absence of trained Christian geologists." In what was apparently the first systematic critique of the "new geology" Kulp concluded that the "major propositions of the theory are contraindicated by established physical and chemical laws." Conservatives within the ASA not unreasonably suspected that Kulp's exposure to "the orthodox geological viewpoint" had severely undermined his faith in a literal interpretation of the Bible.[46]

Before long it became evident that a growing number of ASA members, like Kulp, were drifting from strict to progressive creationism and sometimes on to theistic evolutionism. The transition for many involved immense personal stress, as revealed in the autobiographical testimony of another Wheaton alumnus, J. Frank Cassel:

First to be overcome was the onus of dealing with a "verboten" term and in a "non-existent" area. Then, as each made an honest and objective consideration of the data, he was struck with the validity and undeniability of datum after datum. As he strove to incorporate each of these facts into his Biblico-scientific frame of reference, he found that—while the frame became more complete and satisfying—he began to question first the feasibility and then the desirability of an effort to refute the total evolutionary concept, and finally he became impressed by its impossibility on the basis of existing data. This has been a heart-rending, soul-searching experience for the committed Christian as he has seen what he had long considered the *raison d'être* of God's call for his life endeavor fade away, and as he has struggled to release strongly held convictions as to the close limitations of Creationism.

Cassel went on to note that the struggle was "made no easier by the lack of approbation (much less acceptance) of some of his less well-informed colleagues, some of whom seem to question motives or even to imply heresy."[47] Strict creationists, who suffered their own agonies, found it difficult not to conclude that their liberal colleagues were simply taking the easy way out. To both parties a split seemed inevitable.

CREATIONISM ABROAD

During the decades immediately following the crusade of the 1920s American antievolutionists were buoyed by reports of a creationist revival in Europe, especially in England, where creationism was thought to be all but dead. The Victoria Institute in London, a haven for English creationists in the nineteenth century, had by the 1920s become a stronghold of theistic evolution. When Price visited the institute in 1925 to receive its Langhorne-Orchard Prize for an essay on "Revelation and Evolution," several members protested his attempt to export the fundamentalist controversy to England. Even evangelicals refused to get caught up in the turmoil that engulfed the United States. As historian George Marsden has explained, English evangelicals, always a minority, had developed a stronger tradition of theological toleration than revivalist Americans, who until the twentieth century had never experienced minority status. Thus while the displaced Americans fought to recover their lost position, English evangelicals adopted a nonmilitant live-and-let-live philosophy that stressed personal piety.[48]

The sudden appearance of a small but vocal group of British creationists in the early 1930s caught nearly everyone by surprise. The central figure in this movement was Douglas Dewar (1875–1957), a Cambridge graduate and amateur ornithologist, who had served for decades as a lawyer in the Indian Civil Service. Originally an evolutionist, he had gradually become convinced of the necessity of adopting "a provisional hypothesis of special creation . . . supplemented by a theory of evolution." This allowed him to accept unlimited development within biological families. His published views, unlike those of most American creationists, betrayed little biblical influence. His greatest intellectual debt was not to Moses but to a French zoologist, Louis Vialleton (1859–1929), who had attracted considerable attention in the 1920s for suggesting a theory of discontinuous evolution, which antievolutionists eagerly—but erroneously—equated with special creation.[49]

Soon after announcing his conversion to creationism in 1931, Dewar submitted a short paper on mammalian fossils to the Zoological Society of London, of which he was a member. The secretary of the society subsequently rejected the piece, noting that a competent referee thought Dewar's evidence "led to no valuable conclusion." Such treatment infuriated Dewar and convinced him that evolution had become "a scientific creed." Those who questioned scientific orthodoxy, he complained, "are deemed unfit to hold scientific offices; their articles are rejected by newspapers or journals; their contributions are refused by scientific societies, and publishers decline to publish their books except at the author's expense. Thus the independents are today pretty effectually muzzled." Because of such experiences Dewar and other British dissidents in 1932 organized the Evolution Protest Movement, which after two decades claimed a membership of two hundred.[50]

HENRY M. MORRIS AND THE REVIVAL OF CREATIONISM

In 1964 one historian predicted that "a renaissance of the [creationist] movement is most unlikely." And so it seemed. But even as these words were penned, a major revival was under way, led by a Texas engineer, Henry M. Morris (b. 1918). Raised a nominal Southern Baptist, and as such a believer in creation, Morris as a youth had drifted unthinkingly into evolutionism and religious indifference. A thorough study of the Bible following graduation from college convinced him of its absolute truth and prompted him to reevaluate his belief in evolution. After an intense period of soul-searching he concluded that creation had taken place in six literal days, because the Bible clearly said so and "God doesn't lie." Corroborating evidence came from the book of nature. While sitting in his office at Rice Institute, where he was teaching civil engineering, he would study the butterflies and wasps that flew in through the window; being familiar with structural design, he calculated the improbability of such complex creatures developing by chance. Nature as well as the Bible seemed to argue for creation.[51]

For assistance in answering the claims of evolutionists, he found little creationist literature of value apart from the writings of Rimmer and Price. Although he rejected Price's peculiar theology, he took an immediate liking to the Adventist's flood geology and incorporated it into a little book, *That You Might Believe* (1946), the first book, so far as he knew, "published since the Scopes trial in which a scientist from a secular university advocated recent special creation and a worldwide flood." In the late 1940s he joined the American Scientific Affiliation— just in time to protest Kulp's attack on Price's geology. But his words fell largely on deaf ears. In 1953 when he presented some of his own views on the flood to the ASA, one of the few compliments came from a young theologian, John C. Whitcomb, Jr., who belonged to the Grace Brethren. The two subsequently became friends and decided to collaborate on a major defense of the Noachian flood. By the time they finished their project, Morris had earned a Ph.D. in hydraulic engineering from the University of Minnesota and was chairing the civil engineering department at Virginia Polytechnic Institute; Whitcomb was teaching Old Testament studies at Grace Theological Seminary in Indiana.[52]

In 1961 they brought out *The Genesis Flood*, the most impressive contribution to strict creationism since the publication of Price's *New Geology* in 1923. In many respects their book appeared to be simply "a reissue of G. M. Price's views, brought up to date," as one reader described it. Beginning with a testimony to their belief in "the verbal inerrancy of Scripture," Whitcomb and Morris went on to argue for a recent creation of the entire universe, a Fall that triggered the second law of thermodynamics, and a worldwide flood that in one year laid down most of the geological strata. Given this history, they argued, "the last refuge of the case for evolution immediately vanishes away, and the record of the rocks becomes a tremendous witness . . . to the holiness and justice and power of the living God of Creation!"[53]

Despite the book's lack of conceptual novelty, it provoked an intense debate among evangelicals. Progressive creationists denounced it as a travesty on geology that threatened to set back the cause of Christian science a generation, while strict creationists praised it for making biblical catastrophism intellectually respectable. Its appeal, suggested one critic, lay primarily in the fact that, unlike previous creationist works, it "looked *legitimate* as a scientific contribution," accompanied as it was by footnotes and other scholarly appurtenances. In responding to their detractors, Whitcomb and Morris repeatedly refused to be drawn into a scientific debate, arguing that "the real issue is not the correctness of the interpretation of various details of the geological data, but simply what God has revealed in His Word concerning these matters."[54]

Whatever its merits, *The Genesis Flood* unquestionably "brought about a stunning renaissance of flood geology," symbolized by the establishment in 1963 of the Creation Research Society. Shortly before the publication of his book Morris had sent the manuscript to Walter E. Lammerts (b. 1904), a Missouri-Synod Lutheran with a doctorate in genetics from the University of California. As an undergraduate at Berkeley Lammerts had discovered Price's *New Geology*, and during the early 1940s, while teaching at UCLA, he had worked with Price in the Creation-Deluge Society. After the mid-1940s, however, his interest in creationism had flagged—until awakened by reading the Whitcomb and Morris manuscript. Disgusted by the ASA's flirtation with evolution, he organized in the early 1960s a correspondence network with Morris and eight other strict creationists, dubbed the "team of ten." In 1963 seven of the ten met with a few other likeminded scientists at the home of a team member in Midland, Michigan, to form the Creation Research Society (CRS).[55]

The society began with a carefully selected eighteen-man "innercore steering committee," which included the original team of ten. The composition of this committee reflected, albeit imperfectly, the denominational, regional, and professional bases of the creationist revival. There were six Missouri-Synod Lutherans, five Baptists, two Seventh-day Adventists, and one each from the Reformed Presbyterian Church, the Reformed Christian Church, the Church of the Brethren, and an independent Bible church. (Information about one member is not available.) Eleven lived in the Midwest, three in the South, and two in the Far West. The committee included six biologists, but only one geologist, an independent consultant with a master's degree. Seven members taught in church-related colleges, five in state institutions; the others worked for industry or were self-employed.[56]

To avoid the creeping evolutionism that had infected the ASA and to ensure that the society remained loyal to the Price-Morris tradition, the CRS required members to sign a statement of belief accepting the inerrancy of the Bible, the special creation of "all basic types of living things," and a worldwide deluge. It restricted membership to Christians only. (Although creationists liked to stress the scientific evidence for their position, one estimated that "only about five percent of evolutionists-turned-creationists did so on the basis of the over-whelming evidence for creation in the world of nature"; the remaining 95 percent became creationists because they believed in the Bible.) To legitimate its claim to being a scientific society, the CRS published a quarterly journal and limited full membership to persons possessing a graduate degree in a scientific discipline.[57]

At the end of its first decade the society claimed 450 regular members, plus 1,600 sustaining members, who failed to meet the scientific qualifications. Eschewing politics, the CRS devoted itself almost exclusively to education and research, funded "at very little expense, and . . . with no expenditure of public money." CRS-related projects included expeditions to search for Noah's ark, studies of fossil human footprints and pollen grains found out of the predicted evolutionary order, experiments on radiation-produced mutations in plants, and theoretical studies in physics demonstrating a recent origin of the earth. A number of members collaborated in preparing a biology textbook based on creationist principles. In view of the previous history of creation science, it was an auspicious beginning.[58]

While the CRS catered to the needs of scientists, a second, predominantly lay organization carried creationism to the masses. Created in 1964 in the wake of interest generated by *The Genesis Flood*, the Bible-Science Association came to be identified by many with one man: Walter Lang, an ambitious Missouri-Synod pastor who self-consciously prized spiritual insight above scientific expertise. As editor of the widely circulated *Bible-Science Newsletter* he vigorously promoted the Price-Morris line—and occasionally provided a platform for individuals on the fringes of the creationist movement, such as those who questioned the heliocentric theory and who believed that Einstein's theory of relativity "was invented in order to circumvent the evidence that the earth is at rest." Needless to say, the pastor's broad-mindedness greatly embarrassed creationists seeking scientific respectability, who feared that such bizarre behavior would tarnish the entire movement.[59]

SCIENTIFIC CREATIONISM

The creationists revival of the 1960s attracted little public attention until late in the decade, when fundamentalists became aroused about the federally funded Biological Sciences Curriculum Study texts, which featured evolution, and the California State Board of Education voted to require public-school textbooks to include creation along with evolution. This decision resulted in large part from the efforts of two southern California housewives, Nell Segraves and Jean Sumrall, associates of both the Bible-Science Association and the CRS. In 1961 Segraves learned of the U.S. Supreme Court's ruling in the Madalyn Murray case protecting atheist students from required prayers in public schools. Murray's ability to shield her child from religious exposure suggested to Segraves that creationist parents like herself "were entitled to protect our children from the influence of beliefs that would be offensive to our religious beliefs." It was this line of argument that finally persuaded the Board of Education to grant creationists equal rights.[60]

Flushed with victory, Segraves and her son Kelly in 1970 joined an effort to organize a Creation-Science Research Center (CSRC), affiliated with Christian Heritage College in San Diego, to prepare creationist literature suitable for adoption in public schools. Associated with them in this enterprise was Henry Morris, who resigned his position at Virginia Polytechnic Institute to help establish a center for creation research. Because of differences in personalities and objectives, the Segraveses in 1972 left the college, taking the CSRC with them; Morris thereupon set up a new research division at the college, the Institute for Creation Research (ICR), which, he announced with obvious relief, would be "controlled and operated by scientists" and would engage in research and education, not political action. During the 1970s Morris added five scientists to his staff and, funded largely by small gifts and royalties from institute publications, turned the ICR into the world's leading center for the propagation of strict creationism.[61] Meanwhile, the CSRC continued campaigning for the legal recognition of special creation, often citing a direct relationship between the acceptance of evolution and the breakdown of law and order. Its own research, the CSRC announced, proved that evolution fostered "the moral decay of spiritual values which contribute to the destruction of mental health and . . . [the prevalence of] divorce, abortion, and rampant venereal disease."[62]

The 1970s witnessed a major shift in creationist tactics. Instead of trying to outlaw evolution, as they had done in the 1920s, antievolutionists now fought to give creation equal time. And instead of appealing to the authority of the Bible, as Morris and Whitcomb had done as recently as 1961, they consciously downplayed the Genesis story in favor of what they called "scientific creationism." Several factors no doubt contributed to this shift. One sociologist has suggested that creationists began stressing the scientific legitimacy of their enterprise because "their theological legitimation of reality was no longer sufficient for maintaining their world and passing on their world view to their children." But there were also practical considerations. In 1968 the U.S. Supreme Court declared the Arkansas antievolution law unconstitutional, giving creationists reason to suspect that legislation requiring the teaching of biblical creationism would meet a similar fate. They also feared that requiring the biblical account "would open the door to a wide variety of interpretations of Genesis" and produce demands for the inclusion of non-Christian versions of creation.[63]

In view of such potential hazards, Morris recommended that creationists ask public schools to teach "only the scientific aspects of creationism," which in practice meant leaving out all references to the six days of Genesis and Noah's ark and focusing instead on evidence for a recent worldwide catastrophe and on arguments against evolution. Thus the product remained virtually the same; only the packaging changed. The ICR textbook *Scientific Creationism* (1974), for example, came in two editions: one for public schools, containing no references to the Bible, and another for use in Christian schools that included a chapter on "Creation According to Scripture."[64]

In defending creation as a scientific alternative to evolution, creationists relied less on Francis Bacon and his conception of science and more on two new philosopher-heroes: Karl Popper and Thomas Kuhn. Popper required all scientific theories to be falsifiable; since evolution could not be falsified, reasoned the creationists, it was by definition not science. Kuhn described scientific progress in terms of competing models or paradigms rather than the accumulation of objective knowledge. Thus creationists saw no reason why their flood-geology model should not be allowed to compete on an equal scientific basis with the evolution model. In selling this two-model approach to school boards, creationists were advised:

> Sell more SCIENCE. . . . Who can object to teaching more science? What is controversial about that? . . . do not use the word "creationism." Speak only of science. Explain that withholding scientific information contradicting evolution amounts to "censorship" and smacks of getting into the province of religious dogma. . . . Use the "censorship" label as one who is against censoring science. YOU are for science; anyone else who wants to censor scientific data is an old fogey and too doctrinaire to consider.

This tactic proved extremely effective, at least initially. Two state legislatures, in Arkansas and Louisiana, and various school boards adopted the two-model approach, and an informal poll of school-board members in 1980 showed that only 25 percent favored teaching nothing but evolution. In 1982, however, a federal judge declared the Arkansas law, requiring a "balanced treatment" of creation and evolution, to be unconstitutional.[65] Three years later a similar decision was reached regarding the Louisiana law.

Except for the battle to get scientific creationism into public schools, nothing brought more attention to the creationists than their public debates with prominent evolutionists, usually held on college campuses. During the 1970s the ICR staff alone participated in more than a hundred of these contests and, according to their own reckoning, never lost one. Although Morris preferred delivering straight lectures—and likened debates to the bloody confrontations between Christians and lions in ancient Rome—he recognized their value in carrying the creationist message to "more non-Christians and non-creationists than almost any other method." Fortunately for him, an associate, Duane T. Gish, holder of a doctorate in biochemistry from the University of California, relished such confrontations. If the mild-mannered, professorial Morris was the Darwin of the creationist movement, then the bumptious Gish was its Huxley. He "hits the floor running" just like a bulldog, observed an admiring colleague; and "I go for the jugular vein," added Gish himself. Such enthusiasm helped draw crowds of up to five thousand.[66]

Early in 1981 the ICR announced the fulfillment of a recurring dream among creationists: a program offering graduate degrees in various creation-oriented sciences. Besides hoping to fill an anticipated demand for teachers trained in scientific creationism, the ICR wished to provide an academic setting where creationist students would be free from discrimination. Over the years a number of creationists had reportedly been kicked out of secular universities because of their heterodox views, prompting leaders to warn graduate students to keep silent, "because if you don't, in almost 99 percent of the cases you will be asked to leave." To avoid anticipated harassment, several graduate students took to using pseudonyms when writing for creationist publications.[67]

Creationists also feared—with good reason—the possibility of defections while their students studied under evolutionists. Since the late 1950s the Seventh-day Adventist Church had invested hundreds of thousands of dollars to staff its Geoscience Research Institute with well-trained young scientists, only to discover that in several instances exposure to orthodox science had destroyed belief in strict creationism. To reduce the incidence of apostasy, the church established its own graduate programs at Loma Linda University, where George McCready Price had once taught.[68]

It is still too early to assess the full impact of the creationist revival sparked by Whitcomb and Morris, but its influence, especially among evangelical Christians, seems to have been immense. Not least, it has elevated the strict creationism of Price and Morris to a position of apparent orthodoxy. It has also endowed creationism with a measure of scientific respectability unknown since the deaths of Guyot and Dawson. Yet it is impossible to determine how much of the creationists' success stemmed from converting evolutionists as opposed to mobilizing the already converted, and how much it owed to widespread disillusionment with established science. A sociological survey of church members in northern California in 1963 revealed that over a fourth of those polled—30 percent of Protestants and 28 percent of Catholics—were already opposed to evolution when the creationist revival began.[69] Broken down by denomination, it showed:

Liberal Protestants (Congregationalists, Methodists, Episcopalians, Disciples)	11%
Moderate Protestants (Presbyterians, American Lutherans, American Baptists)	29%
Church of God	57%
Missouri-Synod Lutherans	64%
Southern Baptists	72%
Church of Christ	78%
Nazarenes	80%
Assemblies of God	91%
Seventh-day Adventists	94%

Thus the creationists launched their crusade having a large reservoir of potential support.

But has belief in creationism increased since the early 1960s? The scanty evidence available suggests that it has. A nationwide Gallup poll in 1982, cited at the beginning of this paper, showed that nearly as many Americans (44 percent) believed in a recent special creation as accepted theistic (38 percent) or nontheistic (9 percent) evolution. These figures, when compared with the roughly 30 percent of northern California church members who opposed evolution in 1963, suggest, in a grossly imprecise way, a substantial gain in the actual number of American creationists. Bits and pieces of additional evidence lend credence to this conclusion. For example, in 1935 only 36 percent of the students at Brigham Young University, a Mormon school, rejected human evolution; in 1973 the percentage had climbed to 81. Also, during the 1970s both the Missouri-Synod Lutheran and Seventh-day Adventist churches, traditional bastions of strict creationism, took strong measures to reverse a trend toward greater toleration of progressive creationism. In at least these instances, strict creationism did seem to be gaining ground.[70]

Unlike the antievolution crusade of the 1920s, which remained confined mainly to North America, the revival of the 1960s rapidly spread overseas as American creationists and their books circled the globe. Partly as a result of stimulation from America, including the publication of a British edition of The Genesis Flood in 1969, the lethargic Evolution Protest Movement in Great Britain was revitalized; and two new creationist organizations, the Newton Scientific Association and the Biblical Creation Society, sprang into existence.[71] On the Continent the Dutch assumed the lead in promoting creationism, encouraged by the translation of books on flood geology and by visits from ICR scientists. Similar developments occurred elsewhere in Europe, as well as in Australia, Asia, and South America. By 1980 Morris's books alone had been translated into Chinese, Czech, Dutch, French, German, Japanese, Korean, Portuguese, Russian, and Spanish. Strict creationism had become an international phenomenon.[72]

NOTES

I would like to thank David C. Lindberg for his encouragement and criticism, Rennie B. Schoepflin for his research assistance, and the Graduate School Research Committee of the University of Wisconsin-Madison for financial support during the preparation of this paper. An abridged version appeared as "Creationism in 20th-Century America" in Science 218 (1982): 538–544.

1. "Poll Finds Americans Split on Creation Idea," New York Times, 29 Aug. 1982, p. 22. Nine percent of the respondents favored an evolutionary process in which God played no part, 38 percent believed God directed the evolutionary process, and 9 percent had no opinion. Regarding Dawson and Guyot, see Edward J. Pfeifer, "United States," in The Comparative Reception of Darwinism, ed. Thomas F. Glick (Austin: Univ. of Texas Press, 1974), p. 203; and Asa Gray, Darwiniana: Essays and Reviews Pertaining to Darwinism, ed. A. Hunter Dupree (Cambridge: Harvard Univ. Press, 1963), pp. 202–203. In The Darwinian Revolution: Science Red in Tooth and Claw (Chicago: Univ. of Chicago Press, 1979), Michael Ruse argues that most British biologists were evolutionists by the mid-1860s, while David L. Hull, Peter D. Tessner, and Arthur M. Diamond point out in "Planck's Principle," Science 202 (1978): 721, that more than a quarter of British scientists continued to reject the evolution of species as late as 1869. On the acceptance of evolution among religious

leaders see, e.g., Frank Hugh Foster, *The Modern Movement in American Theology: Sketches in the History of American Protestant Thought from the Civil War to the World War* (New York: Fleming H. Revell Co., 1939), pp. 38–58; and Owen Chadwick, *The Victorian Church*, Part 2, 2d ed. (London: Adam & Charles Black, 1972), pp. 23–24.

2. William G. McLoughlin, Jr., *Modern Revivalism: Charles Grandison Finney to Billy Graham* (New York: Ronald Press, 1959), p. 213. In *Protestant Christianity Interpreted through Its Development* (New York: Charles Scribner's Sons, 1954), p. 227, John Dillenberger and Claude Welch discuss the conservatism of the common people. On the attitudes of premillennialists see Robert D. Whalen, "Millenarianism and Millennialism in America, 1790–1880" (Ph.D. diss., State University of New York at Stony Brook, 1972), pp. 219–229; and Ronald L. Numbers, "Science Falsely So-Called: Evolution and Adventists in the Nineteenth Century," *Journal of the American Scientific Affiliation* 27 (March 1975): 18–23.

3. Ronald L. Numbers, *Creation by Natural Law: Laplace's Nebular Hypothesis in American Thought* (Seattle: Univ. of Washington Press, 1977), pp. 89–90; Bernard Ramm, *The Christian View of Science and Scripture* (Grand Rapids, Mich.: Wm. B. Eerdmans, 1954), pp. 195–198. On the influence of the *Scofield Reference Bible* see Ernest R. Sandeen, *The Roots of Fundamentalism: British and American Millenarianism, 1800–1930* (Chicago: Univ. of Chicago Press, 1971), p. 222.

4. Charles F. O'Brien, *Sir William Dawson: A Life in Science and Religion* (Philadelphia: American Philosophical Society, 1971). On Guyot and his influence see Numbers, *Creation by Natural Law*, pp. 91–100. On the popularity of the Guyot-Dawson view, also associated with the geologist James Dwight Dana, see William North Rice, *Christian Faith in an Age of Science*, 2d ed. (New York: A. C. Armstrong & Son, 1904), p. 101; and Dudley Joseph Whitney, "What Theory of Earth History Shall We Adopt?" *Bible Champion* 34 (1928): 616.

5. H. L. Hastings, preface to the 1889 edition of *The Errors of Evolution: An Examination of the Nebular Theory, Geological Evolution, the Origin of Life, and Darwinism*, by Robert Patterson, 3d ed. (Boston: Scriptural Tract Repository, 1893), p. iv. On the Darwinian debate see James R. Moore, *The Post-Darwinian Controversies: A Study of the Protestant Struggle to Come to Terms with Darwin in Great Britain and America, 1870–1900* (Cambridge: Cambridge Univ. Press, 1979).

6. G. L. Young, "Relation of Evolution and Darwinism to the Question of Origins," *Bible Student and Teacher* 11 (July 1909): 41. On anti-Darwinian books see "Evolutionism in the Pulpit," in *The Fundamentals*, 12 vols. (Chicago: Testimony Publishing Co., 1910–1915), 8:28–30. See also Peter J. Bowler, *The Eclipse of Darwinism: Anti-Darwinian Evolution Theories in the Decades around 1900* (Baltimore: Johns Hopkins Univ. Press, 1983).

7. Philip Mauro, "Modern Philosophy," in *The Fundamentals* 2:85–105; and J. J. Reeve, "My Personal Experience with the Higher Criticism," ibid., 3:98–118.

8. G. Frederick Wright, *Story of My Life and Work* (Oberlin, Ohio: Bibliotheca Sacra Co., 1916); idem, "The First Chapter of Genesis and Modern Science,"

Homiletic Review 35 (1898): 392–399; idem, introduction to *The Other Side of Evolution: An Examination of Its Evidences*, by Alexander Patterson (Chicago: Winona Pub. Co., 1902), pp. xvii–xix.

9. George Frederick Wright, "The Passing of Evolution," in *The Fundamentals* 7:5–20. The Scottish theologian James Orr contributed an equally tolerant essay, "Science and Christian Faith," ibid., 4:91–104.

10. Lawrence W. Levine, *Defender of the Faith—William Jennings Bryan: The Last Decade, 1915–1925* (New York: Oxford Univ. Press, 1965), p. 277.

11. Ibid., p. 272. The quotation about America going mad appears in Roland T. Nelson, "Fundamentalism and the Northern Baptist Convention" (Ph.D. diss., University of Chicago, 1964), p. 319. On antievolution legislation see Maynard Shipley, *The War on Modern Science: A Short History of the Fundamentalist Attacks on Evolution and Modernism* (New York: Alfred A. Knopf, 1927); and idem, "Growth of the Anti-Evolution Movement," *Current History* 32 (1930): 330–332. On Bryan's catalytic role see Ferenc Morton Szasz, *The Divided Mind of Protestant America, 1889–1930* (University: Univ. of Alabama Press, 1982), pp. 107–116.

12. Levine, *Defender of the Faith*, pp. 261–265.

13. Ibid., pp. 266–267. Mrs. Bryan's statement appears in Wayne C. Williams, *William Jennings Bryan* (New York: G. P. Putnam, 1936), p. 448.

14. William Jennings Bryan, *In His Image* (New York: Fleming H. Revell Co., 1922), pp. 94, 97–98. "The Menace of Darwinism" appears in this work as chap. 4, "The Origin of Man." Bryan apparently borrowed his account of the evolution of the eye from Patterson, *The Other Side of Evolution*, pp. 32–33.

15. Paolo E. Coletta, *William Jennings Bryan*, vol. 3, *Political Puritan, 1915–1925* (Lincoln: Univ. of Nebraska Press, 1969), p. 230.

16. "Progress of Anti-Evolution," *Christian Fundamentalist* 2 (1929): 13. Bryan's reference to a "scientific soviet" appears in Levine, *Defender of the Faith*, p. 289. Bryan gives the estimate of nine-tenths in a letter to W. A. McRae, 5 Apr. 1924, box 29, Bryan Papers, Library of Congress.

17. "Fighting Evolution at the Fundamentals Convention," *Christian Fundamentals in School and Church* 7 (July–Sept. 1925): 5. The best state histories of the antievolution crusade are Kenneth K. Bailey, "The Enactment of Tennessee's Antievolution Law," *Journal of Southern History* 16 (1950): 472–510; Willard B. Gatewood, Jr., *Preachers, Pedagogues and Politicians: The Evolution Controversy in North Carolina, 1920–1927* (Chapel Hill: Univ. of North Carolina Press, 1966); and Virginia Gray, "Anti-Evolution Sentiment and Behavior: The Case of Arkansas," *Journal of American History* 57 (1970): 352–366. Ferenc Morton Szasz stresses the urban dimension of the crusade in "Three Fundamentalist Leaders: The Roles of William Bell Riley, John Roach Straton, and William Jennings Bryan in the Fundamentalist-Modernist Controversy" (Ph.D. diss., University of Rochester, 1969), p. 351.

18. George Herbert Betts, *The Beliefs of 700 Ministers and Their Meaning for Religious Education* (New York: Abingdon Press, 1929), pp. 26, 44; Milton L. Rudnick, *Fundamentalism and the Missouri Synod: A Historical Study of Their Interaction and Mutual Influence* (St. Louis: Concordia Publishing House, 1966), pp. 88–90; Sandeen, *Roots of Fundamentalism*, pp. 266–268, which discusses

the premillennialists. Lutheran reluctance to join the crusade is also evident in Szasz, "Three Fundamentalist Leaders," p. 279. For examples of prominent fundamentalists who stayed aloof from the antievolution controversy see Ned B. Stonehouse, *J. Gresham Machen: A Biographical Memoir* (Grand Rapids, Mich.: Wm. B. Eerdmans, 1954), pp. 401–402, and William Bryant Lewis, "The Role of Harold Paul Sloan and his Methodist League for Faith and Life in the Fundamentalist-Modernist Controversy of the Methodist Episcopal Church" (Ph.D. diss., Vanderbilt University, 1963), pp. 86–88.

19. Edward Lassiter Clark, "The Southern Baptist Reaction to the Darwinian Theory of Evolution" (Ph.D. diss., Southwestern Baptist Theological Seminary, 1952), p. 154; James J. Thompson, Jr., "Southern Baptists and the Antievolution Controversy of the 1920's," *Mississippi Quarterly* 29 (1975–1976): 65–81; Lefferts A. Loetscher, *The Broadening Church: A Study of Theological Issues in the Presbyterian Church since 1869* (Philadelphia: Univ. of Pennsylvania Press, 1954), p. 111; John L. Morrison, "American Catholics and the Crusade against Evolution," *Records of the American Catholic Historical Society of Philadelphia* 64 (1953): 59–71. Norman F. Furniss, *The Fundamentalist Controversy, 1918–1931* (New Haven: Yale Univ. Press, 1954), includes chapter-by-chapter surveys of seven denominations.

20. T. T. Martin, *Hell and the High School: Christ or Evolution, Which?* (Kansas City: Western Baptist Pub. Co., 1923), pp. 164–165. On Riley see Marie Acomb Riley, *The Dynamic of a Dream: The Life Story of Dr. William B. Riley* (Grand Rapids, Mich.: Wm. B. Eerdmans, 1938), pp. 101–102; and Szasz, *The Divided Mind of Protestant America*, pp. 89–91. George M. Marsden, *Fundamentalism and American Culture: The Shaping of Twentieth-Century Evangelicalism, 1870–1925* (New York: Oxford Univ. Press, 1980), pp. 169–170, stresses the interdenominational character of the antievolution crusade. On the expansion of public education see Kenneth K. Bailey, *Southern White Protestantism in the Twentieth Century* (New York: Harper & Row, 1964), pp. 72–73.

21. Both quotations come from Howard K. Beale, *Are American Teachers Free? An Analysis of Restraints upon the Freedom of Teaching in American Schools* (New York: Charles Scribner's Sons, 1936), pp. 249–251.

22. [William B. Riley], "The Evolution Controversy," *Christian Fundamentals in School and Church* 4 (Apr.–June 1922): 5; Bryan, *In His Image*, p. 94.

23. L. S. K[eyser], "No War against Science—Never!" *Bible Champion* 31 (1925): 413. On the fundamentalist affinity for Baconianism see Marsden, *Fundamentalism and American Culture*, pp. 214–215.

24. William Bateson, "Evolutionary Faith and Modern Doubts," *Science* 55 (1922): 55–61. The creationists' use of Bateson provoked the evolutionist Henry Fairfield Osborn into repudiating the British scientist; see Osborn, *Evolution and Religion in Education: Polemics of the Fundamentalist Controversy of 1922 to 1926* (New York: Charles Scribner's Sons, 1926), p. 29. On proevolution resolutions see Shipley, *War on Modern Science*, p. 384.

25. Heber D. Curtis to W. J. Bryan, 22 May 1923, box 37, Bryan Papers, Library of Congress. Two physicians, Arthur I. Brown of Vancouver and Howard A. Kelly of Johns Hopkins, achieved prominence in the fundamentalist movement, but Kelly leaned toward theistic evolution.

26. William D. Edmondson, "Fundamentalist Sects of Los Angeles, 1900–1930" (Ph.D. diss., Claremont Graduate School, 1969), pp. 276–336; Steward G. Cole, *The History of Fundamentalism* (New York: Richard R. Smith, 1931), pp. 264–265; F. J. B[oyer], "Harry Rimmer, D.D.," *Christian Faith and Life* 45 (1939): 6–7; "Two Great Field Secretaries—Harry Rimmer and Dr. Arthur I. Brown," *Christian Fundamentals in School and Church* 8 (July–Sept. 1926): 17. Harry Rimmer refers to his medical vocabulary in *The Harmony of Science and Scripture*, 11th ed. (Grand Rapids, Mich.: Wm. B. Eerdmans, 1945), p. 14.

27. Edmondson, "Fundamentalist Sects of Los Angeles," pp. 329–330, 333–334. Regarding the $100 reward, see "World Religious Digest," *Christian Faith and Life* 45 (1939): 215.

28. This and the following paragraphs on Price closely follow my account in "'Sciences of Satanic Origin': Adventist Attitudes toward Evolutionary Biology and Geology," *Spectrum* 9 (Jan. 1979): 22–24.

29. George McCready Price, *Illogical Geology: The Weakest Point in the Evolution Theory* (Los Angeles: Modern Heretic Co., 1906), p. 9. Four years earlier Price had published his first antievolution book, *Outlines of Modern Science and Modern Christianity* (Oakland, Calif.: Pacific Press, 1902).

30. David Starr Jordan to G. M. Price, 5 May 1911, Price Papers, Andrews University Library.

31. George McCready Price, *The New Geology* (Mountain View, Calif.: Pacific Press, 1923), pp. 637–638. Price first announced the discovery of his law in *The Fundamentals of Geology and Their Bearings on the Doctrine of a Literal Creation* (Mountain View, Calif.: Pacific Press, 1913), p. 119.

32. Charles Schuchert, review of *The New Geology*, by George McCready Price, *Science* 59 (1924): 486–487; Harry Rimmer, *Modern Science, Noah's Ark and the Deluge* (Los Angeles: Research Science Bureau, 1925), p. 28.

33. *Science* 63 (1926): 259.

34. Howard A. Kelly to W. J. Bryan, 15 June 1925; Louis T. More to W. J. Bryan, 7 July 1925; and Alfred W. McCann to W. J. Bryan, 30 June 1925, box 47, Bryan Papers, Library of Congress. Regarding Price, see Numbers, "'Sciences of Satanic Origin,'" p. 24.

35. Numbers, "'Sciences of Satanic Origin,'" p. 24; Levine, *Defender of the Faith*, p. 349.

36. W. J. Bryan to Howard A. Kelly, 22 June 1925, box 47, Bryan Papers, Library of Congress. In a letter to the editor of *The Forum* 70 (1923): 1852–1853, Bryan asserted that he had never taught that the world was made in six literal days. I am indebted to Paul M. Waggoner for bringing this document to my attention.

37. W. B. Riley and Harry Rimmer, *A Debate: Resolved, That the Creative Days in Genesis Were Aeons, Not Solar Days* (undated pamphlet); [W. B. Riley], "The Creative Week," *Christian Fundamentalist* 4 (1930): 45; Price, *Outlines*, pp. 125–127; idem, *The Story of the Fossils* (Mountain View, Calif.: Pacific Press, 1954), p. 39. On Rimmer's acceptance of a local flood see Robert D. Culver, "An Evaluation of *The Christian View of Science and Scripture* by Bernard Ramm from the Standpoint of Christian Theology," *Journal of the American Scientific Affiliation* 7 (Dec. 1955): 7.

38. Shipley, "Growth of the Anti-Evolution Movement," pp. 330–332; Szasz, *The Divided Mind of Protestant America*, pp. 117–125.

39. Beale, *Are American Teachers Free?* pp. 228–237; Willard B. Gatewood, Jr., ed., *Controversy in the Twenties: Fundamentalism, Modernism, and Evolution* (Nashville: Vanderbilt Univ. Press, 1969), p. 39. The quotation comes from Shipley, "Growth of the Anti-Evolution Movement," p. 330. See also Judith V. Grabiner and Peter D. Miller, "Effects of the Scopes Trial," *Science* 185 (1974): 832–837; and Estelle R. Laba and Eugene W. Gross, "Evolution Slighted in High-School Biology," *Clearing House* 24 (1950): 396–399.

40. Joel A. Carpenter, "Fundamentalist Institutions and the Rise of Evangelical Protestantism, 1929–1942," *Church History* 49 (1980): 62–75, provides an excellent analysis of this trend. For a typical statement of creationist frustration see George McCready Price, "Guarding the Sacred Cow," *Christian Faith and Life* 41 (1935): 124–127. The title for this section comes from Henry M. Morris, *The Troubled Waters of Evolution* (San Diego: Creation-Life Publishers, 1974), p. 13.

41. "Announcement of the Religion and Science Association," Price Papers, Andrews University; "The Religion and Science Association," *Christian Faith and Life* 42 (1936): 159–160; "Meeting of the Religion and Science Association," ibid., p. 209; Harold W. Clark, *The Battle over Genesis* (Washington: Review & Herald Publishing Association, 1977), p. 168. On the attitude of the Price faction see Harold W. Clark to G. M. Price, 12 Sept. 1937, Price Papers, Andrews University.

42. Numbers, "'Sciences of Satanic Origin,'" p. 26.

43. Ben F. Allen to the Board of Directors of the Creation-Deluge Society, 12 Aug. 1945 (courtesy of Molleurus Couperus). Regarding the fossil footprints, see the *Newsletters* of the Creation-Deluge Society for 19 Aug. 1944 and 17 Feb. 1945.

44. Numbers, "'Sciences of Satanic Origin,'" p. 25.

45. On the early years of the ASA see Alton Everest, "The American Scientific Affiliation—The First Decade," *Journal of the American Scientific Affiliation* 3 (Sept. 1951): 33–38.

46. J. Laurence Kulp, "Deluge Geology," ibid., 2, no. 1 (1950): 1–15; "Comment on the 'Deluge Geology' Paper of J. L. Kulp," ibid., 2 (June 1950): 2. Kulp mentions his initial skepticism of geology in a discussion of "Some Presuppositions in Evolutionary Thinking," ibid., 1 (June 1949): 20.

47. J. Frank Cassel, "The Evolution of Evangelical Thinking on Evolution," ibid., 11 (Dec. 1959): 26–27. For a fuller discussion see Ronald L. Numbers, "The Dilemma of Evangelical Scientists," in *Evangelicalism and Modern America*, ed. George M. Marsden (Grand Rapids, Mich.: Wm. B. Eerdmans, 1984), pp. 150–160.

48. Numbers, "'Sciences of Satanic Origin,'" p. 25; George Marsden, "Fundamentalism as an American Phenomenon: A Comparison with English Evangelicalism," *Church History* 46 (1977): 215–232; idem, *Fundamentalism and American Culture*, pp. 222–226.

49. Douglas Dewar, *The Difficulties of the Evolution Theory* (London: Edward Arnold & Co., 1931), p. 158; Arnold Lunn, ed., *Is Evolution Proved? A Debate*

between Douglas Dewar and H. S. Shelton (London: Hollis & Carter, 1947), pp. 1, 154; *Evolution Protest Movement Pamphlet No. 125* (Apr. 1965). On Vialleton see Harry W. Paul, *The Edge of Contingency: French Catholic Reaction to Scientific Change from Darwin to Duhem* (Gainesville: University Presses of Florida, 1979), pp. 99–100.

50. Douglas Dewar, "The Limitations of Organic Evolution," *Journal of the Victoria Institute* 64 (1932): 142; "EPM—40 Years On; Evolution—114 Years Off," supplement to *Creation* 1 (May 1972): no pagination.

51. R. Halliburton, Jr., "The Adoption of Arkansas' Anti-Evolution Law," *Arkansas Historical Quarterly* 23 (1964): 283; interviews with Henry M. Morris, 26 Oct. 1980 and 6 Jan. 1981. See also the autobiographical material in Henry M. Morris, *History of Modern Creationism* (San Diego: Master Book Publishers, 1984).

52. Interviews with Morris; Henry M. Morris, introduction to the revised edition, *That You Might Believe* (San Diego: Creation-Life Publishers, 1978), p. 10.

53. John C. Whitcomb, Jr. and Henry M. Morris, *The Genesis Flood: The Biblical Record and Its Scientific Implications* (Philadelphia: Presbyterian & Reformed Pub. Co., 1961), pp. xx, 451.

54. Henry M. Morris and John C. Whitcomb, Jr., "Reply to Reviews in the March 1964 Issue," *Journal of the American Scientific Affiliation* 16 (June 1964): 60. The statement regarding the appearance of the book comes from Walter Hearn, quoted in Vernon Lee Bates, "Christian Fundamentalism and the Theory of Evolution in Public School Education: A Study of the Creation Science Movement" (Ph.D. diss., University of California, Davis, 1976), p. 52. See also Frank H. Roberts, review of *The Genesis Flood,* by Henry M. Morris and John C. Whitcomb, Jr., *Journal of the American Scientific Affiliation* 16 (Mar. 1964): 28–29; J. R. Van de Fliert, "Fundamentalism and the Fundamentals of Geology," ibid., 21 (Sept. 1969): 69–81; and Walter E. Lammerts, "Introduction," Creation Research Society, *Annual,* 1964, no pagination. Among Missouri-Synod Lutherans, John W. Klotz, *Genes, Genesis, and Evolution* (St. Louis: Concordia Publishing House, 1955), may have had an even greater influence than Morris and Whitcomb.

55. Walter E. Lammerts, "The Creationist Movement in the United States: A Personal Account," *Journal of Christian Reconstruction* 1 (Summer 1974): 49–63. The first quotation comes from Davis A. Young, *Creation and the Flood: An Alternative to Flood Geology and Theistic Evolution* (Grand Rapids, Mich.: Baker Book House, 1977), p. 7.

56. Names, academic fields, and institutional affiliations are given in *Creation Research Society Quarterly* 1 (July 1964): [13]; for additional information I am indebted to Duane T. Gish, John N. Moore, Henry M. Morris, Harold Slusher, and William J. Tinkle.

57. *Creation Research Society Quarterly* 1 (July 1964): [13]; [Walter Lang], "Editorial Comments," *Bible-Science Newsletter* 16 (June 1978): 2. Other creationists have disputed the 5-percent estimate.

58. Lammerts, "The Creationist Movement in the United States," p. 63; Duane T. Gish, "A Decade of Creationist Research," *Creation Research Society*

Quarterly 12 (June 1975): 34–46; John N. Moore and Harold Schultz Slusher, eds., *Biology: A Search for Order in Complexity* (Grand Rapids, Mich.: Zondervan Publishing House, 1970).

59. Walter Lang, "Fifteen Years of Creationism," *Bible Science Newsletter* 16 (Oct. 1978): 1–3; "Editorial Comments," ibid., 15 (Mar. 1977): 2–3; "A Naturalistic Cosmology vs. a Biblical Cosmology," ibid., 15 (Jan.–Feb. 1977): 4–5; Gerald Wheeler, "The Third National Creation Science Conference," *Origins* 3 (1976): 101–102.

60. Bates, "Christian Fundamentalism," p. 58; "15 Years of Creationism," *Five Minutes with the Bible and Science,* supplement to *Bible-Science Newsletter* 17 (May 1979): 2; Nicholas Wade, "Creationists and Evolutionists: Confrontation in California," *Science* 178 (1972): 724–729. Regarding the BSCS texts see Gerald Skoog, "Topic of Evolution in Secondary School Biology Textbooks: 1900–1977," *Science Education* 63 (1979): 621–640; and "A Critique of BSCS Biology Texts," *Bible-Science Newsletter* 4 (15 Mar. 1966): 1. See also John A. Moore, "Creationism in California," *Daedalus* 103 (1974): 173–189; and Dorothy Nelkin, *The Creation Controversy: Science or Scripture in the Schools* (New York: W. W. Norton, 1982).

61. Henry M. Morris, "Director's Column," *Acts & Facts* 1 (June–July 1972): no pagination; Morris interview, 6 Jan. 1981.

62. Nell J. Segraves, *The Creation Report* (San Diego: Creation-Science Research Center, 1977), p. 17; "15 Years of Creationism," pp. 2–3.

63. Bates, "Christian Fundamentalism," p. 98; Henry M. Morris, "Director's Column," *Acts & Facts* 3 (Sept. 1974): 2. See also Edward J. Larson, "Public Science vs. Popular Opinion: The Creation-Evolution Legal Controversy" (Ph.D. diss., University of Wisconsin-Madison, 1984).

64. Morris, "Director's Column," p. 2; Henry M. Morris, ed., *Scientific Creationism,* General Edition (San Diego: Creation-Life Publishers, 1974).

65. The quotation comes from Russel H. Leitch, "Mistakes Creationists Make," *Bible-Science Newsletter* 18 (Mar. 1980): 2. Regarding school boards, see "Finding: Let Kids Decide How We Got Here," *American School Board Journal* 167 (Mar. 1980): 52; and Segraves, *Creation Report,* p. 24. On Popper's influence see, e.g., Ariel A. Roth, "Does Evolution Qualify as a Scientific Principle?" *Origins* 4 (1977): 4–10. In a letter to the editor of *New Scientist* 87 (21 Aug. 1980): 611, Popper affirmed that the evolution of life on earth was testable and, therefore, scientific. On Kuhn's influence see, e.g., Ariel A. Roth, "The Pervasiveness of the Paradigm," *Origins* 2 (1975): 55–57; Leonard R. Brand, "A Philosophic Rationale for a Creation-Flood Model," ibid., 1 (1974): 73–83; and Gerald W. Wheeler, *The Two-Taled Dinosaur: Why Science and Religion Conflict over the Origin of Life* (Nashville: Southern Publishing Association, 1975), pp. 192–210. For the judge's decision see "Creationism in Schools: The Decision in McLean versus the Arkansas Board of Education," *Science* 215 (1982): 934–943.

66. Henry M. Morris, "Two Decades of Creation: Past and Future," *Impact,* supplement to *Acts & Facts* 10 (Jan. 1981): iii; idem, "Director's Column," ibid., 3 (Mar. 1974): 2. The reference to Gish comes from an interview with Harold Slusher and Duane T. Gish, 6 Jan. 1981.

67. "ICR Schedules M.S. Programs," *Acts & Facts* 10 (Feb. 1981): 1–2. Evidence for alleged discrimination and the use of pseudonyms comes from: "Grand Canyon Presents Problems for Long Ages," *Five Minutes with the Bible and Science,* supplement to *Bible-Science Newsletter* 18 (June 1980): 1–2; interview with Ervil D. Clark, 9 Jan. 1981; interview with Steven A. Austin, 6 Jan. 1981; and interview with Duane T. Gish, 26 Oct. 1980, the source of the quotation.

68. Numbers, "'Sciences of Satanic Origin,'" pp. 27–28; Molleurus Couperus, "Tensions between Religion and Science," *Spectrum* 10 (Mar. 1980): 74–88.

69. William Sims Bainbridge and Rodney Stark, "Superstitions: Old and New," *Skeptical Inquirer* 4 (Summer, 1980): 20.

70. "Poll Finds Americans Split on Evolution Idea," p. 22; Harold T. Christensen and Kenneth L. Cannon, "The Fundamentalist Emphasis at Brigham Young University: 1935–1973," *Journal for the Scientific Study of Religion* 17 (1978): 53–57; "Return to Conservatism," *Bible-Science Newsletter* 11 (Aug. 1973): 1; Numbers, "'Sciences of Satanic Origin,'" pp. 27–28.

71. Eileen Barker, "In the Beginning: The Battle of Creationist Science against Evolutionism," in *On the Margins of Science: The Social Construction of Rejected Knowledge,* ed. Roy Wallis, Sociological Review Monograph 27 (Keele: University of Keele, 1979), pp. 179–200, who greatly underestimates the size of the E.P.M. in 1966; [Robert E. D. Clark], "Evolution: Polarization of Views," *Faith and Thought* 100 (1972–1973): 227–229; [idem], "American and English Creationists," ibid., 104 (1977): 6–8; "British Scientists Form Creationist Organization," *Acts & Facts* 2 (Nov.-Dec. 1973): 3; "EPM—40 Years On; Evolution—114 Years Off," supplement to *Creation* 1 (May 1972): no pagination.

72. W. J. Ouweneel, "Creationism in the Netherlands," *Impact,* supplement to *Acts & Facts* 7 (Feb. 1978): i–iv. Notices regarding the spread of creationism overseas appeared frequently in *Bible-Science Newsletter* and *Acts & Facts.* On translations see "ICR Books Available in Many Languages," *Acts & Facts* 9 (Feb. 1980): 2, 7.

*Robert Clouse claims that Jerry Falwell and Tim LaHaye, taken as repre-
sentatives of the New Christian Right, attempt unsuccessfully to combine two views of the millenium, both of which have been influential in
American history. Mr. Clouse is professor of history at Indiana State
University.*

By Robert G. Clouse

The New Christian Right, America, and the Kingdom of God*

O NE OF THE MAJOR CONCERNS of the American evan-
gelicals has been the coming of the Kingdom of God to earth. Beginning with the
colonial era in the writings of individuals such as Jonathan Edwards, it is possible to trace the currents of millennial thought to the present time. It is apparent
as one does this that there are two divergent views on the matter, often labelled
postmillennialism and premillennialism. At times the distinction between these
interpretations has not been clearly maintained, and both sides have adopted a
type of postmillennialism which teaches that the United States has a divine
mission as the last best hope of humanity. A recent example of an attempt to
merge these millennial interpretations is the New Christian Right and its interest
in the return of Christ and the millennium. Such eschatological confusion has
led to other contradictions in the thinking of the new group. To understand the
matter more fully it is necessary to describe the two major millennial views in
greater detail, to trace the development of millennialism in America, and to
illustrate the problems of confusing the two millennial outlooks from the teaching of two leading New Christian right representatives, Jerry Falwell and Tim
LaHaye.[1]

I. Definitions of the two major varieties of millennialism

Two basic understandings of the Second Coming, postmillennialism and
premillennialism, have been influential among believers in America. These la-

*This is a slightly revised form of a paper presented at a meeting on "The Public Face of Evan-
gelicalism" held at Huntington College, November 13, 1981. The author wishes to thank Professor Jack
Barlow of Huntington College for arranging the Conference.

[1]Jerry Falwell and Tim LaHaye have been chosen as examples of the New Christian Right because
of their reputations and because they have published several books on eschatology. Falwell has become
such a media favorite that he needs little introduction, and for LaHaye see "Door Interview, Tim
LaHaye," *The Wittenburg Door* (June–July 1980): 8–12.

bels, although helpful, are in a sense misleading, because the distinction involves much more than merely whether Christ returns before or after the millennium. The kingdom expected by the postmillennialist is quite different from that anticipated by the premillennialist, not only with respect to the time and manner in which it will be established but also with regard to the nature of the kingdom and the way Christ excercises his control over it. The postmillenarian believes that the Kingdom of God is extended through Christian preaching and teaching as a result of which the world will be Christianized and will enjoy a long period of peace and righteousness called the millennium. This new age will not be essentially different from the present and it emerges gradually as an ever-larger share of the world's population is converted to Christianity. Evil is not eliminated but is reduced to a minimum as the moral and spiritual influence of Christianity is heightened. During this age the church assumes a greater importance, and many social, economic, and educational problems are solved. The period closes with the second coming of Christ, the resurrection of the dead, and the final judgment.

The other major interpretation, premillennialism, claims that the Lord's return will be followed by a period of peace and righteousness before the last judgment, during which Christ will reign as king in person or through a select group of people. This kingdom is not to be established by the conversion of individual souls over a long period of time, but suddenly and by overwhelming power. The new age will be characterized by the conversion of the Jews and the reign of harmony in nature to such an extent that the desert will blossom like a rose and even ferocious beasts will be tame. Evil is held in check during this period by Christ, who rules with a rod of iron. Despite these idyllic conditions men are not satisfied and launch one last rebellion against God and his followers. This final exposure of evil is crushed by Christ and then the judgment is held. Many premillennialists have believed that during this golden age believers who have died will be resurrected with glorified bodies to help Christ rule the earth. Usually they also teach that the return of Christ will be preceded by certain signs such as a great tribulation, apostasy, wars, famines, earthquakes, and the appearance of the Antichrist.[2]

II. The development of postmillennialism and premillennialism in American thought

Although many colonial American clergymen were premillennialists, this view was eclipsed by postmillennialism during the eighteenth century. Postmillennialism, when coupled with the idea of the new Republic, would have serious implications for the growth of the nation's opinion of itself. One of the most influential American theologians, Jonathan Edwards, adopted this out-

---

[2]For some of the variety of millennial thought see Robert G. Clouse, ed., *The Meaning of the Millennium: Four Views* (Downers Grove, Ill.: InterVarsity Press, 1977).

look.[3] Millennial considerations were more important to Edwards than has been realized. In fact, he kept a notebook on the apocalypse which included the signs of the times that he believed were leading to the millenium. Edwards confessed his belief that there will be a golden age for the church on earth achieved through the ordinary process of preaching the gospel in the power of the Holy Spirit. He might have held his millennial views in an academic way had it not been for the Great Awakening of 1742, which filled him with excitement about the prospect of the coming of the new age. Edwards suggested that the revival then going on in New England surpassed any that had ever been seen and would lead to the establishment of the millennium in America. But when the revival died out a problem was posed for Edwards and those who followed his views. The failure of the Great Awakening made it difficult to explain how America was to lead the world to the new age. Among the solutions suggested for this dilemma, the one advanced by Edwards was a concern for revival on an international scale. However, his method did not prevail as a vehicle for continuing the millennial tradition. It was war with France which injected new meaning into eschatology, as the conflict gave the New England ministers a more appealing basis for millenialism than that which had been supplied by the revival.

In the course of the French and English wars the colonialists identified the defense of their land with the cause of God. Sermons reminded the faithful that the conduct of the French "bespeaks them the offspring of the Scarlet Whore, the Mother of Harlots, who is justly the abomination of the Earth."[4] If France would win the war, "cruel Papists would quickly fill the British Colonies, seize our Estates, abuse our Wives and Daughters, and barbarously murder us; as they have done the like in France and Ireland."[5] By 1760 many colonial pastors lost a clear distinction between the millennial reign of Christ and the future of their own political community. Military victories over the French seemed to be preparing the way for a millennium of civil and religious liberty.

After the French were defeated and forced to leave North America the colonialists anticipated the total destruction of Antichrist. This did not happen, however, but rather the power of tyranny appeared in a new form. The ministers applied millennialist ideology to the English when the Royal Government attempted to create an American Bishopric and enforce the Stamp Act. They taught that the prophecies of Revelation could be applied continuously throughout the Christian era. Consequently, prophecy could refer to the French, but also to the British if they acted in a 'tyrannical way. The cause of America was

---

[3]Jonathan Edwards, *Apocalyptic Writings*, ed. Stephan J. Stein (New Haven: Yale University Press, 1977). For the history of millennialism see Le Roy Edwin Froom, *The Prophetic Faith of Our Fathers* (Washington: Review and Herald, 1946-54), 4 vols. This work is probably the most complete treatment of millennial thought but it must be used carefully because it is biased in approach and perhaps of necessity rather fragmentary in its handling of many subjects.

[4]John Burt, *The Mercy of God to His People* (Newport, 1759), p. 4.

[5]Thomas Prince, *The Salvations of God in 1746* (Boston, 1746), p. 12.

identified with the preservation of civil and religious liberty against the forces of tyranny from Britain.

Freedom and faith flowed together in this outlook, twin ideas which America was destined by God to spread throughout the entire world. Consequently, at a time when the national consciousness was being formed, patriotic preachers and spokesmen were giving the new nation a transcendent postmillennial purpose. Despite the fact that this teaching was rooted in the work of scholars like Jonathan Edwards, a fundamental change had come over millennialism. The earlier teaching that God's people in the New World might be granted the spiritual power to prepare the world for Christ's kingdom through prayer and revival preaching was transformed into a conviction of the high political and moral destiny of the United States.

The millennium was related not only to contemporary events but also to the national structure of the American republic.[6]

During the nineteenth century postmillennial views of the destiny of America played a vital role in justifying national expansion.[7] Although there were other explanations for the nation's growth, the idea of a Christian republic progressing toward the golden age appealed to many people. Millennial nationalism was attractive because it harmonized the Republic with religious values, and it was fostered by ministers, teachers, and writers who played an important role in the intellectual life of the developing nation. They believed that millennialism could check the dangers to faith implicit in republican ideology. Yet, once America was identified with God's purpose on earth, believers were obligated to give it a place of great religious importance. They were forced to continue the teaching that the American achievement was God's special work and that the system was chosen to spread the same blessing to the rest of the world, thus preparing the way for the millennial reign of Christ.

The nineteenth century saw America develop from an idea to a reality. The westward movement coupled with the growth of population fostered an optimistic expansive attitude. As the settlers found new homes they identified with the land and became more self-consciously American. The frontier experience led to the development of a brash, exuberant, undisciplined outlook. One of the first expressions of this spirit was an aggressive nationalism directed against any foreigners who blocked the path of westward movement.

Later in the century, western expansion resulted in a war with Mexico and the elaboration of a doctrine called Manifest Destiny. This phrase, although not used until mid-century, was based upon ideas that were prevalent much earlier. As defined by a party journalist in 1845 it indicated "the fulfillment of our manifest destiny to overspread the continent allotted by Providence for the free development of our yearly multiplying millions."[8] What the writer meant was

---

[6]Early identification of the American cause with the millennium is capably treated in Nathan O. Hatch, *The Sacred Cause of Liberty, Republican Thought and the Millennium in Revolutionary New England* (New Haven: Yale University Press, 1977.)

[7]See Ernest L. Tuveson, *Redeemer Nation: The Idea of America's Millennial Role* (Chicago: University of Chicago Press, 1968).

[8]Richard N. Current et al., *American History, a Survey* (New York: Alfred A. Knopf, 1975), p. 341.

that the United States had a divine mission to take the whole of North America, by force if needed, and make room for its growing population while carrying the blessings of democracy to these areas.

With such driving energy the nation doubled in size by 1830 and achieved its present continental limits at the time of the Civil War. Despite the vastness of the land and the seeming diversity of the population, there was an underlying unity to the American people to which postmillennialism appealed. Their religion was Protestant and most were of British descent. Even when immigration introduced many new people and languages, the economic and social origin of the inhabitants remained fairly uniform with most of them coming from farming classes.

One of the most complete statements of nineteenth-century American postmillennialism comes from Hollis Read (1802–1887). Educated at Williams College and Princeton Seminary, Read was ordained at Park Street Church, Boston in 1829. He spent the years 1830–1835 as a missionary in India under the American Board of Foreign Missions and was prevented from continuing his foreign service because of his wife's poor health. After returning to the United States he preached, wrote several books, and taught. His most ambitious work was a two-volume analysis, *The Hand of God in History*, which attempted to demonstrate that world history was culminating in America.[9] Read believed that his age was so advanced in learning, in the arts, in morality and religion, and in the science of government that the millennium was coming to earth. As he looked back over history he could discern the hand of God in the discovery and settlement of America. Among the reasons that he advances to support this contention is the time of the discovery of the new land. When America was discovered it was needed to help the Reformed church triumph over papal errors. Many providential events such as epidemics among the Indians, confusion of the French fleet, and the fact that the settlements were made under Stuart rule when opposition to the crown was great aided the growth of the colonies. The character of the settlers also contributed to the growth of the elect nation.[10]

Read could see God's blessing on America in the favorable geographical position and the abundant resources of the land. The broad ocean separated the country from Europe with its age-old tensions and despotism, thus providing a natural environment for the growth of civil and religious freedom. The resources of the United States consisting of land, population, learning, enterprise, and the wise application of steam technology seemed to be boundless. As he expands upon these points he gives a spiritual blessing upon imperialist forces. He

[9]Hollis Read, *The Hand of God in History* (Hartford, 1856), vol. 1, pp. 380f. This work was first published in 1849 and was republished in 1851, 1856, and 1870.

[10]A contemporary book that echoes many of Read's ideas and his interpretation is Peter Marshall and David Manuel, *The Light and Glory* (Old Tappan, New Jersey: Fleming H. Revell, 1977). The advertisement for this work stated: "Never before has the saga of the founding of America been told from the point of view that God had a plan for this country, and that He intervened repeatedly to ensure that it had a chance to come to pass. This is the story of the discovery of that plan, and how it dramatically affects the conventional view of our nation's history and purpose."

looked on the extension of European, especially Anglo-Saxon control of other nations, as an opportunity to spread the gospel in these areas. The prevalence of the English language made it easier to preach the Word and teach the native peoples the rudiments of culture. Modern conveyances such as the locomotive and the steamship also helped to bring enlightenment and the gospel to the ends of the earth.

Read was excited about political events which were taking place in Europe at the time he wrote his book. The Revolution of 1848 and the Crimean War indicated to him that the war of Gog and Magog, which would usher in the millennium, was on its way.

In a later book he shifts the focus of the conflict to America during the Civil War. After the struggle was over the millennium, a period consisting of the triumph of Protestant Christianity, would arrive. Although not marked by a literal resurrection of the dead, the era would be characterized by peace, health and long life, increased knowledge, and cultural advance. Using Isaiah's prophecies as a text Read pointed out that "the improved moral character of man—the vastly increased industry, enterprise, and public spirit . . . will do wonders to overcome the physical ruins of the fall, and to renovate, beautify, and fertilize the whole face of the earth."[11] Rather than an age of rest, however, the millennium was to be characterized by great advancement in industry, inventions, public works, and a continued application of technology to human problems. Read's optimistic vision continued as the belief of many Americans in the twentieth century. Although secularized to a greater extent than he would have liked, the identification of progress with the government of the United States and free enterprise capitalism was adopted as a consensus ideology by immigrants who would never accept his interpretation of the gospel of Christ.

Despite the popularity of postmillennialism in the eighteenth and early nineteenth centuries there were always some individuals who preached premillennialism, and by the early nineteenth century their number increased because of a renewal of interest in prophecy fostered by the French Revolution. When the French overthrew their monarch, Europe was plunged into turbulence that encouraged apocalyptic thinking. Many Bible scholars in Britain came to the conclusion that the end of the age was near. Most of these interpreters, like their American counterparts, believed that the papacy had to be destroyed before the millennium would come. The Revolution caused the destruction of papal power in France, the seizure of the property of the church, the founding of a religion of reason, and even for a time the banishment of the Pope from Rome. The new prophetic movement centered in Britain, where a vast literature on millennial themes developed in the first half of the nineteenth century.

British premillennial writings became well known in the United States beginning in 1820 and encouraged American millennialism. An outstanding example of the new premillennialism, David Nevins Lord (1792–1880), was born in Connecticut and educated for the Christian ministry at Yale, but was prevented

---

[11]Hollis Read, *The Coming Crisis of the World* (Columbus, 1861), p. 20.

from fulfilling his vocational goal by poor health.[12] After his business failed in the panic of 1837 he dedicated himself to theological studies of a premillennialist variety. He mastered nearly all of the contemporary Anglo-American work on millennialism. He was an able writer and scholar with an excellent knowledge of Greek and a thorough acquaintance with English literature and the classics.

Lord was the editor of a periodical, *The Theological and Literary Review*, which appeared quarterly from 1848 to 1861 featuring articles of interest to premillennial scholars. He also systematized premillennialism to an extent never before attempted. He set forth rules for the literal method of interpretation so that among many there was an agreed-upon standard for prophetic analysis. Using these principles, Lord proceeded to elaborate a premillennial system based upon the historicist interpretation of the Book of Revelation. He states flatly at the beginning of one of his books that mankind as a whole is not to be redeemed under the present dispensation. Rather it is a period of trial when men choose between good and evil and show whether they follow God or not. Mocking the idea of worldwide revival by tracing the history of spiritual awakenings through the centuries, he demonstrated to his readers that every period of revival has been followed by a time of backsliding and trouble. The optimism of the postmillennialists, he felt, could be disproven from Scriptures such as John 16:31–33, Acts 14:22, and 1 Thessalonians 3:3–4, which picture the present age as a time of trial and discipline in which evil and good are tested and made to reveal themselves. The purpose of the present age was to prepare the way for another dispensation under which the world could be redeemed and salvation extended to all nations.

Christ is to return, Lord taught, to inaugurate the millennium and to reign in person during this period. His throne is to be on Mount Zion (Isa. 9:6–7, Psalm 2:6–10) and he is to reign on earth with the risen saints (Rev. 20:6). Lord consistently bolstered his views with Scripture references and appeals to the original languages of the Bible. When one reads his work he feels that here is an individual who consciously tried to follow the philological approach of the Protestant Reformers, while the postmillennial interpreters put a great stress upon a philosophical understanding of the Christian message. Lord's system gained many followers who believed in the literal interpretation of the Scriptures, two resurrections, and the restoration of the Jews to the Holy Land. These premillennialists delivered a minority report on much of American society dissenting from the normal enthusiasm for democracy, patriotism, technology, and social change. America, they taught, was not a special recipient of God's favor but was rather just another Gentile world power.

Since Christianity was always a persecuted minority, faith and missions were a witness against an evil world and there was no reason to assign a special role in evangelization to any country. What was true of the individual was true of a group, and so Americans could not enter the millennium by their own

---

[12]Robert K. Whalen, "Millenarianism and Millenialism in America, 1790-1880" (Unpublished Ph.D. Thesis, State University of New York at Stony Brook, 1972), pp. 41ff.

efforts any more than sinners could save themselves. Democracy would join the list of governmental schemes defeated by an evil world.

Such pessimistic attitudes doomed premillennialism to be the faith of a small group. A condemnation of national pride did not make for popularity in the antebellum United States. It was not just the political system and patriotism of their fellow Americans that upset the premillennarians, but also the fact that so many nineteenth century people were supremely confident of the positive benefits of technological improvement. David Lord reminded his readers that evil "derives as powerful aid for the spread and propagation of its empire, as pure Christianity does, from the improvements of the age in the methods of communicating knowledge through the press, and the union of numbers in the profession and dissemination of opinion."[13]

Nineteenth century premillennialism experienced considerable change upon the introduction of dispensationalism. This view describes the premillennial coming of Christ in two stages: the first, a secret rapture when the church is removed from earth; and the second, Christ's coming with his saints to set up the kingdom after a period of great tribulation. Originally propounded by J. N. Darby (1800–1882), this teaching was accepted and preached by many American interdenominational evangelists. It was also made an integral part of the notes in the *Scofield Reference Bible* and consequently became normative for much of American fundamentalism. It often led to a decline in social involvement on the part of those who followed its teaching. Among many dispensationalists the main activity became soul-winning, and those who wished to engage in acts of physical compassion toward their fellow human beings were dismissed as followers of the social gospel.[14]

III. The millennial views of the new Christian right

There has always been inconsistency on the part of premillennial dispensationalists with regard to the interpretation of world events and their desire to be patriotic Americans. On the one hand they were forced to admit that America was just another secular power, but on the other hand they wanted to preserve their country as the unique expression of God's purpose in a sinful world. As long as they maintained a strict separation from politics this contradiction in their attitude toward America and God's kingdom was not so apparent. However, with the recent political involvement of the New Christian Right the problem has been made more acute. Individuals such as Jerry Falwell and Tim LaHaye have felt called to enter the social and political arena, but they do not have a consistent eschatological base for such activities. In essence they want to support a certain type of postmillennial vision for America while maintaining a premillennial eschatology. Both Falwell and LaHaye have written books that teach the dispensational view of Bible prophecy. One of the more popular of

---

[13]David Lord, *The Coming and Reign of Christ* (New York: 1858), p. 52.

[14]For dispensationalism see Ernest R. Sandeen, *The Roots of Fundamentalism: British and American Millenarism, 1800–1930* (Chicago: University of Chicago Press, 1970).

these presentations, *Dr. Jerry Falwell Teaches Bible Prophecy*,[15] is a Bible study course which has lessons on such familiar dispensationalist themes as the signs of tbe last days, the Rapture, the tribulation, and the millennial reign of Christ. Falwell includes a list of books for further study which features the writings of dispensational stalwarts such as Charles Feinberg, J. Dwight Pentecost, Herman Hoyt, and John Walvoord. LaHaye has written a commentary on the book of Revelation as well as a volume entitled *The Beginning of the End*. In the latter work he explains the Rapture with great detail. As he puts it:

> The Rapture of the Church will be an event of such startling proportions that the entire world will be conscious of our leaving. . . . There will be airplane, bus, and train wrecks throughout the world. Who can imagine the chaos on the freeways when automobile drivers are snatched out of their cars! One cannot help but surmise that many strangers will be in churches the first Sunday after the Rapture. . . . Liberal churches, where heretics in clerical garb have not preached the Word of God and the need for a new-birth experience, may be filled to capacity with wondering and frantic church members. Many a minister will have to "explain it away" in some fantastic manner or seriously alter his theology.[16]

Among the signs that he feels point to the imminence of this event are worldwide war, famine and disease, earthquakes in various places simultaneously, the restoration of Israel, the growing power of the Soviet Union, the struggle between labor and capital, the increase of knowledge and travel, apostasy in the churches, the growth of witchcraft and astrology, growing wickedness and social decay, the ecumenical movement in the churches, and the activities of the United Nations. In most respects LaHaye's work along with that of Falwell could be called consensus thinking among dispensationalists. The New Christian Right's attitude toward Israel is an example of how premillennialism has shaped their political and social thinking. Jerry Falwell has been extremely outspoken in supporting the Jewish nation. He is convinced that the continued existence of the Jewish people and the birth of modern state of Israel are miracles of God. Divine care gave the Jews victories in the Six Day War and the Yom Kippur Conflict. The future, he claims, holds even more trouble for Israel as a great coalition led by Russia and her Arab allies will attack the land. But on the basis of Ezekiel 38 and 39 he believes that "Russia will be defeated and Israel will again be spared by the hand of God." He cautions America not to be neutral in war involving Israel because

> God is not finished with the nation Israel. . . . Every nation that has ever persecuted the Jews has felt the hard hand of God on them. Every nation that has ever stood with the Jews has felt the hand of God's blessing on them. I firmly believe God has blessed America because America has blessed the Jew. If this nation wants her fields to remain white with grain, her scientific achievements to remain notable, and her freedom to remain intact, America must continue to stand with Israel.[17]

[15]This study booklet and tapes are published by The Old-Time Gospel Hour, Inc., Lynchburg, Virginia, 1979.

[16]Tim LaHaye, *The Beginning of the End* (Wheaton, Ill.: Tyndale House Publishers, 1972), p. 28.

[17]Jerry Falwell, *Listen, America!* (New York: Bantam Books, 1981), p. 98.

In a recent interview Falwell urged his listeners to support Israel "because they are the only true friends America has in the Middle East."[18] The tendency to identify God's cause with Israel leads the New Right to part company with much of the anti-Semitism of other conservative groups.

The dispensationalist theology that the New Christian Right desires to follow teaches that God's purpose is not centered in America but in Israel. Despite this belief Falwell and LaHaye speak of America in terms that would fit into postmillennialist expositions, even though they both condemn this outlook. Falwell states: "Finally the postmillennial theory was quietly laid to rest amid Hitler's gas ovens during the Second World War! Today a postmillennialist is harder to find than a 1940 Wendell Willkie button."[19] LaHaye also adds: "Although this view was quite popular before the turn of the century and was given some impetus during the great revival movement of the Wesleys, Finney, Moody and others, it has been almost eliminated as a result of the two great world wars, the great depression, and an overwhelming rise in moral evil."[20] Both of the New Right spokesmen fail to realize that they have adopted a postmillennial view of their homeland.

Falwell and LaHaye attempt to base their patriotic millennialism on an interpretation of American history. Falwell believes that his country was founded as a Christian nation and that the typical early settlers of the land were the Pilgrims. They signed the Mayflower Compact indicating that they came to America to found a godly nation and to advance the Christian faith. The settlers in Pennsylvania and Virginia shared the same holy purpose. Those who formed the New England Confederation of 1643 confessed "we all came into these parts of America with one and the same aim, namely, to advance the Kingdom of our Lord Jesus Christ."[21] Whenever Americans lost this vision great revival preachers led the people back to God. Outstanding examples of these individuals were "John Wesley and George Whitefield [who] came to America preaching the Gospel of Jesus Christ and spreading revival throughout the land."[22]

When the colonists rebelled against England they depended upon God for leadership. The phrasing of the Declaration of Independence, the numerous references to the deity by colonial leaders, and the frequent recourse to prayer by the Founding Fathers all indicate that they "were putting together God's country, God's republic, and for that reason God has blessed her for two glorious centuries."[23] Falwell's list of the great Christian patriots includes Benjamin Franklin, George Washington, Patrick Henry, John Quincy Adams, Daniel Webster, Abraham Lincoln, and Woodrow Wilson, He cites the Constitution of the United States, the inscriptions on public buildings, and the inaugural

[18]Jerry Falwell, "An Interview with the Lone Ranger of American Fundamentalism," *Christianity Today* 25 (Sept. 4, 1981): 25.

[19]*Dr. Jerry Falwell Teaches Bible Prophecy*, p. 68.

[20]Tim LaHaye, *Revelation, Illustrated and Made Plain* (Grand Rapids: Zondervan, 1973), p. 290.

[21]Falwell, *Listen, America!* p. 29.

[22]*Ibid.*

[23]Jerry Falwell, *America Can Be Saved* (Murfreesboro, Tenn.: Sword of the Lord Publishers, 1979), p. 23.

addresses of the Presidents as further evidence of the Christian foundations of the nation. The continued national and spiritual blessings of God on America also testify to the special place she has in his purpose. LaHaye explains:

> Have you ever tried to imagine what America would be like if all Christian and Bible influence were removed? What would we have left? The ugliest kind of twentieth century barbarism, resembling the inhumanity perpetuated in Nazi Germany or Russia and China since the war. . . . Two-thirds of the world today is enslaved in communism and socialism. America is the human hope of the world, and Jesus Christ is the hope of America. Our present weakness, confusions, bureaucracy, immorality, and other national evils cannot be traced to the Bible or Christians but to the subversive erosion of basic Christian principles that have made this the greatest nation under God that the world has ever known. Unless we return to the Bible principles that provided our nation's greatness, we will pass like others before us.[24]

The New Right leaders identify one of the historic strengths of America as her economic development. The free enterprise system is the key element in America's rise to riches. Falwell assures his readers:

> The free enterprise system is clearly outlined in the Book of Proverbs in the Bible. Jesus Christ made it clear that the work ethic was a part of His plan for man. Ownership of property is Biblical. Competition in business is Biblical. Ambitious and successful business management is clearly outlined as a part of God's plan for His people.[25]

He does not support his point with arguments from Scripture but rather from the writings of Milton Friedman, Jesse Helms, William E. Simon, and Robert Tinger. His opinions are based on certain misconceptions about the economy. For example, he believes that there are plenty of jobs for all and that wives work so that they can have several cars, beautiful houses, and television sets.[26] One of Falwell's favorite arguments against social welfare is based on an illustration about two beautiful dogs given to his family by a friend. The giver recommended that they be fed meat but instead Falwell fed them dry dog food. Four days passed and finally the spoiled animals ate the food. The same tactics, he feels, would apply to "that lazy, trifling bunch lined up in unemployment offices who would not work in a pieshop eating the holes out of donuts."[27]

The government, in the New Right's opinion, once it is freed from heavy welfare burdens could balance the federal budget and support a larger military establishment. Falwell states plainly: ". . . I am a superpatriot. If you would like to know where I am politically, I am to the right of wherever you are. I thought Goldwater was too liberal!"[28] His list of the seven greatest Americans consists of Patrick Henry, Thomas Jefferson, Abraham Lincoln, Robert E. Lee, Teddy Roo-

---

[24]Tim LaHaye, *The Bible's Influence on American History* (San Diego, Cal.: Master Books, 1976), pp. 58–59.

[25]Falwell, *Listen, America!* p. 12. For a thoughtful critique and response to the New Christian Right from an evangelical perspective see Robert E. Webber, *The Moral Majority: Right or Wrong?* (Westchester, Ill.: Cornerstone Books, 1981).

[26]To counter Falwell's point see the U.S. Dept. of Labor Bulletin *20 Facts on Woman Workers*, August, 1979.

[27]Falwell, *America Can Be Saved*, p. 35.

[28]*Ibid*, pp. 97–98

sevelt, Douglas McArthur, and J. Edgar Hoover.[29] He believes that we need a strong authoritarian president who will lead the nation in cutting welfare spending at home and abroad and increase "our defense budget to whatever it takes to put us solidly back to No. 1 for good."[30]

The reason that the United States must have large military forces is to combat communism. The danger is very great because currently the Soviet Union is the world's leading power. As Falwell explains:

Ten years ago we could have destroyed much of the population of the Soviet Union had we desired to fire our missiles. The sad fact is that today the Soviet Union would kill 135 million to 160 million Americans, and the United States would kill only 3 to 5 percent of the Soviets because of their antiballistic missiles and their civil defense. Few people today know that we do not have one antiballistic missile. We once had $5.1 billion worth of them, but Ted Kennedy led a fight in the Senate and had them dismantled and removed. From 1971 to 1978 the Soviets outspent the United States by $104 billion for defense and an additional $40 billion for research.

It is not too late, however, for America to rearm and once again gain the military advantage. This is necessary to protect the gospel in America and to save our land from the ravages that communism inevitably brings.

Social changes are also necessary if America is to be spared. There are opponents to these moral moves, namely, the secular humanists. These individuals have a great deal of control through organizations such as the American Civil Liberties Union, American Humanist Association, National Education Association, National Organization for Women, and the labor unions. They exercise their power through the news media, government, public education, and the foundations. With divine aid the fundamental churches and Christian school movement can reverse the evil humanistic tide and save America. The process of salvation will restore the traditional family, defeat feminism, guarantee child discipline, stop abortion, restrict homosexuals, stamp out pornography, stop drug abuse, and suppress rock music. In short, the millennial vision of the New Christian Right for America would make it a very conservative place where one group's traditional values would hold sway. The nation would then be prepared to launch an international soul-winning movement and be a true friend to Israel.

Contrary to the usual dispensationalist pessimism, there is a hopeful tone to the preaching of the New Christian Right. Tim LaHaye observes enthusiastically: "Never have I seen a better time for effectively sharing one's faith than today. By the year 1990, I expect 40 percent of all America to profess a personal faith in Christ and over 51 percent of all educable schoolchildren to be attending Christian or private schools."[32]

The writers of the *Liberty Bible Commentary* concur with his assessment as they declare:

But how does one explain the dynamic resurgence of the impact of the gospel in the last few years? Where does this fit into the scheme of the seven churches of Revelation?

---

[29]*Ibid.*, p. 141.
[30]*Ibid.*, p. 35.
[31]Falwell, *Listen, America!* pp. 84–85.
[32]Tim LaHaye, *The Battle for the Mind* (Old Tappan, N.J.: Fleming H. Revell, 1980), p. 222.

Perhaps the answer is that today in the waning years of the twentieth century, examples of each type of the seven churches can be found. In the United States alone there are Smyrnean churches, Philadelphian churches and Laodicean churches. Perhaps, too, God is causing more pastors and churches to see the "open door" today, more so than even during the eighteenth and nineteenth centuries, so that one great final outbreak of revival may bring glory to His name. Perhaps we are presently experiencing a post-seven-church revival. . . . We must become keenly aware that whoever is to receive the message of salvation prior to the awful Tribulation Period must receive it now.[33]

One can detect that at times the spokesmen for the New Christian Right perceive some inconsistency between their dispensationist theology and their postmillennial vision for America. Recently Jerry Falwell responded to an interviewer: "I think America is great, but not because it is a Christian nation: it is not a Christian nation, it has never been a Christian nation, it is never going to be a Christian nation."[34] On another occasion when asked if there was a conflict between being pro-Israel and pro-American, he declared that God would bless America because it supports Israel.[35] Tim LaHaye has tried to merge the two millennial views by proposing the existence of two tribulation periods. As he explains:

The seven year tribulation period . . . is predestined and will surely come to pass. But the pretribulation tribulation—that is, the tribulation that will engulf this country if liberal humanists are permitted to take total control of our government—is neither predestined nor necessary. But it will deluge the entire land in the next few years, unless Christians are willing to become much more assertive in defense of morality and decency than they have been during the past three decades.[36]

If Christians will make America the kind of land that LaHaye and his friends want there will be a great international soul harvest.

Much of the New Christian Right's program seems to be contradictory and inconsistent. Perhaps this is due to a confused eschatology. A further problem with their millennialism is its encouragement of the renewal of what many scholars and pastors have called "civil religion."[37] The New Christian Right's postmillennial vision for America has given great encouragement to civil millennialism. One can do no better than to caution them with the words of an earlier advocate of civil religion:

---

[33]*Liberty Bible Commentary on the New Testament*, ed. Jerry Falwell *et al.* (Nashville, Tenn.: Thomas Nelson, 1978), p. 757.

[34]Falwell, *Christianity Today* interview, p. 24.

[35]Jerry Falwell, "The Nutshell Interview, Jerry Fallwell," *Nutshell*, 1981–1982, p. 37.

[36]LaHaye, *The Battle For The Mind*, pp. 218–219.

[37]For the problem of American Civil Religion see the following volumes: *American Civil Religion*, ed. Russell E. Richey and Donald G. Jones (New York: Harper & Row, 1974); Robert T. Handy, *A Christian America: Protestant Hopes and Historical Realities* (Oxford: Oxford University Press, 1971); *The Religion of the Republic*, ed. Elwyn A. Smith (Philadelphia: Fortress Press, 1971); *God's New Israel: Religious Interpretations of American Destiny*, ed. Conrad Cherry (Englewood Cliffs: Prentice-Hall, 1971); William A. Clebsch, *From Sacred to Prophane America: The Role of Religion in American History* (New York: Harper & Row, 1968); Richard J. Neuhaus, *Time Toward Home: The American Experiment as Revelation* (New York: Seabury Press, 1975); Sidney E. Mead, *The Nation With The Soul of a Church* (New York: Harper & Row, 1975); and Robert D. Linder and Richard V. Pierard, *Twilight of the Saints: Biblical Christianity and Civil Religion in America* (Downers Grove, Ill.: InterVarsity Press, 1978).

[It is an] error to identify the Gospel with any one particular system or culture. This has been my own danger. When I go to preach the Gospel, I go as an ambassador for the Kingdom of God—not America. To tie the Gospel to any political system, secular program, or society is wrong and will only serve to divert the Gospel. The Gospel transcends the goals and methods of any political system or any society, however good it may be.[38]

[38]Billy Graham, "Why Lausanne?" *Christianity Today* 18 (Sept. 13, 1974): 7. For Graham's struggles with the ideas of the millennium see William D. Apel, "The Lost World of Billy Graham," *Review of Religious Research* 20 (Spring 1979): 138–49.

Religious Broadcasting and the Mobilization of the New Christian Right*

JEFFREY K. HADDEN†

INTRODUCTION

Evangelical religious broadcasters are amassing a power base which has the potential of changing American society in ways that are revolutionary in character. This address seeks to shed light on how this has happened. The argument which will be pursued can be succinctly summarized as follows:

First, the charismatic leaders of religious broadcasting are the principle actors in a social movement of monumental importance in the late twentieth century. Second, the ideological origins of this social movement are deeply grounded in the long held view of America as a "New Israel," a land providentially endowed by God with a special mission in world history. Third, the organization resources and managerial techniques of modern televangelism grew out of nineteenth century urban revivalism and they are now being applied to fuel a social movement. Fourth, by stereotyping fundamentalism as a backwater anti-intellectual reaction to the modern world, scholars and the mass media alike have seriously misunderstood the complexity, diversity, and strength of evangelical faith in America. Fifth, the collapse of the liberal vision now provides the opportunity for evangelical Christians to reassert their influence and reshape American culture.

Pursuit of this argument requires some minimal conceptual framework. I begin with a brief critique of secularization theory. This critique is necessary because secularization, the dominant social science theory of modern societies, offers a model of change which renders religion incapable of mustering sufficient strength to have revolutionary consequences in the modern world. The critique of secularization theory is followed by a presentation of a few key concepts from the resource mobilization theory of social movements. Here I also offer a brief delineation of how my own work departs from the work of others who are utilizing this approach.[1]

*Presidential Address delivered to the Society for the Scientific Study of Religion, October 26, 1985, Savannah, Georgia. The author wishes especially to thank Razelle Frankl for the critical insights that helped this analysis fall in place and Anson Shupe for helpful comments on an early draft.

†Jeffrey K. Hadden is professor of sociology at the University of Virginia.

1. This is not the occasion for a detailed critique of secularization theory, or for an elaboration of the conceptual underpinnings of resource mobilization theory. My task, rather, is to provide the minimum conceptual critique and tools for the analysis which follows.

This conceptual section is followed by an effort to trace the historical roots of the contemporary New Christian Right social movement. Basically, I argue that in both form and substance, the roots of the electric church are firmly anchored in urban revivalism. The third major section of the paper demonstrates how the contemporary New Christian Right social movement reemerges out of its roots in urban revivalism and its prodigy fundamentalism. By tracing these historical links and placing them in the resource mobilization theoretical framework, I try to demonstrate why the present religiously led social movement is a major development in American history.

CONCEPTUAL MATTERS

A Note on Secularization Theory

The Sacred in a Secular Age (Hammond, 1985), published in 1985 under the imprimatur of the Society for the Scientific Study of Religion, stands as a landmark in challenging the linear image of history that is implicit in secularization theory, and in reassessing the viability not only of this theoretical perspective but also of a much broader set of presuppositions that have been central to modern social science thought.

"Modernization," "rationalization," "bureaucratization," "industrialization," and "urbanization," Hammond notes, are but a few of the concepts social scientists have developed to analyze and describe social change. Every one of these concepts carries an implicit one-dimensional view of history which "[helps] to maintain the notion that social life is systematically 'coming from' somewhere and 'going' elsewhere" (1).

At the risk of offending many scholars who have contributed to a rich and complex literature, I would reduce my critique of secularization theory to this simple proposition: Secularization theory infers a cognitive result (unbelief) from a structural process (differentiation) and then concludes that an institution with only a remnant of believers must necessarily be politically impotent. To confirm this faulty reasoning, scholars examine the European experience without recognizing that Europe constitutes a highly particularistic and non-representative case in the secularization process.

Those who are not yet ready to reassess secularization theory argue that both the persistence and intermittent revival of religion are to be expected, and that the theory implies no timetable. But even a casual effort to be informed about world affairs brings one again and again into dramatic contact with the entanglement of religion and politics around the globe. Whereas the imprint of secularization is clearly apparent everywhere, I do not see compelling evidence for secularization theory's fundamental proposition that religion is receding from the public arena. Equally important, religious institutions possess resources independent of other institutions which they may deploy independently or in concert with other social movement organizations.

Some years ago I proposed that the concept secularization could be freed of the linear assumptions of the erosion of religious influence in culture if we were to radically restrict the meaning of the concept to the simple process by which the historical yoke between church and state is broken (Hadden, 1980a). I still think this is a reasonable idea. More recently, Randall Collins (1986) has helped me see how the process of secularization, so conceived, is useful for much broader historical and cross-cultural research. Challenging the classicial sociological assertion that all states must be religiously legitimated, Collins identifies three broad relationships between religion and regime: 1) identical religious and political structure and activity; 2) religiously-legitimated states; and 3) secularly-legitimated states. Societies may proceed from the first to the second to the third structural relationship between religion and regime, but this pattern is not inevitable nor is a relationship, having been achieved, irreversible. Collins' work develops examples of secularly-legitimated societies temporally and geographically outside our general conceptual notions of modernization. These general types should not be viewed as absolute conditions, but as ideal-types.

Within this framework, we can define *secularization* as the process by which religiously-legitimated states are transformed into secularly-legitimated states. Secularization, thus, does not refer to some transformation of beliefs or behavior, but *only* to the *legal and quasi-legal institutional relationships between religion and regime.*[2]

If this narrower conception of secularization is less aesthetically pleasing, its analytical potential may be much richer. First of all, it is much easier to assess empirically the structural relationships between religion and regime than to evaluate the religious beliefs, practices and institutional strength of religion at some point in time. The former is not dependent upon cross-cultural and cross-faith measures of belief and practice; the latter is. Second, analyzing the structural relationship between religion and regime may suggest hypotheses that are obscured by the old way of thinking about secularization. For example, the high levels of religious practice and institutional viability in the United States, especially in comparison with Europe, have always appeared incongruous with similar levels of secularization.

The secularization concept I am recommending leads us immediately to a critical difference. Whether we view Europe as generally less secularized, or if you prefer to view the secularization process as having taken a different structural route, the focus on

---

2. In communist nations we have witnessed abrupt transformations to secularly-legitimated states, but in some communist countries we have also seen quasi-legal understandings between church and state such that religion continues to legitimate the regime (Pankhurst, 1986). Hence, we can speak of a process and not a simple administrative act.

In the United States the process has been gradual. The First Amendment to the Constitution formally differentiated church from the national state, but several states retained formally established churches for three or four decades. The Fourteenth Amendment was the milestone that made explicit the application of the First Amendment to the states. During the second half of the twentieth century, we have experienced a long string of court decisions which have progressively separated church and state. As a result, the state must of necessity rely upon other measures for legitimacy. The Supreme Court's decisions during its last session failed to realize the predicted shift in the direction of "accommodation." But the prospect of President Ronald Reagan's "packing" the Court during his second term in office suggests the possibility of the secularization process being reversed in this nation.

The European experience with secularization, in this sense of the concept, is highly varied. Generally, however, except in communist Eastern Europe, the process of secularization has not gone as far as it has in the North America. Or, perhaps some would wish to argue that it has merely gone in a different direction. *Pluralism*, that is, the formal recognition of the legitimacy of many churches exists along side the retention of a state church. Like old soldiers, the institutional church and her leaders pick up their pensions (i.e., state support) and hardly bother anybody at all. Their symbols may be too old and their leaders too tired to provide the state a great deal of legitimacy. Nevertheless, if secularization is to be understood as a general process of transition from total religious-legitimation to complete secular-legitimation, most of Europe has opted for hanging on to the vestments of religious-legitimation of a secular state.

In still another part of the world, Islamic countries have witnessed numerous examples of change toward secular-legitimation of the state followed by abrupt retreat. Iran is the spectacular example, but many Islamic countries have backed away from earlier moves toward secularization.

structure provides the key to understanding the difference. By granting no establishment, conceding no privilege, and permitting no disability, the climate of religion in the United States approximated the free enterprise economic structure of the new nation. Hussle and product differentiation in response to market demand created a diverse and vital religious landscape.

The relatively high level of structural differentiation between religion and regime in the United States obviously does not account fully for religious vitality, but it is an important element which opens other analytical doors. Other nations, with different religious compositions, will respond differently to explicit efforts to shift the legitimizing base of the state from religious to secular foundations. But given that nation states are constantly in the process of forming and dissolution, it seems quite possible that we might develop propositions of the "If X, then Y" variety.

Permit me to conclude these observations about secularization with a proposition: *The greater the degree to which modern states legitimize their existence in secular rather than religious foundations, the greater the autonomy of religious institutions to pursue their own interests vis-a-vis the state.* This involves a good deal more than religious institutions joining many other subsystems (cf. Luhmann in Dobbelaere, 1984: 204). I agree with Collins that "this very separation of organizational resources is what gives religion the possibility of mobilizing rebellions against an existing regime" (Collins, 1986). Religious institutions are unique among all other institutions in their potential to provide transcendental legitimacy, as well as other unique resources to social movements.

Thus, the tremendous surge of religious energy in the world today is not an aberration, an intermittent revival of a soon-to-be archaic institutional form, or a "transitional phenomenon on the road to 'modernity' " (Collins, 1986). Rather, we are looking at a process which has become a permanent feature of modern politics. The potential of religious or parareligious organizations to mobilize social movements in pursuit of reform, rebellion, or revolution will vary from culture to culture depending upon the numerical strength of the group and the form of government.[3]

Resource Mobilization Theory

As an alternative to secularization theory, I purpose that the "resource mobilization theory" of social movements holds much greater promise for understanding the bouyant quality of religion in the modern world. Social movements may take the form of reform, rebellion, or revolution. Religion is a resource which will be used both to legitimize and to repress social movements.

Traditionally, the concept social movement has been subsumed under the broader concept of collective behavior. Deeply felt and broadly held grievances coupled with an ideological rationale for change have been viewed as necessary conditions for the emergence of social movements. The research mobilization approach to the study of social movements has shifted the focus of inquiry away from grievances and ideology to look at the structure

---

3. Pankhurst (1986) properly observes that the potential to effect change appears rather more limited under regimes that have an absolute monopoly over the means of violence and no predilection against its use to suppress opposition.

of movements and how resources are mobilized and managed to achieve goals.[4]

From the onset, my own interest in the rise of the New Christian Right has been centrally informed by resource mobilization theory. What Charles Swann and I argued in *Prime Time Preachers* (1981) is that media access is a critical *resource* in a social movement and that the televangelists have greater unrestrained access to media than any other interest group in America.

For the most part, the vocabulary of resource mobilization analysis does not depart from the general lexicon of social science in ways that require detailed presentation of concepts. Just a few terms may be helpful for those not familiar with the perspective. Also, I need to call attention to how my own thought departs in important ways from other resource mobilization scholars.

The general phenomenon of a *social movement* (SM) may be differentiated from *social movement organization* (SMO), and a *social movement industry* (SMI). Following McCarthy and Zald (1977), a *social movement organization* is "a complex, or formal, organization which identifies its goals with the preferences of a social movement . . . and attempts to implement these goals" (1218). As a social movement grows, so also will the number of SMOs multiply. The *social movement industry* constitutes "all SMOs that have as their goal the attainment of the broader preferences of a social movement" (1219).

It is also useful to differentiate between *adherents*, "those individuals and organizations that believe in the goals of the movement," and *constituents*, "those providing resources for it" (1221). The *resources* of social movements include economic assets, leadership, personnel, and legitimacy conferring powers. As I have noted elsewhere, a unique role of religious institutions is their capability to confer legitimacy both on the movement itself and upon specific activities pursued (Hadden, 1980a: 104). But this is by no means the only contribution of religious groups. Religious organizations have regular meetings which can be used to educate and recruit SMO volunteers. Furthermore, religious leaders tend to have greater elasticity in their work schedules. When a cause arises, they can redirect their time away from their professional activities toward the social movement cause. They may even be able to define the SMO activity as their work.

When one recalls the civil rights movement of the 1960s, one remembers vast numbers of people assembling for demonstrations and marches. But social movements almost always become complex and highly diverse entities as they grow. Some SMOs may specialize in certain types of activities, e.g., protest demonstrations, legal defense, publication, and education, etc. While major social movements tend to engage in at least some mass activities, this is not always the case. The movement to curb drunken driving has been highly effective with virtually no mass demonstrations. Some movements, like the Trilateral Commission, are quite elitist.

A relatively new feature of social movements, in broad historical perspective, is what I have called the *"living room social movement."* Direct mail and telephone banks work to transform adherents into constituents. Constituents, in turn, are alerted to the needs

---

4. In a recent H. Paul Douglass lecture before this assembled body, Mayer Zald, a leading exponent of this perspective, demonstrated its utility for the study of intergroup religious conflict (Zald, 1982). While less formally developed, Zald also argued that resource mobilization is a useful approach for studying the involvement of religious organizations in social movements aimed at changing the broader society.

of movements, and the postoffice carries their responses — a petition signature, a letter to an elected official, or money to the SMO. And the evening news delivers evidence of the SMOs effectiveness back to the constitutents' living room.

I also find it useful to differentiate between social movements, social problems and social issues.[5] A *social problem* exists when a collectivity of people or institutions identifies some social issue as intolerable *and*, as a result of that labeling, endeavors to mobilize resources to effect change. A *social movement* consists of the organized activities of individuals and institutions directed at changing social conditions they have defined as intolerable. The *problem* side of the equation is the act of labeling a condition intolerable, while the *movement* consists of the activities aimed at attempts to effect change. Movements may seek change through reform (utilizing both legitimate and illegitimate measures), through rebellion (*coup d'etat*), or by revolution. Rebellion and revolution are activities which follow the labeling of the social system incapable of change via reform. *Countermovements* are the organized activities of those who define the proposed changes as intolerable. Their presence and level of activity can be expected to increase in proportion to the effectiveness of the social movement.

An additional distinction should be made between social issues and social problems. *Social issues* are social phenomena which have at least some minimal collective recognition as being harmful to individuals, some segment of society, or society as a whole.[6]

In free societies, mass media inform the public of many harmful and potentially harmful social phenomena and, as a result, the number of social issues multiplies greatly. By calling attention to a large number of social issues, the media is tacitly encouraging people to transform social issues into social problems. Just beneath the surface of much "objective" reporting, there is the unspoken cry that "somebody ought to do something about this." And in some instances, media actors consciously agitate and encourage people to take up causes. This is a fairly common occurrence on news magazine programs like *60 Minutes* and *20/20*. It is nearly a daily occurrence on the Christian Broadcasting Network's *700 Club*.[7]

Finally, some observations about the life span of social movements. First of all, the vast majority of efforts to define a social issue as a social problem are short lived and do not really generate much of a social movement. But when movements do get off the

5. There is an inseparable link between social problems and social movements. Armand Mauss has developed this principle in his social problems textbook, but the utility of seeing this interaction is yet to be picked up by resource mobilization theorists (Mauss, 1975). Social movements, in this conceptual framework, are the overt activities which follow the labeling of a social phenomenon as problematic. Social problems and social movements are not coterminous but, rather, separate moments of the same process.

6. This recognition gives cause for the expression of grievances, but grievances do not automatically or necessarily follow. Furthermore, grievances may be expressed and responded to in many ways short of the mobilization of a social movement. Natural disasters (e.g., hurricanes and droughts) may be viewed as beyond human control. Some religious perspectives and folk philosophies teach that any relief from suffering and hardship will have to come in another world. For many reasons, thus, most social issues are not and do not become social problems.

7. Thus, the media function as a conduit for agitation. By defining protests and demonstrations, as well as controversial speakers and authors, as newsworthy, the media provide an open stage for social movement activists to plead their cases to large publics, thus seeking to move public opinion in the direction of their goals. Effective media communication has become as important to social movements as it is for political candidates and holders of public office.

ground, they may be thought of as having a natural life-cycle.

A few problems can be thought of as macro or master social problems, because they recur over an extended time span albeit in different forms and with different emphases on goals. Humanity's despoiling of the environment and the vexing problem of how to control war are two examples of macro social problems that have generated several cycles of social movements in the twentieth century.

We can identify at least a few social problems and social movements that have traversed several centuries. Certainly the democratic and egalitarian instinct, which has addressed the problem of social inequality, is one such movement. During the first half of this century, the thrust of this movement focused on the plight of industrial workers, and the results were collective bargaining and Social Security. During the second half of the century the movement has focused first on the rights of blacks and then on women and other minorities.

The major thesis of this inquiry is that what we are today experiencing, with the emergence of the New Christian Right, is another cycle of a centuries' old social movement. This centuries' old social problem has no formal name because its existence has not been recognized as such. Its social movement manifestation corresponds in a rough, but by no means precise way to what historians have identified as "great awakenings."

THE ROOTS OF A SOCIAL MOVEMENT

The Crisis of Dominion

If a single concept can describe what the problem and social movement is all about, I would say it is a recurring *crisis of dominion*. It is undergirded by a creation myth of America and is periodically fueled when Christians perceive God's dominion over this land to be threatened. Such is the case today. The creation myth can be roughly recounted as follows:

> After God had created the heavens and the earth he made Adam and Eve and told them to multiply and have dominion over the earth. They sinned and fell short of the glory of God's commandments, as did subsequent generations, but each time man fell, God provided a pathway for repentence and redemption. And in the course of time, God showed man this promised land and taught him of unalienable rights. And when man stumbled, God raised up a great leader who stood in the blood-stained battlefield of Gettysburg and renewed our pledge that this nation, under God, shall not perish from the earth.

While there are many variations on this creation myth it is the central motif in the New Christian Right's image of contemporary America. At the heart of its proponents' anguish is the belief that America, this special place in God's divine plan, has stumbled again. Their rhetoric resonates with an imagery of God's dominion, humanity's unfaithfulness to stewardship, the call for repentance, and the promise of redemption.

Their sermons intertwine the Old and New Testaments with American history and the contemporary malaise as if it were all a continuous experience. They proclaim: "Righteousness exalteth a nation, but sin is a reproach to any people" (Proverbs 14:34). We have "forgotten God," wrote Abraham Lincoln in the anguish of our Civil War. "It behooves us, then," he wrote in proclaiming a National Day of Humiliation, Prayer, and

Repentance, "to humble ourselves before the offended Power, to confess our national sins, and to pray for clemency and forgiveness."

A call for renewed repentance was sounded at the Washington for Jesus rally on the Mall, April 29, 1980 before a quarter of a million faithful:

"If my people, which are called by my name, shall humble themselves, and pray, and seek my face, and turn from their wicked ways; then will I hear from heaven, and will forgive their sin, and will heal their land" (2 Chronicles 7:14).

Some of the particulars of this creation myth are unique to fundamentalism, but the broad contours have been a part of American history from the outset. From the days of the earliest settlers in North America through the mid-twentieth century, the idea that this land and its people had a very special relationship to God was never very far from center stage. "Throughout their history," wrote Conrad Cherry, "Americans have been possessed by an acute sense of divine election. They have fancied themselves a New Israel, a people chosen for the awesome responsibility of serving as a light to the nations, a city set upon a hill" (Cherry, 1971: vii). Each generation has embellished the theme with particulars which gave special meaning to the events of their time — the birth of the nation, westward expansion, the Civil War, the industrial revolution, the journeys to Europe and Asia to defend freedom. Each epoch was interpreted to have a special meaning and purpose in God's divine plan.

Urban Revivalism As Precursor to the New Christian Right

So pervasive and persistent is this dominion theme that one could tell the story of the rise of the New Christian Right from the first permanent settlement in Jamestown in 1607. Indeed, M. G. "Pat" Robertson finds providential meaning in the location of his Christian Broadcasting Network.[8]

If locating the roots of this social movement is somewhat arbitrary, I think we appropriately set the stage if we look back to the early nineteenth century. In 1800, the United States was still a small nation of a mere five million people, and approximately ninety-five percent of them lived in rural areas. In 1820, even as great numbers were packing their wagons and moving westward to explore and settle the vast territory on the other side of the Allegheny and Blue Ridge Mountains, the number of new immigrants to the nation was still at a trickle of about seven hundred per month.

But then this changed. At first, it was like opening a valve just a little. Then the trickle became a steady stream, and the stream became a flood. By the 1880s the flow of immigrants was approaching a half-million a year. By the end of the century there were seventy-six million Americans, a fifteen fold increase. And while the westward settlement increased throughout the century, most newcomers took up residence in cities. And those cities, which by our present standards were only towns in the early 1800s,

---

8. In *The Secret Kingdom*, Robertson writes: "At a point just ten miles from the site I had purchased [for CBN], the first permanent English settlers in America had planted a cross on the sandy shore and claimed the land for God's glory and for the spread of the gospel. After 370 years, the ultramodern television facility with worldwide capabilities began to fulfill their dreams" (1982: 196). Further providential significance is found in his creation of CBN University. These early Virginia settlers had planned Henrico College, a school to train young persons to teach and preach the gospel.

grew into industrial metropolitan areas.

Between westward resettlement and the millions of urban newcomers, America was fast becoming an unchurched nation. The vast population increase included millions who were not Protestants, and this threatened the hegemony that had characterized this land from its beginning.

The revival emerged as a response to the migration flows which had left whole new settlements unchurched. The agents of revival were itinerant Baptist and Methodist preachers. They were, literally, socializing agents taming the violent frontier (Hofstadter, 1963). But the great waves of immigrants were also making frontiers of the cities. So revival found its way to the city as well. By mid-century "the cutting edge of American Christianity . . . was the revival, adopted and promoted in one form or another by major segments of all denominations" (Smith, 1957: 45).

We face here a question of what to call this revival phenomenon. Sociologists of religion have differentiated between *religious movements,* which are aimed at changing religious beliefs, values, or practices, and *social movements,* which aim to change some broader aspect of society (Stark & Bainbridge, 1979: 124). While revivalism clearly sought to change religious practice, I think its broader objective was to change society. The intolerable social problem was unbelief and its concomitant "uncivilized" behavior. Perhaps we can think of the religious problem as an inadequate number of churches and preachers to launch an adequate assault on the social problem. Revivalism, thus, is partly a religious movement, but those who fueled the movement were fundamentally committed to changing society through changing religious behavior. If more people were religious, then society would be a more civilized and tolerable.

When the revival went to the city its social movement character, in contrast to its religious movement character, became more pronounced. The goal of the urban revivalists was not simply to make Christians of the unwashed masses. Rather, they had in mind a more or less explicit image of how Christianizing people transformed their social character. Salvation was viewed as a solution to the problem of urban poverty. Save souls and people will lift themselves out of poverty. Some revivalists argued that the way to win souls was first to care for physical needs so that people might then be better prepared to understand their spiritual condition. But that view did not prevail for long among the big-time urban revivalists. For them, salvation begat motivation to work — it was straight out of Andrew Carnegie's "gospel of wealth" (Carnegie, 1949).

The most important thing that happened in the life of urban revivals is that revivalism became a *discrete institutional form.* The process by which urban revivalism became differentiated as an institution apart from the denominational structures of American Protestantism has recently been analyzed by Razelle Frankl (1984).[9] Frankl shows how these autonomous structures provide, in both form and content, the organizational model for the contemporary electric church. Others have traced the roots of modern televangelism to urban revivalism, but the importance of this observation is not clear until one notes the significance of revivalism as an autonomous institutional form.

With no obligation to answer to any Protestant denomination or ecclesiastical

---

9. I am endebted to Frankl for her brilliant analysis of how urban revivalism became an autonomous institution form. A revision of her doctoral dissertation (Frankl, 1984) will be published by Southern Illinois University Press (Frankl, 1987).

authority, urban revivalism took on a life and form all its own. Out of this autonomous structure emerged other detached structures which contributed to private or parachurch structures, most notably Bible institutes and conferences, and independent missionary societies.

These structures, in turn, provided environments which encouraged and fostered much of the rich variety of religious activity in the late nineteenth and early twentieth century. Pentecostalism, the holiness movement, and fundamentalism were all nourished within the broad tenets of urban revivalism.

Frankl singles out three figures who made marked contributions to the development of urban evangelism as an institutional form: Charles G. Finney, Dwight L. Moody and Billy Sunday (Frankl, 1984; see also Frankl, 1985 and Frankl, 1987). I shall briefly identify the contributions of each, although it is perhaps best to think of a cumulative process, with each building upon his predecessors.

Charles G. Finney (1792-1875), widely acclaimed as "the father of modern revivalism," stands as transitional figure between the agrarian camp meeting, which characterized the Second Great Awakening in the West, and the urban revival. Finney established the principles of modern evangelism. Well ahead of Frederick Winslow Taylor's development of the "principles of scientific management" (1911), Finney pioneered the development of principles of scientific revivalism. Systematic and comprehensive planning, creation of a proper environment, development of rational arguments, playing on emotions, and utilization of theatrics were all critical elements of Finney's approach to evangelism.

Undergirding Finney's innovative techniques, "New Measures" he called them, were important transformations in theological thought which were popularized by the spread of Methodism. John Wesley's doctrines of "salvation by grace through faith" and "individual perfection" cut a deep rift with the Calvinist doctrine of predestination. Responsibility for salvation was shifted from the whims of a God of wrath to the individual who had only to reach out and accept a merciful and loving God's grace. Finney believed it was the responsibility of the evangelist to do everything he could to confront the individual with that choice. Here too, the process of revival was turned on its head: we are responsible for producing revivals rather than waiting upon God to send revival.

It was Dwight L. Moody (1837-1899) who advanced urban revivalism from emerging principles to established and routinized techniques. Moody's legacy is the institutionalization of revivalism and the autonomously standing religious institutions which it spawned. Moody was a businessman merchandising salvation. His was a simple business proposition, eternal life in exchange for accepting Christ. Most importantly, Moody rationalized the organization of revivals, creating a complex division of labor with specialized roles and expertise to assure the smooth execution of every detail in the planning and execution of a revival (Frankl, 1984).

Of particular interest is the manner in which he was able simultaneously to engage local ministers in very elaborate planning and participation, while keeping some considerable social distance from denominational and other institutionalized religious authorities. This he achieved, in considerable measure, by the utilization of the business community to raise money for the revivals. With their leading laity enthusiastic about the impending revival, local pastors were in no position to be indifferent, but neither were they in a position to control the organization of the revival.

Thus, Moody developed the autonomous institution of religious revival not by schism, as had been the pattern since the Reformation, but by innovation. There was, to be sure, plenty of antagonism toward this autonomous parachurch structure, but neither denominational nor local ecclesiastical authorities were in a position to block the development of this new institutional form. Moody's revivals provided surplus resources for creating Moody Bible Institute which, in turn, became a model for other autonomous Bible schools and institutes. It was, to be sure, an institution formed solely on the principles of free enterprise (Marsden, 1980: 34). This was a lesson contemporary televangelists learned well.

Moody also left an important theological legacy. In some ways the principal progenitor of fundamentalism, Moody "lacked the one trait that was essential to a 'fundamentalist' — he was unalterably opposed to controversy" (Marsden, 1980: 33). He was not interested in theology and, in fact, claimed not to have a theology (Hudson, 1973: 233). Still, his theological views were not hard to discern. He was a premillenialist and accepted biblical infallibility (Marsden, 1980: 33). His reluctance to pursue theological issues publicly no doubt contributed to opening even wider the door of privatized faith. Thus, Moody may be seen as a key progenitor of fundamentalism, while also tacitly endorsing the proliferation of privatized faith.

If Moody was not openly political, the underlying political implications of his preaching were clear; "a closer reading of the sermons showed that he was a thoroughgoing conservative" (Weisberger, 1958: 224). One of the reasons he got on so well with businessmen was because he believed and preached a doctrine that was in total accord with their worldview.

Billy Sunday (1862-1935) built upon and refined Moody's organizational skills, but he added also the roles of entertainer and celebrity to the evangelist's repertoire. Both would prove to be important components of the recipe for success when revivalism found a new form of expression in television. At the organizational level, William McLoughlin notes some twenty specialized roles, and cites the claim of an economics professor that Sunday's organization was one of the five most efficient businesses in the country (cited in Frankl, 1985: 7.3; 7.1). Weisberger notes that "There was never a machine better designed for publicizing 'the Lord's work' " (1958: 251). Entertainment had been creeping into evangelism for a while, but Sunday greatly magnified that trend. Wrote Weisberger, "He synthesized and magnified to the hundredth degree the tendencies towards big-time religious showmanship begun by those before him" (1958: 231).

No preacher before or since has commanded so much media attention and celebrity status as did Billy Sunday. His flamboyant antics were criticized as vulgar and an abomination of religious leadership. But Sunday also had many supporters. After his New York crusade in 1918, which claimed nearly 100,000 souls for the Lord, Billy Sunday's supercharged evangelistic career began to wind down. The New York crusade was not only his last great hurrah, but the last hurrah for revivalism for three decades (Weisberger: 267). If ever the revivals had brought in the unwashed masses in need of being snatched from the grasp of Satan, increasingly the audiences were there for a peculiar kind of amusement and entertainment. The revival in the end, writes Weisberger, "was not even so much . . . a ritual as a spectator sport with religious overtones" (1958: 272).

In the later years of his ministry, Billy Sunday became a transition figure to a new role that would become important to televangelism. Sunday was long a staunch supporter of the laissez-faire capitalists he golfed with, but when the U.S. belatedly entered World War I, Sunday became a super-patriot.

Fundamentalism As A New But Short-Lived Social Movement.

Among the several religious movements which were spawned by urban revivalism, fundamentalism no doubt grew strongest. The term fundamentalism derives from a series of paperback volumes published between 1910 and 1915 called *The Fundamentals.* Conceived as a testimony to biblical literalism, in contraposition to "higher criticism," they served the cause of pulling together a diverse coalition of similar minded conservative Christians. *The Fundamentals* were clearly an important resource in a religious movement. The free distribution of three million copies of the booklets went a long way toward creating a communications infrastructure.

Fundamentalism as a critique of culture and, hence, a social movement, did not appear until later. When it appeared, its life as a social movement was short-lived and stormy. Billy Sunday was an important figure. Initially fundamentalists stood in staunch opposition to America's becoming involved in the war effort, in clear opposition to the support lent by the "modernists" (Marsden: 141-153). The transition to super-patriots resulted, in part, because the modernists accused them of being sympathetic to Germany. But fundamentalists associated Germany with Neitzsche's "might makes right" philosophy, which they in turn associated with "the survival of the fittest" and "evolutionism." They came to view the war effort, thus, as a struggle between civilization and barbarism.

Marsden notes Sunday's role in whipping up fundamentalist patriotism:

> As the war effort accelerated [Sunday] used the rhetoric of Christian nativism to fan the fires of anti-German furor and was famous for sermons that ended with his jumping on the pulpit waving the flag. "If you turn hell upside down," he said, "you will find 'Made in Germany' stamped on the bottom (142).

This step into the political arena, once taken, gave fundamentalism a much more decidedly political character. But as a nascent social movement, it was not well focused. The enemy was omnipresent in the form of "barbarism," "Bolshevism," "religious modernism," "evolutionism," and a dozen other 'isms. As fundamentalist leaders came to see themselves as locked in a struggle for the survival of God's dominion over the affairs of man, their premillenialist view of the futility of involvement in this world, which had earlier kept them at arm's length from worldly affairs, was overshadowed. They had to get involved.

Diffuse, poorly organized, and inexperienced in politics, the fundamentalists enlisted William Jennings Bryan to lead the movement. Fighting the sinister evolutionists became their main crusade. When John Scopes decided to challenge a Tennessee statute against the teaching of evolution in 1925, Bryan felt that this was an opportune time to square off and do battle. Scopes was found guilty of teaching evolution, but Bryan's knowledge of the Bible was no match for the sharp cross-examination of Clarence Darrow. Victorious, but humiliated, Bryan took ill and died five days later in Dayton, Tennessee. And with the passing of Bryan, the burgeoning social movement collapsed. What remained became more and more like the stereotyped image the media had delighted in portraying during the eleven day trial — a backwater remnant of fools comically battling the inexorable forces of modernity.

The "Monkey Trial" became an epoch-ending event from which fundamentalism might never regain respectability or power. The demise of urban revivalism parallels this devestating blow to conservative religion in America. After Billy Sunday, there were no more great evangelists until Billy Graham. After William Jennings Bryan, there was Billy James Hargis, Carl McIntire, Dr. Frederick Schwarz, Edgar Bundy and a whole host of lesser luminaries who probably contributed more to reinforcing the stereotypes of evangelists offered by Sinclair Lewis and H. L. Mencken than they did to saving America from communism and other evils. Each of these persons attracted followings, but none ever managed to develop sufficient resources to challenge seriously any aspect of American life. Thus, while they represent surviving residuals of the movement, we can conclude that the fundamentalist social movement, which had gained considerable strength during the first half of the 1920s, lost its momentum and then fizzled after 1925.

In retrospect it is fairly obvious that the fundamentalist social movement was neither well defined nor its attention well focused. Also, with the benefit of hindsight, the movement was probably doomed when they picked Williams Jennings Bryan to be its leader. Three times the Democratic Party's standard-bearer for the presidency, Bryan had never won any office for which he was nominated. His oratory skills were not matched with organizational acumen.

When Jerry Falwell came upon the national political scene in 1980, proud to be a fundamentalist and professing to speak for millions who were going to reclaim America in the name of God, few political analysts were prepared to take him seriously. Contemporary political analysts had come to see fundamentalism through the eyes of secularization theorists — as intermittant residual noise from an archaic religious form.

UNDERSTANDING THE RE-EMERGENCE OF A SOCIAL MOVEMENT

To understand how and why fundamentalism eventually reappears as a social movement in the late 1970s, and why it is destined to become the major social movement in America during the last quarter of the twentieth century, we need to pull together four separate threads or developments. The first is the link of televangelism to urban revivalism. This is important because religious broadcasting is the critical resource in mobilizing the social movement.

The second thread involves the process by which the religious airwaves became the near exclusive domain of evangelical and fundamentalist Christians. This dominance was not always the case. In fact, for a long time in the history of radio, and during the first major phase of television, the liberal church tradition had a near monopoly of the airwaves. Without this very considerable media access, the New Christian Right would be a much less formidable social movement.

The third thread involves the role of fundamentalism in preserving the dominion creation myth as the central motif in interpreting American history. As the liberal dream seemed to be collapsing during the 1960s and 70s, the fundamentalists had an alternative vision of a better America, along with the motivation to press their grievances against what they saw as the causes of a failing social order.

A final thread involves the restoration of the Christian worldview of dominion. Dispensational premillenialism, with its pessimistic view of this world, was just plain bad luck for evangelicals. A critical step in the transformation of the New Christian Right into a world transforming social movement is the junking of premillenialism and the restoration of a post millenial worldview. Restoration of post millenialism can provide the social movement with the conceptual, or theoretical, rationale for resuming the historic quest for dominion. This process has already begun.

The Link to Urban Revivalism

Urban revivalism as an important cultural form never completely disappeared, but it achieved a high profile once again in 1949 when William Randolph Hearst delivered his celebrated two word command, "Puff Graham," to his newspaper empire. The following year, after an incredible flurry of media attention, Billy Graham decided to do a weekly radio program, *Hour of Decision.* That decision firmly linked nineteenth century urban revivalism to modern religious broadcasting.

Billy Graham's decision to go on radio was even more momentous to his career and the future of religious broadcasting than the great boost he got from Hearst's patronage. About a year after going on radio, Graham made another important decision. In 1951, the Billy Graham Evangelistic Association began packaging his crusades for television. This gave Graham even greater visibility and success.

Like Moody and Sunday before him, Graham relished rubbing shoulders with the rich and powerful. He particularly liked presidents until he got soiled by the carnage of Watergate. His role as the "preacher of presidents" no doubt rendered legitimacy to the political status quo. Graham's sermons have always had a ring of patriotism, but never the bellicose "100 percent Americanism" of Billy Sunday in his later days. Still, while Graham would eventually repudiate his own involvement in politics, he set the stage for others to become more deeply involved than he.

Excluding Graham, the early days of televangelism could be characterized as apolitical. At about the same time Billy Graham decided to go on television, two itinerate evangelists from Oklahoma and Arkansas also recognized the potential of television for saving souls. Oral Roberts brought the television cameras into his revival tent; Rex Humbard sold his tent and built a cathedral especially equipped for broadcasting. A new era was born.

These three men play roles in the development of the electric church which parallel Finney, Moody, and Sunday in the development of urban evangelism. Building on the organizational principles which resulted in the institutionalization of urban revivalism, Graham, Roberts and Humbard created yet another institution — the electric church.

The Billy Graham Evangelistic Association modeled its crusades after the methods of Finney, Moody, and Sunday: the engagement of local pastors and churches before the decision to conduct a crusade, advanced planning activities to arouse interest, top flight entertainment — albeit in a much more subdued form than Sunday's vaudeville antics — , guest celebrity appearances, appeals to the emotions, and the urgency of a decision for Christ, etc. The great boost Graham received from Hearst gave him a competitive edge in access to evening prime time television which Roberts and Humbard were never able to overcome. Roberts and Humbard became innovators in the structure of programming and the development of communication feedback loops with their audiences.

Of the two, Roberts has been much more innovative. He hired top flight secular entertainers to appear on his programs as a way of hooking audiences. Through rather crude studies of audience response, he guaged audience size and aggressively bought the best time slots. He learned early that people will give for bricks and mortar while they do not get excited about paying for air time. With this knowledge, he built a university. That he eventually overextended himself with his hospital does not detract from the principle he established, and almost every successful televangelist has copied: major projects excite audience response.

Humbard has been most successful in mastering the art of parapersonal communication. When he looked into the camera, people thought he was talking directly to them. From the beginning, his viewers were participants in the program, and audiences came to feel that they too were members of the Humbard family, or vice versa. Humbard developed audience loyalty rarely achieved in television and they stuck with him for a very long while.

What all of these ministries have in common, which would in time revolutionize religious broadcasting, is their relationship to their audience. Whereas commercial broadcasting sells advertising to support programming, the electric church sells itself and its projects. They solicit support for general and specific projects from their audiences and offer premiums in exchange for donations. Over time, their solicitation tactics have become extremely sophisticated.

All but one of the major televangelists have used their broadcasting to build substantial off-camera empires. The limits of growth potential for existing operations is still unknown. With increased competition, air time has become very expensive. There can be no question that it is a precarious operation. Humbard's empire collapsed in 1985 and Roberts has appeared on the brink of disaster since the raising of the medical complex. If Roberts survives, the implication of the Humbard fate is that the dangers of *failing* to build a bricks and mortar empire are more hazardous than trying to sustain a broadcasting enterprise without projects.

If one can use a television ministry to build colleges, cathedrals, hospitals and spiritual Disneylands, it seems likely that it can also be used to pursue projects that are not necessarily direct derivatives of the religious broadcasting. Jerry Falwell's decision to create the Moral Majority was the first bold attempt to test this proposition.

In a more subtle, and in the long run more effective way, Pat Robertson too has used his religious television role to demonstrate to the world his political acumen and to build a following for his political views. Whether or not he eventually chooses to run for political office, his blending of religion, politics, and economic analysis on *The 700 Club* has elevated his personal status as a respected conservative spokeperson. His potential to capitalize upon this status is considerable.

Other televangelists, e.g., Jimmy Swaggart and D. James Kennedy, are well positioned to channel their audiences toward explicit political projects. So long as the Federal Communication Commission, the Congress, and the Courts do not change broadcasting rules, the potential for using religious broadcasting as a base for building social movement-political organizations seems quite likely.

In summary, the institution of urban revivalism has been resurrected in the form of the electric church. Like revivalism, the electric church functions autonomously from other religious institutions. Two unique qualities are 1) the potential to generate surplus resources; and 2) the capability of directing those surplus resources to other projects. Other projects may include building colleges, hospitals, cathedrals, etc. What has happened over the past decade, with increasing velocity, is that resources have been directed toward social movement activities. Resources are here understood to include exposure (advertisement of the social movement cause), recruitment of constituents, and transfer of fiscal and leadership resources from "conventional" electric church activities to social movement activities.

Control of Airwaves

A second important thread for understanding how the New Christian Right has become an important social movement is the process whereby religious broadcasting in America came almost exclusively into the hands of evangelicals and fundamentalists. From the beginning of broadcasting, the liberal church traditions associated with the Federal Council of Churches (which became the National Council of Churches in the 1950s) held a privileged position in access to network air time. This advantage increased over time so that by the early 1940s evangelical and fundamentalist broadcasters were virtually squeezed off the air.

How the transition from liberal dominance to virtual monopoly by the conservatives occured is a fascinating and basically untold story — and, regrettably, beyond the scope of this paper. It is possible only to identify the most critical variables which brought about this revolution.

The Communications Act of 1934 authorized the FCC to license individual stations. The granting of a license, in effect, constitutes a monopoly to use a scarce commodity, namely a specific airwave, to transmit messages. Without specifying precise details, it has always been presumed that stations "owe" some proportion of their broadcast time to general "public interest" in exchange for this monoply right. From the beginning, religious broadcasting has been designated as public interest broadcasting. Both network and local broadcasters preferred to allocate this time to the "mainline" religious traditions on a "sustaining" (free) basis. Among the networks, only the Mutual Broadcasting System offered commercial air time and, eventually, they severely restricted the amount of time and the conditions for access.

In 1960, the FCC released a policy directive specifing that no important public interest was served by differentiating between gratis airtime and commercially sponsored programming. The implication of this ruling was that local stations could sell airtime for religious programs and still get "public interest credit" in the eyes of the FCC overseers.

Several important developments followed. First, the evangelical and fundamentalist syndicators rushed in to compete with one another for the airtime. Their competition enhanced the value of the time slots. As a result, many local stations which had previously followed network policies of not selling air time for broadcasting decided to cash in on the new demand.

The implications of this process for sustaining-time programs was devastating. Local stations dropped sustaining-time programs produced both by their network and individual denominations. This effect was dramatically demonstrated in a report by the Communications Committee of the U.S. Catholic Conference to the FCC in 1979 (Horsfield, 1984: 89). In 1959, fifty-three percent of all religious broadcasting in America was paid-time programming. Following the FCC ruling, that proportion increased to ninety-two percent in 1977. For all intents and purposes, thus, religious broadcasting has become a commercial enterprise in America on both radio and television. The *de facto* implication of this is that the "mainline" or liberal denominations of Protestantism, as well as Catholics and Jews, have been virtually squeezed off the air.

Mainline denominations have not been able to compete in commercial religious broadcasting for a variety of reasons (for details, see Hadden, 1980b). Among other things, liberal church traditions are less comfortable in asking people to give money or give their soul to Jesus. By contrast, the confluence of evangelical proselytizing zeal and the commercial free-enterprise system go together well. And the oligarchically structured bureaucracies of the electric church permit rapid adaptation and prompt execution of decisions in response to perceived opportunities.

Other FCC policies have a significant bearing on the preponderance of commercial time religious broadcasting. It is conceivable that these policies might be altered significantly in the future. But it is also unlikely that this will happen without confrontation. The National Religious Broadcasters have proved themselves to be effective lobbyists in Washington. Their chief legal counsel is a former chairman of the FCC. At the annual meeting of the NRB, a luncheon is held to honor the Commissioners.

Whereas the founding of the National Religious Broadcasters is grounded in a religious conflict between evangelicals and "modernists," it has become an important resource for the New Christian Right social movement. NRB retains top legal counsel to lobby for the group's interests and protect against any attempt to assault the favored position it occupies. It has over 1,100 member radio and television stations.

The Preservation and Restoration of the Dominion Creation Myth

A third thread which provides insight into why a fundamentalist social movement of the first quarter of twentieth century reappears during the last quarter involves the peculiar and ironic circumstances by which the fundamentalists became the custodians of the dominion creation myth.

There was also, of course, a certain irony to the political engagement of fundamentalists in the first quarter of the century. The irony results from the fact that premillenialism is an eschatology of defeat and despair. Things are getting so bad, the doctrine holds, that only Christ's return can stay the tide against Satan. It invites retreat from the world, not hand-to-hand combat with the minions of Satan.

However one may view the inconsistency, the fundamentalists of the 1920s, perceiving Christian civilization to be crumbling around them, could not content themselves with waiting for Christ's return. At least some of them were not able to disengage from the world. The reason why this was so in the '20s and again in the '70s is that premillenialism runs in sharp contradiction to the dominion creation myth.

For nearly the whole of American history, its people have believed that God had special providential plans for this land. However compelling the premillenialist doctrine may be for those who see the modern world as sinful and corrupt beyond redemption, premillenialism has never completely defeated the dominion doctrine. The latter was reinforced throughout the nineteenth century by the doctrine of free will. If the individual is compelled to choose between good and evil and if God has a plan for America, then it falls to each individual to join in the struggle.

The premillinealist doctrine focuses attention on saving as many souls as possible before it is too late. The doctrine of dominion, in sharp contrast, would transform society. The tension between engagement and disengagement is a struggle that goes on not only between groups within the fundamentalist and evangelical sectors, but it also involves personal struggle to discern God's will.

In 1965, Jerry Falwell preached that he could not imagine ever turning his attention from preaching the "pure saving gospel . . . [which] . . . does not clean up the outside but rather regenerates the inside" (Hadden & Swann: 160). That was premillenialist Falwell. A decade later, as America prepared its bicentennial, Falwell was feeling the tension of the "broken covenant" as he organized "I Love America" rallies. In 1980, after he had organized the Moral Majority, he wrote "If Americans will face the truth, our nation can be turned around and can be saved from the evils and the destruction that have fallen upon every other nation that has turned its back on God" (Falwell, 1980: 18). Falwell is still preaching premillenialist theology, but the dominion covenant is tugging at his soul.

The dominion creation myth with its recurring cycles of sin, repentance, and redemption is an inextricable part of the American conscience. About the same time Falwell began sounding the trumpet for America to repent, sociologist Robert Bellah wrote: "Once in each of the last three centuries America has faced a time of trial, a time of testing so severe that not only the form but even the existence of our nation have been called in question" (1975: 1). Bellah's understanding of why the covenant has been broken and what must be done to repent is at odds with Falwell's. The important observation is that two persons, so different in their intellectual understanding of America, agree that America is broken and in need of fixing.

This consensus is the result of the survival of the dominion creation myth, and it is widely shared in sacred and secular form across American culture. Harvey Cox's scenario for the return of the sacred to the secular city is a liberal church vision of repentence and redemption (1984). Neo-conservative Richard John Neuhaus would return God to the "naked public square" as a first step in restoration of the covenant (1984), and even Catholic atheist Michael Harrington agonizes about "the politics at God's funeral" and pleads for rational re-creation of something like the creation myth (1983).

Secularization fosters pluralism which, in turn, de-monopolizes religious dogma (Berger, 1967: 134). This is true of both formal theological doctrine and religious mythology. The dominion creation myth has survived in America, but the tremendously diverse array of variations on the theme cited above bear testimony to the reality of cultural pluralism. If there is a general sense that America needs to mend its ways so that it can be healed and once again return to greatness, how that is to be achieved is a matter of deep division.

The critical question from the perspective of resource mobilization analysis is whether any group or coalition of groups possesses resources sufficient to generate a major social movement. To address this question, we have to move from the generalized perception that the society faces serious problems to more precise identification of those problems.

This is not the occasion to analyze the conservative mood in America today vis-a-vis the more liberal mood of previous decades. But the liberal social problems agendas of the previous two decades do have important implications regarding the options for working through a crisis of dominion. Two factors are of particular importance.

First, social movements are not easily sustained for long periods of time. Maintaining momentum requires more and more resources. The capacity of social structures (public and private) to respond is not without limits. Movements tend to become bureaucratized and, with this, they develop survival imperatives independent of the movements. In brief, movements have cycles. Over time, they loose momentum and must either be regenerated or they fade.

Second, movements do not necessarily give way to counter-movements, but it is not unusual that new problems are identified as a consequence of solving others problems. Thus, addressing the problems of inequality and human rights during the '60s and '70s is now seen as a source of some of the problems of the 1980s. Whereas most movement leaders would not wish their movements to be judged in terms of the goals and activities of the most extreme elements of the movement, there is a tendency to do so. Counter-movements will seek to highlight fringe groups and characterize them as exemplary of the entire movement. Thus did a violent element of the black community in America hasten the conclusion of a formal civil rights movement. And radical feminists and gays have provided highly visible targets for those who seek to organize counter-movements against them.

These more or less natural movement tendencies have important implications for understanding (and misunderstanding) the present conservative tendencies in America. First, a variety of liberal social movement activities dominated American politics for over two decades. Whatever the empirical links may be between the liberal social movement agendas and the current array of intolerable social issues, there is an appearance of causal relationship — an appearance that the New Christian Right (NCR) has not hesitated to exploit.

Viewed as a countermovement to the liberal social agenda of the 1960s, the New Christian Right bears a lot of similarities to the fundamentalist movement of the 1920s. And viewed from this assumptive framework, it is easy to dismiss it as yet another backlash by persons who have been left behind and are out of step with the mainstream of American culture. From this perspective, it seems terribly significant that "creationism" should reappear as a serious issue on the NCR agenda.

One cannot escape an element of backlash in the NCR agenda. But my quarrel with much of the analysis about the NCR to date is 1) it fails to understand the importance of playing on the old themes as an instrument for mobilizing previously passive adherents; and 2) it does not see new elements which have the potential for mobilizing much broader followings.

First, some comments on the value of playing on old themes. To an outsider, the "creationism" issue has the appearance of a *deja vu* of William Jennings Bryan on the witness stand being cross-examined by Clarence Darrow. To fundamentalists, there is a redemptive vindication. The new creationism is not going to put evolution to rest, but it is a far more sophisticated scrutiny of the theory of evolution than the first round of Bible thumping. Meticulous rational scrutiny of the theoretical aspects of the science of evolution has laid bare the "faith" assumptions of secular scientists. Further, the fact that secular scientists are generally not prepared to deal with fundamentalists at this level of analysis has provided a tremendous moral victory and, from their perspective, vindication against the stereotypes of "anti-evolutionists" as "dummies."

But more importantly, the evolution issue provides a springboard for a more general critique of contemporary public education. If evolution can be demonstrated to be a theory, with some gaps in empirical evidence, and if secular education refuses to teach the alternative biblical "theory" of creation, then public education must certainly be guilty of committing "secular humanism," as charged.

In this context, creationism becomes a very important issue because it symbolizes a critique of public education in America which has dismissed God and morality. For a hundred and fifty years the constitutional issue of the relation between church and state was essentially moot. With few notable exceptions, there were no cases in the courts. Not only was God's place in the classroom firmly established, but socializing "barbarians" to godly and moral principles was also an important basis for public education (Collins, 1979: 95-118). After a long history of congenial accommodation of God in public places, the last twenty years have seen the court dockets filled with challenges to the entanglement of religion in public life. An extraordinary number of cases have dealt with religion and public education. And, on balance, the large number of cases points toward exclusion of God from the classroom.

As a springboard to a more general critique of what has happened to public education, the creationism issue may have important indirect implications for a broader support base. If creationism can develop and then retain an aura of respectability, it bolsters the argument that not only has God been excluded from ceremonial matters in public education, but also that public education has developed a biased curriculum which systematically excludes Christian principles.

The school prayer issue demonstrates the vulnerability of the legal and structural barriers to penetration by the NCR. Several recent public opinion polls have demonstrated that significant proportions of Americans do not accept the legitimacy of prohibiting prayer in public schools. For example, a national poll conducted earlier this month revealed that three-quarters of the American population think that prayer in school does not violate the separation of church and state. Sixty-three percent not only think prayer should be permitted, but that it should be encouraged (*The Daily Progress*, 1985: A6). These views, of course, stand in direct contradiction to several Supreme Court decisions, including *Wallace v. Jaffree* (1985) in the most recent court session.

Thus far, the New Christian Right has been held off by the courts and frustrated in its efforts to get a school prayer amendment before the full Congress for consideration. However symbolic the issue may appear, it seems to me to be an issue that stands a chance of eventually winning, either through progressive testing of the Courts or by a Constitutional Amendment.

Abortion is another issue about which the fundamentalists feel very deeply. It has proved not only to be an issue which has motivated action, but has also served to begin the process of building coalitions with other groups who share such concerns. Until only recently, fundamentalists have a long history of animosity toward Roman Catholics and Mormons. On the issue of abortion, they are substantially together.

A critical indicator of developing movement maturity and strength is the ability to put aside ideological differences to work together on at least limited objectives with those who have otherwise been adversaries. Through the abortion issue, evangelicals, Catholics, Mormons and other groups are now finding a broader range of common concerns and, as a result, their organizational alliances are beginning to extend into other areas of cooperation.

In summarizing the early years of the New Christian Right movement, I would offer four observations.

First, this decade has been characterized by a broadening of the base of support among conservative Christians, large numbers of whom have not heretofore been engaged in the political process. We have no sensitive empirical indicators of just how extensive this type of activity has been, but there is an abundance of evidence that more and more people are shifting from indifferent bystanders to the category of adherents, and from adherents to constituents. Second, the number of New Christian Right organizations has increased precipitously, but again I am not aware of any systematic inventories of identity, memberships, or budgets. Third, and equally impressive, is the extent to which the New Christian Right leadership has been building organizational coalitions to groups who share common concerns on both broad fronts and limited social agendas.

Notwithstanding the potential of the New Christian Right to build a broader base of support for its critique of public education, support for its position on abortion, and perhaps several other matters as well, the fact remains that it appears to be swimming against the mainstream of American culture (Cf: Hammond, 1983: 219). It seems out of step with emerging values about interpersonal relations between the sexes. Its long laundry list of personal "vices" covers a large proportion of matters that Americas feel ought to be left to personal choice.

A fourth observation in reviewing the NCR movement to date involves an assessment of their potential to develop issues that command national attention to a degree that requires collective resolution. The Civil Rights movement was of this character. It dramatically presented massive evidence of social inequality and mistreatment which demanded public attention and at least some sense of progress toward resolution.

Is there such an analogue in the burgeoning New Christian Right movement? The strongest critics of the New Christian Right argue that, if left unchecked, these Christian "zealots" will rewrite the laws and the Constitution to impose their pietistic moral code on the rest of the nation. Clearly the New Christian Right stands sharply at odds with the general direction of American culture in some very important ways. My own sense, however, is that such thinking about the NCR locks us into a mindset of the movement as a counter-movement to the social movements of the 1960s. I think this leads both to misunderstandings and to an underestimation of potential strength.

At the beginning of this address, I characterized the heart of the New Christian Right movement as a *crisis of dominion*. Let me return to this theme. The past quarter of a century has sorely tested the American character. We have achieved technological accomplishments which were beyond the human imagination only a generation ago. We have taken giant strides to erase inequities and injustices inherited from our ancestors. And we have grown increasingly conscious of our planet, our place, and our requirement to be a good neighbor.

For all of our intentions at home and abroad, we seem to stumble. We have suffered the pain of assassination, the agony of defeat, the humiliation of impotence, the frustration of the inability to use strength, and the disgrace of failed leadership.

Domestically, the evidence for our having made lasting progress in eliminating discrimination against minorities and women seems ambiguous, poverty stubbornly resists our prescriptions for eradication, drug use and crime seem out of control. The list of ills and frustrations, at home and abroad, goes on and on.

Others, in other times and places, have faced prolonged frustrations and humiliations and learned to live with them. But the events of the past quarter of a century have not gone down well with the American people because we have always thought of ourselves as a good God-fearing people who, with his hand, have the ability to control destiny. As individuals, most of us are doing well. But as we look around at all that has gone wrong, these happenings seem to bear testimony to a much deeper and fundamental cultural malady.

Sociologically speaking, if it were possible to step outside our world and objectively assess our cultural condition, it wouldn't matter. People act upon what they perceive to be real. Whether or not our problems seriously threaten our destiny as a people, we think they do.

The liberal tradition in America does not stand in a strong position to direct us out of this perceived condition of cultural malaise. We stand too close in time to too many liberal programs that were supposed to help solve the very problems that now disturb us so.

The New Christian Right has offered us an old diagnosis — one that we have bought before: Our problems are of our own making. We must repent and make things right with our maker before we can resume our providential role in his divine plan.

What is required of us is not yet clear. Social movements never unfold in conformity with a masterplan or blueprint. They are made up as they go. Jerry Falwell's vision of a repentant America has received a lot of media attention to date. His neighbor on the other end of Virginia now seems positioned to get a lot of visibility in the immediate future. Others, presently just off center stage, could move into the national spotlight.

My political theory leads me to believe that there are powerful forces which pull all movements toward the center of our culture. Within this framework, I find some hope that the movement will not take America to the extremes that many liberal groups fear. I also find hope in what I detect as a distinct turn from premillenial doctrine by some televangelists.[10] On the one hand, a return to postmillenial theology will likely give the social movement a great boost of energy. If this nation and the world is yet to be conquered before Christ's return, that could charge the movement with renewed zeal to restore God's dominion in America. The other side of postmillenialism, however, is that it could spell a much more responsible posture toward our tiny planet and all that dwells therein. And that, too, could be a source of hope. If Jesus isn't coming very soon, then it behooves

10. Since this address was delivered, I have had the opportunity to discuss this issue with several televangelists, including Pat Robertson and Jerry Falwell. Both of these men reject the notion that their thinking is moving toward postmillenialism. However, Robertson did admit to backing away from his sense that the second coming of Christ was imminent. Both assert that eschatology is not a significant variable in explaining the political engagement of the New Christian Right. On the basis of these interviews I have to conclude that explicit and dramatic rejection of premillenialism is not imminent. But I do think that this-worldly political success will gradually lead to a deemphasis on premillenialism.

all who are in positions of leadership and responsibility to recognize and protect our fragile interdependence on this planet.

REFERENCES

Ahlstrom, Sidney E.
1972   *A Religious History of the American People.* New Haven, Connecticut: Yale University Press.
Armstrong, Ben
1979   *The Electronic Church.* Nashville, Tennessee: Thomas Nelson.
Bellah, Robert N.
1970   *Beyond Belief.* New York: Harper and Row.
1975   *The Broken Covenant.* New York: Seabury Press.
Berger, Peter L.
1967   *The Sacred Canopy.* Garden City: New York, Doubleday.
Carnegie, Andrew
1949   "Wealth." Pps. 1-8 in Gail Kennedy (Ed.), *Democracy and the Gospel of Wealth.* Boston: D. C. Heath.
Cherry, Conrad (Ed.)
1971   *God's New Israel: Religious Interpretations of American Destiny.* Englewood Cliffs, New Jersey: Prentice-Hall.
Collins, Randall
1979   *The Credential Society.* New York: Academic Press.
1986   "Historical perspectives on religion and regime: Some sociological comparisons of Buddhism and Christianity." Pps. 254-71 in Jeffrey K. Hadden and Anson Shupe (Eds.), *Prophetic Religions and Politics.* New York: Paragon House.
Cox, Harvey
1984   *Religion in the Secular City.* New York: Simon and Schuster.
Dobbelaere, Karel
1981   *Secularization: A Multi-Dimensional Concept.* Vol. 29, Current Sociology Series. Beverly Hills, California: Sage Publications.
1984   "Secularization theories and sociological paradigms: Convergences and divergences." *Social Compass* 31: 199-219.
Falwell, Jerry
1980   *Listen, America!* Garden City, New York: Doubleday.
Finney, Charles F., Jr.
1960   *Lectures: On Revivals of Religion.* Ed. by William G. McLoughlin. Cambridge, Massachusetts: Belknap Press of Harvard University Press.
Frady, Marshall
1979   *Billy Graham: A Parable of American*

Righteousness. Boston: Little, Brown and Company.
Frankl, Razelle
1984   *Popular Religion and the Imperatives of Television: A Study of the Electric Church.* Ph.D. Dissertation, Bryn Mawr College.
1985   "The Historical Antecedent of the Electric Church." Unpublished paper presented at the Annual Meeting of the Society for the Scientific Study of Religion.
1987   *Televangelism: The Marketing of Popular Religion.* Carbondale, Illinois: Illinois University Press.
Gallup Organization
1983   *Religious Television in America.* Princeton, New Jersey: The Gallup Organization.
Hadden, Jeffrey K.
1980a   "Religion and the construction of social problems." *Sociological Analysis* 41: 99-108.
1980b   "Soul-saving via video." *Christian Century* 97: 609-613.
Hadden, Jeffrey K. and Charles E. Swann
1981   *Prime Time Preachers.* Reading, Massachusetts: Addison-Wesley.
Hammond, Phillip E.
1983   "Another great awakening?" Pp. 207-23 in Robert C. Liebman and Robert Wuthnow (Eds.), *The New Christian Right.* New York: Aldine.
Hammond, Phillip E. (Ed.)
1985   *The Sacred in a Secular Age.* Berkeley, California: University of California Press.
Harrington, Michael
1938   *The Politics at God's Funeral.* New York: Holt, Rinehart and Winston.
Hofstadter, Richard
1963   *Anti-Intellectualism in American Life.* New York: Random House.
Horsfield, Peter G.
1984   *Religious Television.* New York: Longman.
Hudson, Winthrop S.
1973   *Religion in America,* 2nd edition. New York: Charles Scribner's Sons.
Jennings, Ralph M.
1968   *Policies and Practices of Selected National Religious Bodies as Related to Broadcasting in the Public Interest: 1920-1950.* Ph.D. Dissertation, New York University.

Marsden, George M.
    1980   *Fundamentalism and American Culture.*
            New York: Oxford University Press.
Martin, David
    1965   "Towards eliminating the concept of
            secularization." In Julius Gould (Ed.),
            Penguin Survey of the Social Sciences.
            Baltimore: Penguin.
    1978   *A General Theory of Secularization.* New
            York: Harper and Row.
Marty, Martin E.
    1970   *Righteous Empire.* New York: Dial Press.
Mauss, Armand
    1975   *Social Problems as Social Movements.*
            Philadelphia: Lippencott.
McCarthy, John D. and Mayer N. Zald
    1977   "Resource mobilization and social
            movements: A partial theory." *American
            Journal of Sociology* 82: 1212-41.
Neuhaus, Richard John
    1984   *The Naked Public Square.* Grand Rapids,
            Michigan: Eerdmans.
Pankhurst, Jerry
    1968   "Comparative perspectives on religion and
            regime in Eastern Europe and the Soviet
            Union." Pps. 272-306 in Jeffrey K. Hadden
            and Anson Shupe (Eds.), *Prophetic
            Religions and Politics.* New York: Paragon
            House.
Parsons, Talcott
    1963   "Christianity and modern industrial
            society." Pps. 33-70 in Edward Tiryakian
            (Ed.), *Sociological Theory, Values and

            Sociocultural Change.* Glencoe, Illinois:
            Free Press.
Robertson, Pat
    1982   *The Secret Kingdom.* Nashville: Thomas
            Nelson Publishers.
Schiner, Larry
    1967   "The concept of secularization in empirical
            research." *Journal for the Scientific Study
            of Religion* 6: 207-220.
Smith, Timothy L.
    1957   *Revivalism and Social Reform.* New York:
            Harper and Row.
Stark, Rodney and William Sims Bainbridge
    1979   "Of churches, sects and cults: Preliminary
            concepts for a theory of religious
            movements." *Journal for the Scientific
            Study of Religion* 18: 117-31.
Taylor, Frederick Winslow
    1911   *The Principles of Scientific Managements.*
            New York: Harper.
The Daily Progress
    1985   "Prayer should be encouraged in
            classrooms, poll finds." October 14: A6.
Weisberger, Bernard A.
    1958   *They Gathered at the River.* Chicago:
            Quadrangle Books.
Yinger, Milton
    1957   *Religion, Society and the Individual.* New
            York: Macmillan.
Zald, Mayer N.
    1982   "Theological crucibles: Social movements in
            and of religion." *Review of Religious
            Research* 23: 317-36.

Copyright Information

Index

Publisher's notes:
Underlined page numbers indicate illustrations.
Editorial corrections and additions appear in brackets adjacent to the entry
as it appears in the text.